Aughey and Frye's Comparative Veterinary Histology with Clinical Correlates

Organised by body system, the new edition of this highly illustrated textbook covers the normal histological appearance of tissues in a wide range of animals, both domestic and exotic species, with relevant clinical correlates emphasising the need to appreciate the normal in order to recognise the abnormal.

In this update by two experienced veterinary pathologists and histology lecturers, new species, such as other companion mammals, aquatic species, and livestock, are introduced into each chapter along with a wealth of new high-quality images. A new chapter covers epithelial tissue, and new techniques used in histology and histopathology are discussed throughout, including *in situ* hybridisation (ISH) and digital image analysis. Pathogenesis explanations are introduced in the current (and many new) cases of histopathology.

The breadth of coverage – farm animals, dogs, cats, horses, birds, reptiles, amphibians, and fish – and the integration of normal and abnormal tissue provide a reference of lasting value to veterinary students as well as veterinary practitioners and pathologists requiring a quick refresher.

Aughey and Frye's Comparative Veterinary Histology with Clinical Correlates

Second Edition

Francisco Javier Salguero Bodes
Francisco José Pallarés Martínez

CRC Press
Taylor & Francis Group
Boca Raton London New York

CRC Press is an imprint of the
Taylor & Francis Group, an **informa** business

Second edition published 2023
by CRC Press
6000 Broken Sound Parkway NW, Suite 300, Boca Raton, FL 33487-2742

and by CRC Press
4 Park Square, Milton Park, Abingdon, Oxon, OX14 4RN

CRC Press is an imprint of Taylor & Francis Group, LLC

© 2023 Taylor & Francis Group, LLC

Reasonable efforts have been made to publish reliable data and information, but the author and publisher cannot assume responsibility for the validity of all materials or the consequences of their use. The authors and publishers have attempted to trace the copyright holders of all material reproduced in this publication and apologize to copyright holders if permission to publish in this form has not been obtained. If any copyright material has not been acknowledged please write and let us know so we may rectify in any future reprint.

Except as permitted under U.S. Copyright Law, no part of this book may be reprinted, reproduced, transmitted, or utilized in any form by any electronic, mechanical, or other means, now known or hereafter invented, including photocopying, micro-filming, and recording, or in any information storage or retrieval system, without written permission from the publishers.

For permission to photocopy or use material electronically from this work, access www.copyright.com or contact the Copyright Clearance Center, Inc. (CCC), 222 Rosewood Drive, Danvers, MA 01923, 978-750-8400. For works that are not available on CCC please contact mpkbookspermissions@tandf.co.uk

Trademark notice: Product or corporate names may be trademarks or registered trademarks and are used only for identification and explanation without intent to infringe.

ISBN: 978-1-032-36797-2 (hbk)
ISBN: 978-1-032-36448-3 (pbk)
ISBN: 978-1-003-33380-7 (ebk)

DOI: 10.1201/9781003333807

by ITC Garamond Std
Typeset in KnowledgeWorks Global Ltd.

Printed in Great Britain by Bell and Bain Ltd, Glasgow

CONTENTS

PREFACE

The understanding of the normal structure of cells and tissues is crucial to evaluate the pathological lesions observed in diseased animals. The main objective of this book is to stimulate in veterinary undergraduates an appreciation of the relationship between structure and function. Following the steps of Elizabeth Aughey and Frederic L. Frye, we have long experience teaching histology and pathology in different countries. The veterinary degree curriculum is constantly changing and adapting, and the time allocated to histology has been significantly reduced in recent years, being taught on many occasions together with physiology and anatomy as 'structure and function' system modules in the first year. However, students are expected to have a good breath of knowledge of histology to understand the histopathology taught in subsequent years, as well as pathophysiology, oncology, and other clinical disciplines in which a good knowledge of histology is paramount.

In this new edition, we have updated each chapter including new images and clinical correlates, information about new techniques widely used in the histology and histopathology laboratory, and aspects from other species. Each chapter discusses mammalian aspects of the topic first and foremost. However, reptiles, birds, and various other species are kept as pets and included in the undergraduate curriculum with the common domestic animals as well as laboratory animals. Therefore, some relevant histology is included in the text to highlight evolutionary differences and clinical relevance. More detailed coverage can be found in specialised texts.

The importance of the student's knowledge of normal structure and function is emphasised with the inclusion of a series of clinical correlates in each system and a wide variety of species. You cannot recognise the abnormal if you do not know the normal. In this way, we hope to stimulate the student to think about histology as an integral part of biology, with relevance to anatomy, physiology, cell biology, immunology, molecular biology, and most importantly, histopathology. We also expect veterinary professionals, including surgeons, clinicians, and pathologists to find this text valuable as well as nonveterinarians that have to deal with histological and histopathological descriptions in scientific publications.

Francisco Javier Salguero Bodes, DVM PhD DECPHM FHEA FRCPath MRCVS
United Kingdom Health Security Agency (United Kingdom)

Francisco José Pallarés Martínez, DVM PhD DECPHM
University of Córdoba (Spain)

ACKNOWLEDGEMENTS

First, we would like to thank the previous authors, Elizabeth Aughey and Frederic L. Frye, for the original editions of this text that have served as the foundations for contribution to this new edition. We would like to thank many colleagues that have shared with us interesting discussions, samples from uncommon species, and clinical cases throughout our professional careers. Special thanks to our colleagues from the University of Córdoba (Spain) L. Carrasco, M.J. Bautista, I. Rodríguez-Gómez, M.Y. Millán, the University of Murcia (Spain) J.I. Seva, S. Gómez, the University of Surrey (U.K.) D. Grainger, E. May, A. Coppi, and the United Kingdom Health Security Agency (UK) L. Hunter and C. Kennard.

We are also thankful to our families; without their support and love, we would not have been able to devote the time and dedication to our many histology and pathology endeavours, including this textbook. Thank you to Mar, Toñi, Marina, Beatriz, Javier Jr., and Santiago.

ABOUT THE AUTHORS

Francisco Javier Salguero Bodes graduated as a veterinarian in 1997 from the University of Córdoba, Spain, and then went on to gain his PhD in comparative pathology in 2001, studying the pathogenesis of African swine fever. He moved to the Spanish Government National Institute for Research in Animal Health (CISA-INIA) in Valdeolmos, Madrid, as a veterinary researcher, working on the pathogenesis and diagnostics of transboundary and emerging diseases. He was the Head of the Experimental Pathology Unit at CISA-INIA from 2005 until 2007, when he moved to the UK to work for the Veterinary Laboratories Agency (now APHA) as a veterinary research pathologist. He was appointed Reader in Comparative Pathology at the University of Surrey in October 2013 and moved to Public Health England (now United Kingdom Health Security Agency) as a Senior Veterinary Pathologist and Project Manager in September 2018. He has been studying the host–pathogen interaction and working on vaccine development in porcine diseases, tuberculosis, COVID-19, and many other zoonoses. Dr. Salguero has been carrying out numerous consultancy missions on veterinary pathology and infectious diseases for FAO, EU, GTRP, and other international institutions. He is a frequent speaker at scientific conferences and has been lecturing on veterinary histology, comparative pathology, and infectious diseases in numerous universities in different countries.

Francisco José Pallarés Martínez graduated as a veterinarian in 1993 from the University of Murcia, Spain. He received a scholarship from the Ministry of Education and Universities to carry out his PhD in swine pathology to study the effectiveness of a vaccine against swine enzootic pneumonia, which was completed in 1999. In 2000, he won a position as Assistant Professor of Histology and Pathology at the University of Murcia. He completed postdoctoral studies at the Veterinary Diagnostic Laboratory of Iowa State University in the United States for 15 months (2001–2002). After which, he developed his teaching and research work at the University of Murcia, as Associate Professor from 2007 and Full Professor from 2019, until 2020 when he moved to the University of Córdoba, where he is currently based doing his research and teaching in Veterinary Histology and Comparative Pathology. He has focused his research activity on the study of the pathogenicity and immune response against several of the main swine pathogens (PRRS virus, PCV2, and *Mycoplasma hyopneumoniae*). Dr. Pallarés has been carrying out numerous consultancy and training activities on veterinary pathology for different veterinary pharmaceutical companies at national and international levels and is a frequent speaker at scientific conferences.

BIBLIOGRAPHY

Embryology and Anatomy

P. Hyttel, F. Sinowatz, M. Vejlsted. Editorial assistance of Keith Betteridge. Essentials of domestic animal embryology. Saunders Elsevier, London, First published 2010. (printed 2014). ISBN: 978-0-7020-2899-1.

T.A. MacGeady, P.J. Quinn, E.S. Fitzpatrick, M. T. Ryan, D. Kilroy, P. Lonergan. Veterinary Embriology (2nd ed.). Wiley Blackwell, Ames, Iowa, 2017. ISBN: 9781118940617.

V. Aspinall and M. Cappello. Introduction to veterinary anatomy and physiology textbook (3rd ed.). Elsevier, London, 2015.

K.V. Kardong. Vertebrates: Comparative anatomy, function evolution (5th ed.). McGraw Hill Education, New York, 2008.

Veterinary and Comparative Histology

J.A. Eurell and B.L. Frappier. Dellmann's textbook of veterinary histology (6th ed.). Blackwell Publishing, Oxford, U.K., 2006.

L.P. Gartner. Textbook of histology (4th ed.). Elsevier, Philadelphia, Pennsylvania, 2017.

V.P. Eroschenko. Atlas of histology with functional correlations (13th ed.). Wolters Kluwer, Baltimore, Maryland, 2017.

W.J. Bacha and L.M. Bacha. Color atlas of veterinary histology (3rd ed.). Wiley-Blackwell, Oxford, 2012.

D.A. Samuelson. Textbook of veterinary histology. Saunders Elsevier, St Louis, Missouri, 2007.

W. Pawlina. Histology: A text and atlas with correlated cell and molecular biology (7th ed.). Wolters Kluwer, Philadelphia, Pennsylvania, 2015.

F.L. Frye. Biomedical and surgical aspects of captive reptile husbandry (2nd ed.). Krieger Publishing, Malabar, Florida, 1991.

Veterinary and Comparative Pathology

M.G. Maxie. Jubb, Kennedy & Palmer's pathology of domestic animals (6th ed.). Edited by M.G. Maxie. Elsevier, St Louis, Missouri, 2016.

J.F. Zachary. Pathologic basis of veterinary disease (6th ed.). Elsevier, St Louis, Missouri, 2017.

D.J. Meuten. Tumors in Domestic Animals (5th ed.). Wiley Blackwell, Ames, Iowa, 2017.

V. Kumar, A. Abbas, J. Aster. Robbins and Cotran Pathologic basis of disease (10th ed.). Elsevier, Philadelphia, Pennsylvania, 2020.

V. Zappulli, L. Peña, R. Rasotto, M.H. Goldschmidt, A. Gama, J.L. Scruggs, M. Kiupel. Surgical pathology of tumors of domestic animals: Volume 2: mammary tumors. Davis-Thomson DVM Foundation, Gurnee, Illinois, 2008.

J.J. Zimmerman, L.A. Karriker, A. Ramirez, K.J. Schwartz, G.W. Stevenson, J. Zhang (Editors). Diseases of swine (11th ed.). Willey Blackwell, Hoboken, New Jersey, 2019.

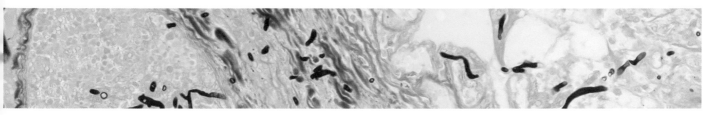

1

INTRODUCTION

Anatomy is the science of the shape and structure of organisms and their parts. Early anatomists recognised that an animal's body is made up of different types of tissues, and with the development of the light microscope, histology – the science that studies the microscopic structure of tissues – became a new field of study. Histology expanded further when varieties of dyes able to stain dissected material specifically were developed. Four basic tissues were described: epithelial tissue, connective tissue, muscle tissue, and nervous tissue. All the various parts of the body are derived from these components, and the distinctive appearance of gross anatomical structures depends on which type of tissue is predominant. Each tissue consists of cells and extracellular components. The balance of cells of different types and their derivatives together with the extracellular matrix and the combination of the different tissues give each part of the body a definitive appearance that can be identified microscopically.

Numerous microscopic techniques are available for studying cells and tissues. The most common of these is the examination of living or fixed dead cells (which can be stained with various dyes) under a light microscope. Fixed dead cells can be examined at a much higher resolution under the transmission electron microscope, and three-dimensional contours of cells can be revealed under the scanning electron microscope.

Before the appearance of the various organs of the body systems can be studied, the four basic tissues must be understood, the embryonic origin identified, and the capacity for growth, regeneration, and repair assessed.

Veterinary science has changed significantly during the past decades. The diverse species examined and cared for by veterinarians has increased from traditional domestic animals bred for food, fibre, work, and human companionship to include many 'exotic' animals, such as ornamental fish, amphibians, and reptiles and laboratory animal species.

Therefore, the variety of tissues that are illustrated and described in this text reflects the diversity of the animals that are now the responsibility of the veterinary profession.

In order to make this text more pertinent within the current clinical milieu, we have added clinical correlate sections (discussed below) that will facilitate comparing normal tissues with diseased tissues and will help students appreciate why it is so important to study and understand histology. Historically, students have wondered why they must learn the myriad number of names and be able to identify the specialised cells and tissue types, but in order to recognise and understand the often subtle changes in tissues that are induced by disease, it is imperative to know what normal tissues look like. Physiological details of some species and the pathophysiology of various conditions are included so that their influence on form and function can be better comprehended. For example, consider the osmoregulatory stresses imposed on teleost fish, which spawn in hypoosmotic fresh water and then must migrate and grow to maturity in hyperosmotic seawater, or the enormous and momentous anatomical, metabolic, and physiological changes that occur during the metamorphosis of amphibian larvae to their adult stage.

It is beyond the scope of this text to cite every abnormal condition known to occur in every organ, in every tissue type, and in every species likely to be examined by a veterinary clinician. Rather, examples of those diseases most likely to be encountered in general and specialised veterinary practices are included.

Embryology – Origin of Tissues

During fertilisation, the secondary oocyte is penetrated by the spermatozoon resulting in the formation of the zygote, which will undergo a series of mitotic divisions until the

DOI: 10.1201/9781003333807-1

blastomeres reach the morula stage. The morula, through the process of blastulation, becomes a blastocyst, which expands to form the bilaminar embryonic disc. In the next phase, the bilaminar embryonic disc is transformed into a trilaminar embryonic disc through the process of gastrulation, which leads to the formation of the three germ layers: ectoderm, mesoderm, and endoderm. The ectoderm will form the central nervous system and the epidermis, the mesoderm will give rise to the majority of the muscles and bones, cartilage, urogenital system, and blood, and the endoderm will form the organs belonging to the digestive and respiratory systems.

Preparation of Tissue Sections

Fixation

Cell and tissues need to be preserved for examination under microscopy. Adequate fixation stops the putrefaction and autolysis of specimens *post mortem* and keeps cells and tissues in a state similar to their living condition. For tissue sections to be evaluated, they must have been fixed or preserved so that their cells and architecture do not decompose after cell death. Generally, a 10% neutral buffered formalin (NBF) solution is employed as a tissue preservative. Formalin is an aqueous-saturated solution of formaldehyde at 40%. For certain tissues, such as adrenal gland, brain, eyes, and a few other structures, special fixative solutions such as Bouin's or Karnovsky's may be preferable. Moreover, tissues from some species like fish show better fixation using alternative fixatives to NBF. Alcohol-based fixatives can be used occasionally depending on the downstream analyses for the samples but are not normally used for histology or histopathology. Usually, specimens of blood and some body fluids containing cells are fixed onto the glass slide with absolute methanol before staining with one of the various dyes. In some cases, such as when supravital staining is used, the stain is applied directly to a specimen without prior fixation.

For fixation, small portions (blocks) of tissue, usually less than 0.5 cm in thickness, are removed from the animal as soon as possible after death and immersed in a special preservative fluid, a fixative (**Figure 1.1**). Delay in fixation after death can lead to serious degenerative changes in the tissue caused by the release of enzymes from the cells. The smaller the sample, the faster the fixative can penetrate the whole block of tissue before degenerative (autolytic) changes occur.

Although many different fixatives are available for different purposes, the most commonly used general-purpose fixative is 10% NBF. It is important to use adequate volumes of fixative: approximately, a minimum of ten volumes of fixative to one volume of sample is enough to ensure good structure preservation. Depending on the temperature, size

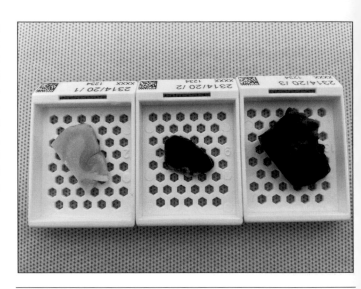

Figure 1.1 Specimens of fixed tissues are placed into disposable plastic cassettes that confine and identify each sample during laboratory processing.

of the specimen, and the type of fixative, fixation time may vary from minutes to days. It is important to note that NBF, the usual fixation solution in the histology and histopathology laboratory, works best at room temperature and the fixation process is compromised at low temperatures (e.g. 4°C).

Fixation kills the cell quickly, stops the *post mortem* degenerative processes, and preserves the structural integrity of the cellular components of the tissue. Soft specimens, with high contents in fat, such as brain, are hardened by fixation, which allows easier manipulation. By coagulating proteins, fixation prevents their leakage from the cells and allows their position to be identified *in situ*. However, fixation can alter the structure of nucleic acid and proteins in the tissues and may affect some staining techniques such as immunohistochemistry (IHC) or *in situ* hybridisation (ISH). Fixation facilitates subsequent processing and staining of the tissue.

Paraffin Embedding

Once the specimen is properly fixed, it is embedded in paraffin wax to support the tissue during the cutting process without altering the morphology of the specimen. The process begins with the removal of the water-based fixative by immersion in a graded series of alcohols of increasing concentration until the tissue is saturated with absolute ethanol (i.e. dehydrated). The specimen is then infiltrated with a clearing agent, such as xylene, that is miscible with both paraffin wax and alcohol. The specimen is infiltrated with warm paraffin wax to replace the xylene (**Figure 1.2a**). The wax hardens as it cools, holding the tissue firmly in place (**Figure 1.2b**). This procedure is usually done automatically in a tissue processor. There are disadvantages to paraffin embedding; it is time consuming and a turnaround time will always be longer than cryopreservation, and the

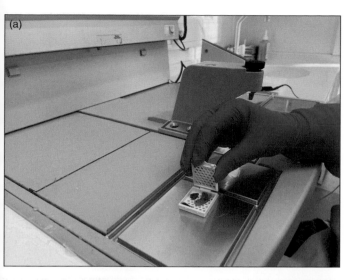

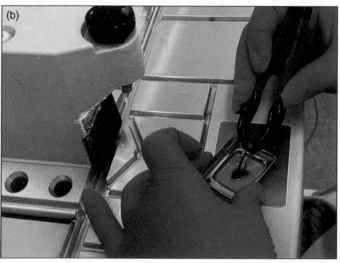

Figure 1.2 (a) Once the tissue has dehydrated, (b) a histology laboratory technician embeds it in a melted paraffin wax–plastic polymer compound.

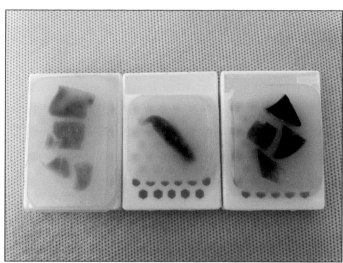

Figure 1.3 Three cassettes containing paraffin-embedded tissue ('tissue blocks') ready for sectioning by a microtome.

necessary to stain the component cells and tissues selectively and make a permanent preparation for examination with the light microscope; a selection of these techniques is described later (*see* the section 'Staining Technique').

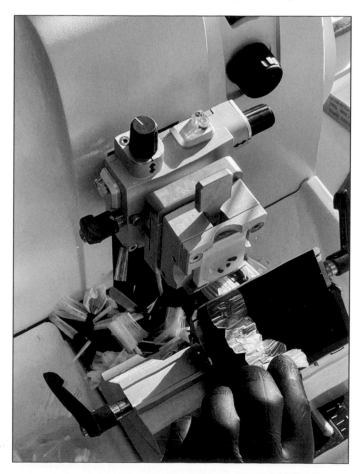

Figure 1.4 Using the adequate microtome blade, the histologist cuts a thin ribbon of paraffin-embedded tissue.

clearing agents are lipid solvents, so this method cannot be used to demonstrate fats (*see* the section 'Freezing').

Once the paraffin-embedded tissue block is cold, excess paraffin is trimmed, and the block is ready for cutting on a microtome (**Figure 1.3**). The block is clamped onto the cutting frame of the microtome and is moved towards the blade of the microtome using an adjustable wheel until the face of the block is against the blade. With each revolution, thin slices (optimally around 4-µm thick) of the block are cut into a ribbon (**Figure 1.4**) and immersed (floated) in a warm water bath. The sections flatten and are floated/fished onto glass slides (**Figure 1.5**). The slides are placed on a warm plate or an oven to dry, and the section adheres to the glass slide (**Figure 1.6**). Removal of the paraffin wax by a suitable solvent, such as xylene, and rehydration allows the tissue to be examined unstained; this has no advantage over the direct examination of living cells. It is

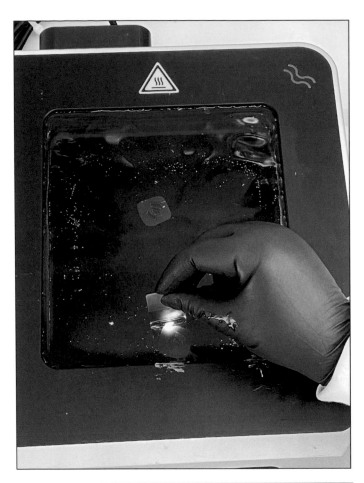

Figure 1.5 The tissue section is transferred to a water bath and is floated onto the glass microscope slide with a fine hair brush. Note the matt black finish of the water bath and retro lighting, which facilitates visualising the nearly transparent tissue section.

Figure 1.6 Slides are dried in an oven at 37°C for a few hours, evaporating the excess water and enhancing the adhesion of the sections onto the glass surfaces, and smooth out irregularities, which is preparatory to xylene clearing.

Decalcification/demineralisation is necessary for tissue with ossified or calcified components before paraffin embedding; otherwise, the hardness of the tissue will result in difficulty in cutting the sections, causing artefacts. Specimens are fixed in formalin or other chemical fixatives and then transferred to the decalcifying solution to allow the removal of the mineral salts. Most of these decalcifying agents contain acids such as formic, glacial acetic, hydrochloric, or nitric or use a chelating agent such as EDTA.

Freezing

A cryostat, a microtome confined to a freezing chamber, is required to cut frozen sections (**Figure 1.7**). These may be from fixed or from unfixed tissue. The advantage of this method is that the time between taking the sample and examining it under the microscope is much reduced. A biopsy may be taken and examined while the patient is still in the operating room. Fat-containing cells retain the lipid content and the tissue is often more lifelike in appearance than non-frozen sections. The disadvantages of this method are tissue distortion, caused by the freezing and thawing, and the necessity of taking thicker sections (normally around 10-μm thick), making them more difficult to interpret. Moreover, the preservation of nucleic acid and protein structures in frozen tissues can be superior to the fixed tissues, making cryosections a good option for downstream analysis included molecular techniques. Once the sections are cut and mounted on glass slides, conventional staining techniques are used.

Consequence of Freezing Unfixed Tissues

When unfixed tissues are frozen and then thawed before being chemically fixed, their delicate cell membranes may become distorted or ruptured, or both, by the forces

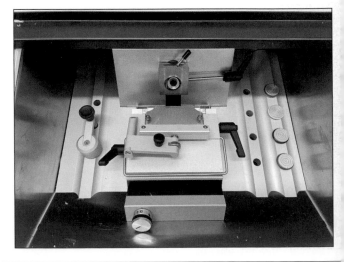

Figure 1.7 Frozen tissue sections are created with the use of a cryostat which is a microtome enclosed within a freezing temperature chamber.

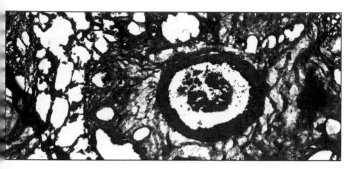

Figure 1.8 Histological section of a lung that was frozen before being fixed in 10% neutral buffered formalin solution. Note the disruption and distortion of the histological architecture and the loss cellular details. H&E ×100.

induced by the expansion and contraction of the intracellular fluid as it freezes and thaws. Therefore, if tissues are to be examined histologically, unfixed specimens must not be frozen. An example of tissue that was frozen before histological fixation and processing is illustrated in **Figure 1.8**.

The rehydrated sections of tissue are now immersed in a solution of one or more stains; any excess stain is removed during this process. The slides are dried again, cleared in xylene, and permanently mounted beneath a glass or plastic coverslip using a mounting medium that is xylene miscible (e.g. DPX). Sections that require special stains are stained and given individual coverslips, as shown in **Figure 1.9**. Sections requiring standard haematoxylin and eosin (H&E) stain are normally stained and coverslipped by automated machines that process the tissues and then dispense an appropriate volume of mounting medium, apply a coverslip, and compress the finished mounted slide to remove any trapped bubbles of air (**Figure 1.10**). The completed stained microsections are placed onto the surface of a warming table beneath a fume hood, where the xylene in the mounting

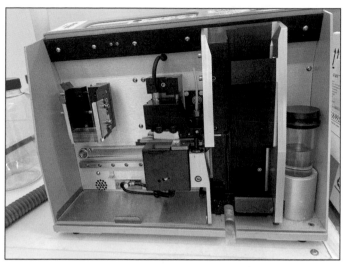

Figure 1.10 When large volumes of slides with standard H&E stain must be coverslipped, an automatic coverslipping machine is normally used.

medium evaporates. This final step fixes the coverslip firmly to the tissue and glass slide, forming a permanent 'sandwich' that can be handled without dislodging any portion of the stained section. Automated slide stainers are frequently used in high-throughput laboratories to decrease the amount of manual work and possible human errors (**Figure 1.11**).

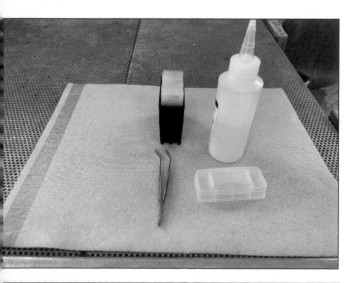

Figure 1.9 Slides that have received special staining are coverslipped manually in a down-draught table or a fume cupboard.

Figure 1.11 Automatic slide stainer used for routine H&E and special stains.

Staining Techniques

Tissue sections are stained with special dyes to enable detailed observations to be made on their structure. The most widely used staining technique is H&E. Haematoxylin stains a deep purple colour and acts as a basic stain (basophilic). Eosin is pink to red in colour and acts as an acid stain (acidophilic or eosinophilic). Haematoxylin reacts with deoxyribonucleic acid and ribonucleic acid, and eosin reacts with cytoplasmic proteins and a variety of extracellular structures. Thus, nuclei and rough endoplasmic reticulum stain blue to purple and cytoplasm stains pink to red depending upon the concentration of the basic and acid components of the cell (**Figure 1.12**).

Numerous different dyes in various combinations are formulated into stains that are used to impart specific and reproducible colouration. Many of these dyes possess positive and negative electrical charges and are attracted or repulsed by electrostatic charges, which are characteristic of certain tissue constituents. In order for some dyes to combine with tissue components, a metallic salt, termed a mordant, is required. The combination of a dye with an appropriate mordant forms a 'lake' and carries a positive electrostatic charge. Dye-mordant combinations with positive charges are cationic and are termed 'basic' stains. These cationic basic lakes combine electrochemically with negatively charged tissue constituents, such as nuclear chromatin, other nucleoproteins, and phosphate groups. Some dyes are inherently basic without requiring the addition of a mordant; they carry their own positive electrostatic charge. Basic fuchsin, toluidine blue, and methylene blue are examples of naturally basic stains. Conversely, anionic or 'acidic' dyes carry a negative or anionic charge

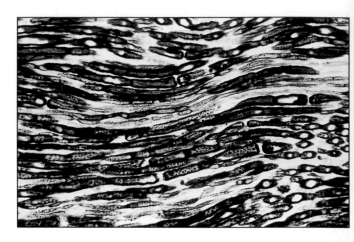

Figure 1.13 Longitudinal section (LS) nerve (dog). The myelin sheath surrounding the nerve fibre reacts with osmic acid and stains black; the supporting connective tissue is unstained. Osmic acid ×250.

and are called 'acidic' because they are attracted to and combine with tissue constituents that possess a positive electrostatic charge. Eosin is an example of an acidic stain. Differential staining is possible because some tissues may be acidic, basic, or amphoteric. Thus, the pH of the extracellular fluid causes their electrostatic charge to vary and, as a result, their acceptance of acidic and basic stains varies.

Many special dye combinations, some requiring rare metallic salts, are used to stain certain tissue types and constituents, microorganisms, metabolic by-products, and so on. Many formularies containing recipe-like staining formulae are available and new staining techniques are continually being developed.

Specialised staining methods are used to illustrate particular features. Osmic acid reacts with fat to give a grey–black colour (**Figure 1.13**), periodic acid-Schiff (PAS) and alcian blue reveal glycosaminoglycans (**Figures 1.14** and **1.15**)

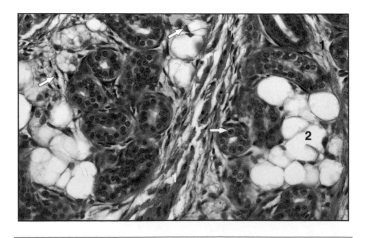

Figure 1.12 Digital pad (dog). The nuclei are stained deep blue (arrows). (1) The cytoplasm and fibres are stained with varying shades of pink with eosin. (2) Adipocytes (fat cells) are unstained as the fat is lost during processing. H&E ×160.

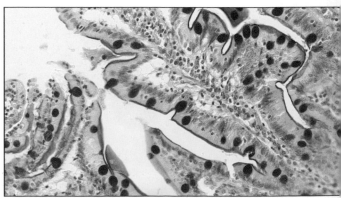

Figure 1.14 Duodenum (dog). The mucus-secreting goblet cells react with PAS (pink). Haematoxylin/PAS ×125.

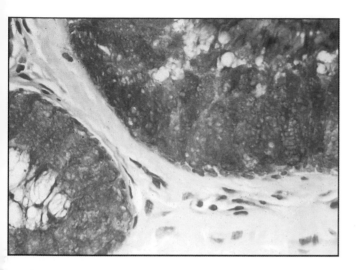

Figure 1.15 Cervix (sheep). The epithelial cells lining the cervix react with either alcian blue or PAS, illustrating chemical differences in the types of mucus secreted. Alcian blue/PAS ×200.

silver impregnation displays reticular fibres and some aspects of nervous tissue (**Figures 1.16** and **1.17**), and Masson's trichrome differentiates between connective tissue and muscle (**Figures 1.18** and **1.19**). Some circumstances require the combination of two or more staining methods to yield the maximum information. These and many other staining methods are available for use in histology and in histopathology to identify pathogens, for example, fungi with PAS or silver impregnation or Feulgen stain to study virus nucleic acid.

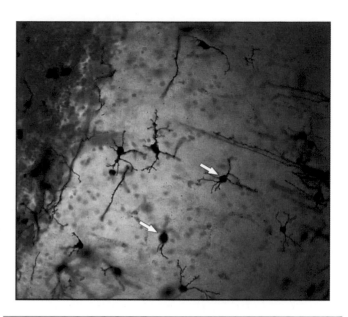

Figure 1.17 Stellate cells in the cerebellum (cat). This method is used specifically to illustrate the cytoplasmic processes of the neurons of the central nervous system (arrowed). Cajal's uranium silver ×250.

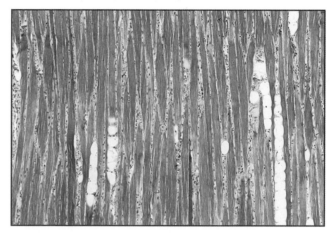

Figure 1.18 Tongue (dog). The muscle is stained red and the connective tissue is stained green. Masson's trichrome ×50.

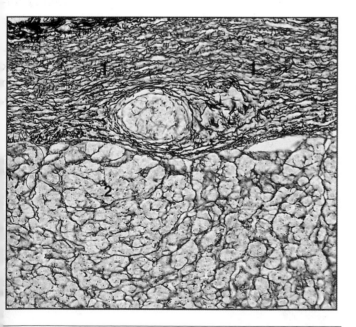

Figure 1.16 Adrenal gland (horse). The reticular fibres form a fine network in (1) the capsule and (2) as a delicate supporting framework for the adrenal secretory cells. The method of Gordon and Sweet for reticular fibres ×125.

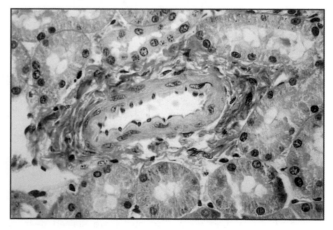

Figure 1.19 Kidney (dog). In this trichrome stain, the connective tissue is stained a blue/green. Gomori's trichrome ×125.

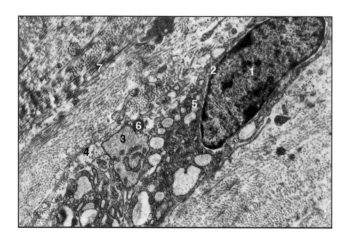

Figure 1.20 Transmission electron micrograph of a fibroblast (sheep). (1) Nucleus, (2) nuclear membrane, (3) cisternae of rough endoplasmic reticulum (RER), (4) plasmalemma, (5) mitochondria, (6) fat droplet, and (7) collagen fibrils ×8000.

Examination of cells and tissues with an electron microscope has necessitated the development of new techniques in preparation procedures to illustrate the arrangement of organelles, membranes, and cell contents (**Figure 1.20**). It has been further refined to provide a three-dimensional picture without distortion (**Figure 1.21**). All of these techniques are now standard tools in histology and have advanced our understanding.

Microscopy

The examination and study of normal cells and tissues by microscopy is called histology or microscopic anatomy. The study of abnormal cells and tissues is histopathology. An understanding of the normal is essential for the recognition of the abnormal. Investigative microscopes range from the simple light microscope to the sophisticated high-resolution electron microscope. In between lies a wide variety of specialised microscopes to meet special needs, such as phase contrast, polarising and fluorescence microscopes, and the scanning electron microscope. Recently, the use of tissue slide scanners has revolutionised how the histological sections are interpreted. Slide scanners convert the glass slides into an electronic file (e-slide) to be viewed with specialised software using a computer and screen, making the process of slide examination more efficient.

Units of Histological Measurement

A micrometre (µm) is equal to a millionth part of a metre and is the unit of measurement of the light microscope; a red blood cell is approximately 8 µm in diameter.

A nanometre (nm) is equal to a billionth part of a metre. The thickness of the basal lamina of an epithelial cell is 70 nm, which can be resolved using the electron microscope.

Light Microscopy

The light microscope is the instrument most commonly used for the visualisation of cells and tissues. With it, magnifications of up to 2000 times are possible. The limit to the size of the structure that can be distinguished with the light microscope is limited by the physical nature of light. The wavelength of visible light ranges from 0.4 to 0.7 µm. Therefore, even with the best optical system available, the resolution, or resolving power, of the light microscope is limited to 0.2 µm, and anything smaller than that will not be clearly distinguished.

In order to achieve the best results, a few basic preliminary checks must be made.

- Ensure that the glass slide is clean and free from dust and smears.
- Ensure that the microscope condenser, objectives, and ocular lenses are clean – take great care to clean the microscope with soft lens tissues.
- Set the microscope up for critical illumination for each objective by:
  1. closing the iris diaphragm (the substage condenser diaphragm),
  2. adjusting the condenser until the circular area of illumination has a sharp edge, and
  3. making sure that the condenser is centred by using the adjusting screws.

Always begin with the lowest objective and increase the magnification slowly.

Digital slide scanners are, in essence, microscopes attached to a computer that converts the optical analogical image as seen with a light microscope into a digital file to be viewed by

Figure 1.21 Scanning electron micrograph of the kidney (dog). (1) Renal tubule, (2) interstitial connective tissue, and (3) free erythrocyte – a biconcave disc with the typical indentation ×675.

a viewing software. The magnification used to digitalise glass slides is similar to that used in light microscopy. However, digital zoom can be used to gain more magnification.

Transmission Electron Microscopy

This microscope uses an electron beam instead of a light source and allows the resolution of structures as small as 1 nm. Small pieces of tissue (cubes of around 1 mm on a side) are fixed rapidly (to avoid artefacts induced by tissue degradation) in a glutaraldehyde-based fixative, dehydrated and embedded in epoxy resins. Sections are cut at around 50 nm on an ultramicrotome using a glass or a diamond knife, mounted on metal grids and stained with heavy metal solutions such as lead citrate and uranyl acetate. The fixative and embedding solutions used in electron microscopy are also hazardous and specialised equipment and cabinets are necessary.

Scanning Electron Microscopy

Solid pieces of tissue fixed in a glutaraldehyde-based fixative are dried, coated with gold, and placed in the microscope. The electron beam scans the specimen, and a three-dimensional representation of the surface is obtained.

Artefacts Induced by Histological Processing

The preparation of tissue sections involves a number of stages during fixing, dehydrating, paraffin embedding, sectioning, deparaffinising, rehydrating, staining, and coverslipping. Each of these processes necessitates the manipulation of tissue specimens and laboratory reagents, thus providing opportunities for artefacts to appear in the final stained slide. Some of the common artefacts are illustrated in **Figures 1.22–1.26**.

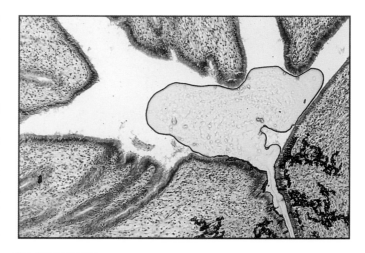

Figure 1.23 Uterus (cat). Shrinkage of the adhesive medium used to mount the coverslip to the slide captures air and causes bubbles. H&E ×65.5.

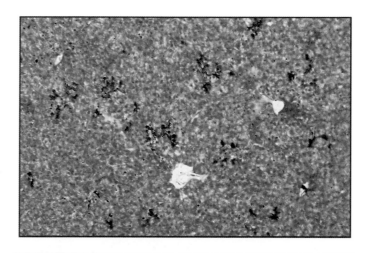

Figure 1.24 Spleen (bird). When crystals accumulate in the stain solutions or are not removed during standard processing, stain deposits precipitate onto the surfaces of the tissue section. H&E ×25.

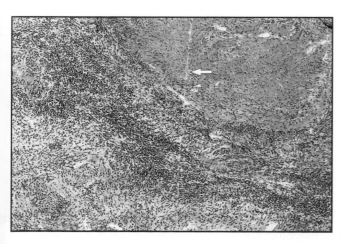

Figure 1.22 Ovary (sheep). A knife mark, caused by a nick in the microtome's cutting edge, leaves a straight line across the section (arrowed). Masson's trichrome ×25.

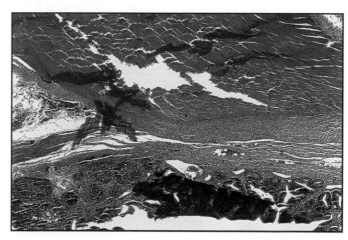

Figure 1.25 Cloacal bursa (bird). Raised areas, overlapping folds, and cracked and separated tissue are present because it is often difficult to flatten the tissue completely, particularly in very thin sections. H&E ×125.

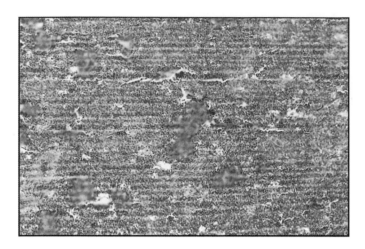

Figure 1.26 Spleen (dog). Compression of the paraffin-embedded tissue causes parallel 'chatter' marks. H&E ×25.

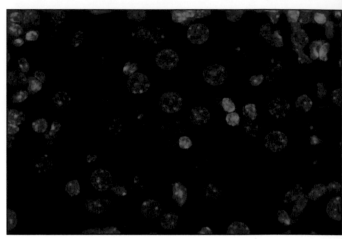

Figure 1.28 Immunofluorescence. Staining of *Leishmania infantum* antigens (red fluorescence) in the liver from an infected dog. The cell nuclei (mostly from hepatocytes in the image) are stained with DAPI (blue fluorescence) ×400.

Immunohistochemistry and *In Situ* Hybridisation

IHC is a special staining technique used to detect specific materials in the tissue section, normally those protein-aceous in nature. This technique uses specific antibodies targeting those proteins in the tissue to be detected and analysed by the histologist or the pathologist. The antigen–antibody formed complex is visualised using an enzymatic reaction and a colour substrate to be detected under light microscopy. It is a widely used technique to study cell populations, using antibodies targeting cell markers (e.g. clusters of differentiation or CDs), protein present in pathogens (e.g. viral or bacterial proteins), and it is very useful in the veterinary pathology laboratory to aid in the final diagnosis of clinical cases (**Figure 1.27**). The immunolabelling can be performed using antibodies tagged with fluorescent dyes (IF) to be analysed under a fluorescence microscope (**Figure 1.28**). Multiplex IHC or IF slides can be analysed using special microscopes or digital scanners (e.g. confocal microscopes).

ISH is another special technique targeting nucleic acids (DNA or RNA) within tissue sections. ISH uses specific probes to reveal a sequence of nucleic acid within the tissue, amplifying the signal to be visualised under a light or fluorescence microscope. This technique is commonly used to detect nucleic acid from pathogens or messenger RNA to perform a transcript analysis *in situ* (**Figure 1.29**).

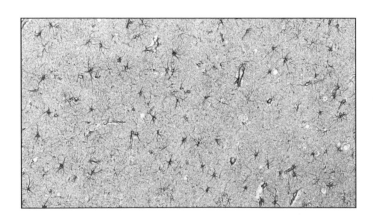

Figure 1.27 Immunohistochemistry. Immunostaining of glial fibrillary acid protein (GFAP) used as a marker of astrocytes in a dog brain. Positive staining is brown, and the tissue is counterstained with haematoxylin ×200.

Figure 1.29 Fluorescent ISH. A specific probe was used to hybridise with *Mycoplasma hyopnemuniae* DNA, tagged with a green fluorescent dye. Positive staining can be observed within the bacteria attached to the apical border of bronchiolar epithelial cells in an infected pig ×300.

Modern Histology and Histopathology Techniques

New techniques have been developed over the past years to study the cell and tissue composition in the normal and the abnormal. Among them, laser-capture microdissection (LCM) is a technique that can harvest cells or tissue structures directly from a tissue section and isolate them into a container to study the molecular composition in downstream analyses like quantitative polymerase chain reaction (PCR) or RNA-sequencing.

Virtual slides (or e-slides) at a very high resolution are often used in the modern histology and histopathology laboratory. Special software is used to drive and view the slides, changing magnification, being able to capture static images, and, more importantly, being able to analyse quantities of cells within a structure, interactions among neighbouring cells, and more complicated analyses involving artificial intelligence, like pattern recognition.

CLINICAL CORRELATES

In order to appreciate the often subtle alterations that accompany disease or other physical abnormalities, it is useful to compare the characteristic changes by which histopathological diagnoses are made and classified. To that end, clinical correlates sections are inserted throughout this text. It is important to note that in many instances the tissues comprising an organ of one species are similar or even identical to those found in a different species.

Generally, there are fewer substantive differences within a phylogenetic group of animals than between different groups of animals. For instance, the livers of sheep, cattle, horses, swine, dogs, and cats are relatively quite similar, although some slight differences are obvious, e.g. the physiological presence of fibrous tissue bridging portal spaces in pig liver, that would be considered pathological in dogs or horses. The liver tissues of many fish resemble the hepatic tissue found in amphibians, and the hepatic tissues of many reptiles resemble those observed in birds. Because of these characteristic similarities and differences, we have selected examples of tissues that are particularly instructive in order to avoid showing repetitively the same tissues for every animal, irrespective of its phylogeny. However, examples from a wide variety of species are included for purposes of comparison.

These correlates are placed where they most readily illustrate specific, clinically significant conditions. Recognising normal tissue facilitates interpreting the often subtle alterations in abnormal tissues. Where appropriate, the physiological attributes or significance, or both, of a particular organ or structure are discussed briefly so that their importance to the survival of the animal becomes apparent.

Regarding staining techniques, many of the stains used to study the normal histological architecture of tissues can be used to study particular alterations in pathological specimens. For example, an increase or decrease in the production of mucous can be studied using PAS stain. Some of the histochemical techniques can be very valuable to identify pathogens, for example, silver impregnation is widely used to identify fungal hyphae in tissue sections (**Figure 1.30**). IHC using antibodies against pathogen antigens (**Figure 1.31**) or ISH using probes targeting pathogen nucleic acid (**Figure 1.32**) is frequently used in the diagnostic and histopathology laboratory.

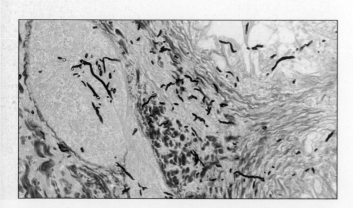

Figure 1.30 Fungal hyphae (black stain over a green counterstain) within the rumen from a cow. Grocott's methenamine silver stain ×300.

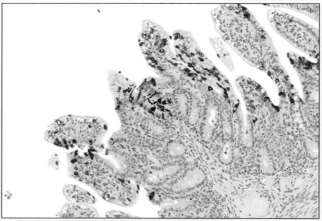

Figure 1.31 Immunohistochemical staining (brown stain over a blue haematoxylin counterstain) of porcine coronavirus antigen in the small intestine from a piglet with acute enteritis ×200.

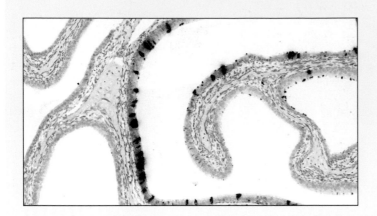

Figure 1.32   *In situ* hybridisation staining (red stain over a blue haematoxylin counterstain) of SARS-CoV-2 RNA in the nasal cavity from a ferret with SARS-Coronavirus-2 ×200.

2

THE CELL

The cell is the basic unit of a living structure, containing proteins, carbohydrates, lipids, nucleic acids, and inorganic material. Cells contain several subcellular structures, or organelles, which possess specific function and structure.

Cells are limited by a cell membrane, the plasmalemma, acting as a barrier between the inside and the outside (**Figure 2.1**). The cell membrane is composed of a double layer of phospholipids with embedded proteins and carbohydrates. Cells can attach to each other by membrane specialisations called junctional complexes, which can be classified into occluding, anchoring, and communicating junctions. Occluding junctions form a barrier preventing materials from crossing the intercellular space (i.e. *zonula occludens*). Anchoring junctions maintain the adherence between cells or between cells and basal lamina (i.e. *fascia adherens*, *zonula adherens*, *macula adherens* or desmosome, and hemidesmosome) (**Figure 2.2**). Communicating junctions allow ions or signalling molecules to pass between adjacent cells (i.e. gap junctions).

Within the cell lies the membrane-bound nucleus (**Figure 2.3**). The nucleus is the fundamental component of the cell, and is surrounded by a double membrane, the nuclear membrane or envelope, and sequesters the cell complement of deoxyribonucleic acid (DNA), which consists of two long strands wound together in a double helix. The DNA is organised into chromosomes, which carry the genetic information: the genes. Chromosomes are rarely visible, except during cell replication or mitosis (**Figure 2.4**) and where protein–DNA complexes are seen as chromatin. The small, darkly staining bodies within the nucleus in an H&E stained section are the

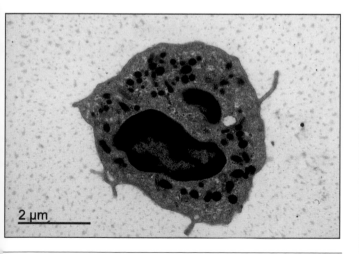

Figure 2.1 Neutrophil (Transmission Electron Microscopy, TEM). The cytoplasm is surrounded by the cell membrane. The nucleus is visible containing dark electrodense chromatin, and different organelles are observed within the cytoplasm. The neutrophil is shrunk due to processing. The typical cell size is 12 microns.

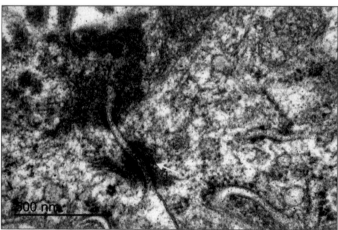

Figure 2.2 Desmosome (TEM) (arrow). Anchoring junction between two epithelial cells observed as dark electrodense material.

DOI: 10.1201/9781003333807-2

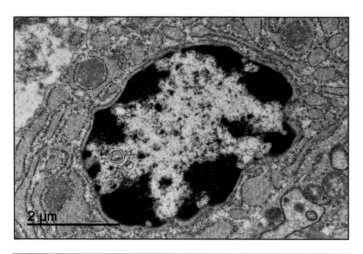

Figure 2.3 Nucleus from a plasma cell (TEM).

nucleoli. Ribonucleic acid (RNA) is present within the nucleoli together with DNA.

Cytoplasm surrounds the nucleus and is bound by the cell membrane. The cytoplasm adjacent to the nucleus is called cytocentrum and is normally free of organelles except for the endoplasmic reticulum. Adjacent to the cytocentrum is the largest part of the cytoplasm, the endoplasm, which contains the majority of cytoplasmatic components. Adjacent to the cell membrane is the narrow cytoplasmatic band, the ectoplasm, with few organelles.

The rough endoplasmic reticulum (rER) is closely associated with the nucleus (**Figure 2.5**), and its membrane is a continuation of the nuclear membrane. rER is full of ribosomes, and the main function is the synthesis and assembly of proteins (**Figure 2.6**). The smooth endoplasmic reticulum (sER) has not got ribosomes attached, and its functions include steroid hormone production, synthesis of

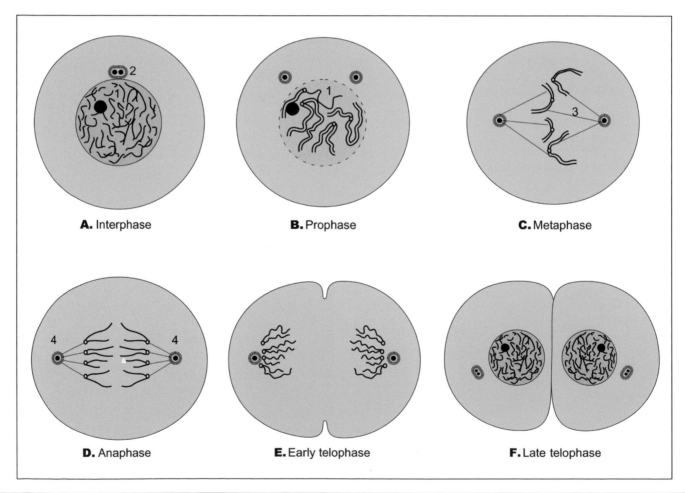

Figure 2.4 The stages of mitotic division. **(A) Interphase.** (1) Nucleus. (2) Centromere. **(B) Prophase.** The granular appearance of the nucleus (1) is the early condensation of the nuclear chromatin in preparation for division. **(C) Metaphase.** The short, compact chromosomes are arranged around a central spindle. **(D) Anaphase.** The centromere has divided and two chromatids for each chromosome have separated and moved towards each centromere (4). **(E) Early telophase.** The separation of the daughter chromosomes is complete and the cytoplasm begins to divide. **(F) Late telophase.** The nucleus of each daughter cell is reconstructed with a nuclear membrane and a nucleolus; the chromosomes are no longer visible. The cytoplasm divides.

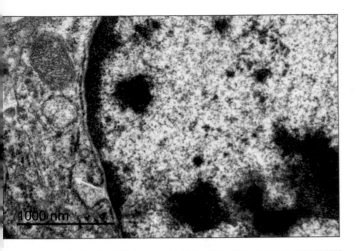

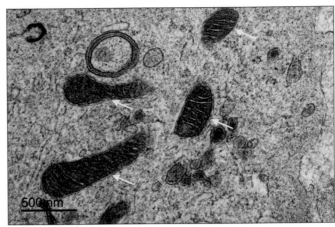

Figure 2.5 Rough endoplasmic reticulum (rER; arrow) adjacent to the nuclear membrane (TEM).

Figure 2.7 Mitochondria (arrows) within the cytoplasm of a macrophage (TEM).

carbohydrates, and lipid complexing from fatty acids. The ER structure costs of interconnecting membrane-bound sacs.

The Golgi apparatus (or Golgi complex) is associated with the ER as further processing of the synthesised material takes place, representing the end of the protein synthesis chain. The Golgi apparatus has a cup-shaped structure with flattened membranes (cisternae) arranged in clusters of parallel sheets. Each cluster is called dictyosome. The distal part of the Golgi apparatus is where the mature secretory vesicles are formed.

There are organelles in charge of degrading material, the lysosomes, normally spherical and bounded by a single membrane. Lysosomes are numerous in phagocytic cells like macrophages. Peroxisomes are also membrane-bound vesicles that contain oxidising enzymes, including peroxidase and catalase involved in the breakdown of fatty acids and the regulation of some oxidising reactions in connection with the mitochondrion.

Mitochondria have a sophisticated function in the cell, providing energy to the rest of the cell components, and are present in all eukaryotic cells except erythrocytes (**Figure 2.7**). The number of mitochondria in a cell depends on the energy needs of the cell. They have a double membrane and small amounts of DNA are also present.

Many cells also store glycogen or lipids as potential energy reserves.

The cytoskeleton is the main support structure of the cell, composed of filaments of various sizes, including (from thinner to thicker) microfilaments (e.g. actin), intermediate filaments (e.g. cytokeratin, desmin, or glial fibrillary acid protein), and microtubules.

There are two major types of cell death, necrosis and apoptosis, being the first normally subsequent to a significant insult and damage, and apoptosis, a result of a well-regulated self-orchestrated process, often called programmed cell death (**Figure 2.8**).

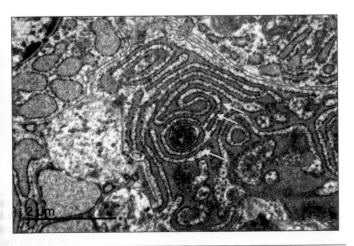

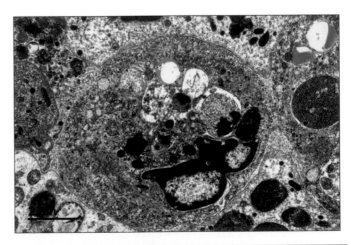

Figure 2.6 Rough endoplasmic reticulum (rER), full of ribosomes (arrows) in a plasma cell producing abundant proteins, including antibodies (TEM).

Figure 2.8 Apoptotic macrophage, showing condensation and fragmentation of the nuclear chromatin, cell shrinkage, and plasma membrane blebbing (TEM).

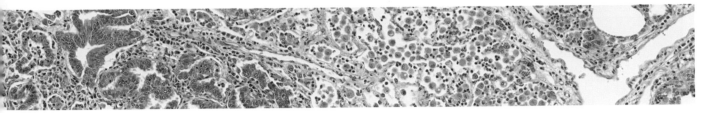

3

EPITHELIAL TISSUE

Epithelial tissue can be divided into two types, surface epithelium and glands (or glandular epithelium). Surface epithelium consists of sheets of cell aggregates covering the internal and external surfaces of the organs. Glandular epithelium consists of specialised secretory (endocrine and exocrine) cells that proliferate into underlying connective tissue.

Epithelium

Epithelium is the term used for all the covering and lining membranes of the body. It is composed of contiguous cells linked by cell junctions and resting on a specialised matrix, the basement membrane. All epithelia are avascular and are supported and nourished by the underlying connective tissue capillary bed via diffusion through the basement membrane. They are derived from all three basic germ layers (ectoderm, endoderm, and mesoderm). The ectoderm provides the nervous system, the outer layer of the skin, and the epidermis, and the endoderm provides the lining of the respiratory and digestive tracts. Both ectoderm and endoderm grow into the underlying embryonic connective tissue (mesenchyme) and form exocrine and endocrine glands. Exocrine glands secrete onto the surface of the epithelial membrane through a system of ducts. Endocrine glands are ductless; islands of secretory cells embedded in connective tissue secrete into the local capillary bed and thus directly into the blood to be carried to the target organ.

Epithelium of mesodermal origin forms a thin squamous membrane lining the pleural, peritoneal, and pericardial cavities of the body. The mesodermally derived epithelium lining the heart, blood, and lymphatic vessels is called endothelium.

The urogenital system is derived from mesoderm, and the epithelial membranes of most of the genital system, the kidneys, and ureters are of mesodermal origin. All epithelial membranes are capable of regeneration and repair.

Damaged and dead cells are replaced by adjoining cells to maintain the cover and the integrity of the membrane.

Epithelium may be either simple, where a single layer of cells is present, or stratified, where a variable number of cell layers are superimposed. There are two other types of epithelia that can be classified as specials by their morphological characteristics: pseudostratified and transitional.

Simple Epithelium

Squamous

Simple squamous epithelium is a single continuous layer of flattened cells, which is often so attenuated that it is difficult to identify the boundaries of individual cells using conventional light microscopy. The nucleus bulges from the thickest part of the cell. This epithelium can be found in the internal lining of blood and lymph vessels, which is called endothelium (**Figure 3.1**), the mesothelium of

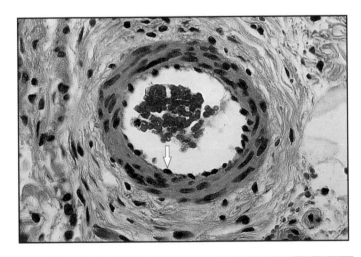

Figure 3.1 Simple squamous endothelium. Uterus (cat). Arteriole. Simple attenuated squamous cells line the lumen; the nucleus of one cell is arrowed. H&E ×250.

DOI: 10.1201/9781003333807-3

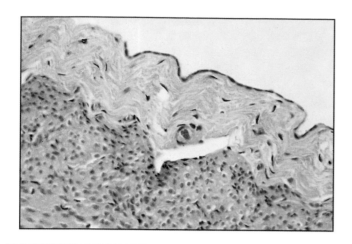

Figure 3.2 Simple squamous mesothelium. Uterus (cat). The simple squamous cells are on the free serous surface of the uterus. H&E ×160.

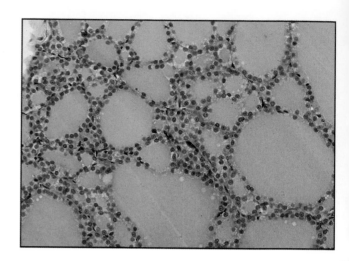

Figure 3.4 Simple cuboidal epithelium. Thyroid (dog). The simple cuboidal epithelium lines the colloid-filled thyroid follicles. H&E ×200.

the body cavities (**Figure 3.2**), pulmonary alveoli, the loop of Henle, or parietal layer of Bowman's capsule in the kidney.

Cuboidal

Simple cuboidal epithelium is a single layer of polygon-shaped cells; each cell is square in cross-section with a central round nucleus. Minor variations in proportion may occur to give short cuboidal and tall cuboidal cells. Examples can be found covering ducts of many glands, kidney tubules, ovary, or thyroid and mammary glands (**Figures 3.3–3.5**).

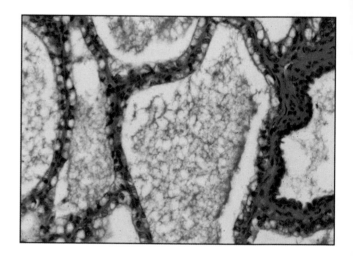

Figure 3.5 Simple cuboidal epithelium. Lactating mammary gland (cow). The secretory alveolus of the mammary gland is lined by simple cuboidal cells. H&E ×125.

Columnar

Simple columnar epithelium is a single continuous layer of tall rectangular cells with a basal ovoid nucleus forming a relatively thick membrane. These cells are often specialised, performing a particular function. In secretory epithelium, the cells secrete mucus and have a lubricant and protective function; examples can be found in the stomach and cervical canal (**Figures 3.6** and **3.7**). In the small intestine, the luminal surface area is markedly increased by microvillous processes to form a striated border, a functional adaptation designed to increase the surface area for absorption. Adjoining goblet cells secrete mucus, keeping the membrane moist and protecting against the luminal contents (**Figures 3.8** and **3.9**). In the uterus, oviducts, efferent ductules, and small bronchi, this epithelium project cilia from the apical surface (**Figures 3.9** and **3.10**).

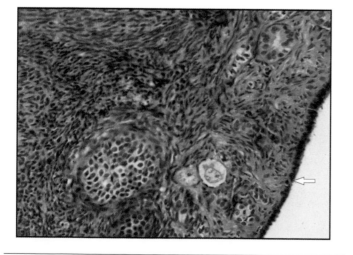

Figure 3.3 Simple cuboidal epithelium. Ovary (sheep). The simple cuboidal epithelium on the free surface of the ovary is arrowed. H&E ×50.

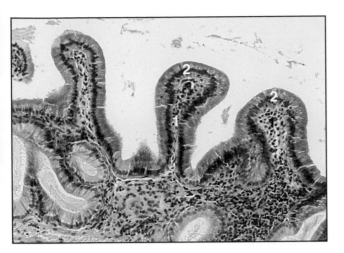

Figure 3.6 Simple columnar epithelium. Gall bladder (dog). (1) Connective core of the lamina propria. (2) Tall epithelial cells with a basal nucleus. H&E ×125.

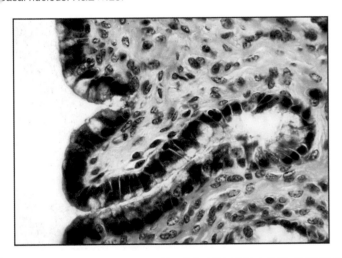

Figure 3.7 Simple columnar epithelium. Cervix (sheep). The epithelium lining the cervix secretes mucus, stained green. Masson's trichrome ×200.

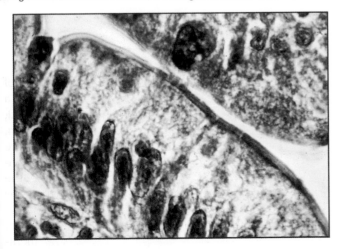

Figure 3.8 Simple columnar epithelium. Duodenum (dog). The brush (striated) border appears as a dark line on the luminal surface; the single mucus-secreting goblet cell is stained purple. Haematoxylin/periodic acid-Schiff (PAS) ×500.

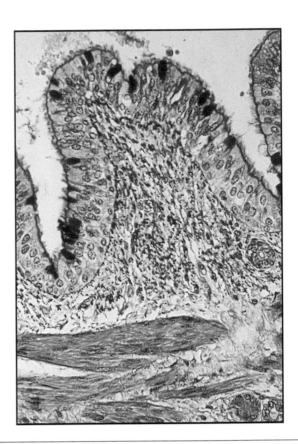

Figure 3.9 Respiratory epithelium. Lung (cow). The mucus-secreting goblet cells are individually stained (blue/purple). Gomori/aldehyde fuchsin ×100.

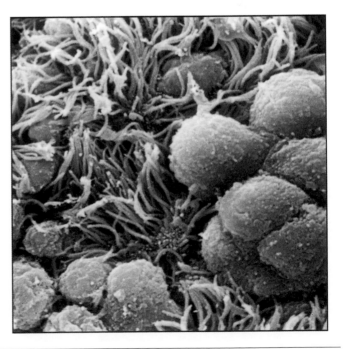

Figure 3.10 Respiratory epithelium. Lung (donkey). In this scanning electron micrograph, the mucus secretion is a bulbous projection surrounded by cilia ×2500.

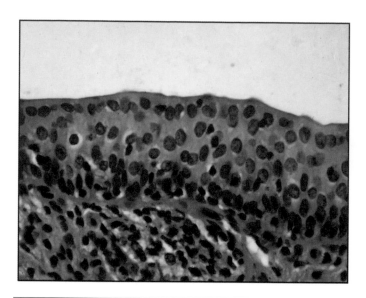

Figure 3.11 Stratified columnar epithelium. Penile urethra (horse). The epithelium is several layers deep; the superficial layer is columnar. H&E ×400.

Stratified Epithelium

Designed to withstand wear and tear, stratified epithelia consist of two or more layers of cells with only the basal layer resting on the basement membrane. Stratified columnar and cuboidal epithelia are found lining large gland ducts and the urethra (**Figures 3.11** and **3.12**). They are usually not suitable for absorption and require gland secretion to keep the surface moist, but the epithelium lining the rumen is absorptive (*see* Chapter 9, Digestive System).

Squamous

Stratified squamous epithelium may be keratinised or non keratinised. The latter is common on surfaces subject to wear and tear, where the secretions necessary to keep the surface wet come from associated glands and is found

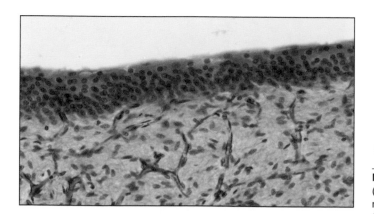

Figure 3.12 Stratified cuboidal epithelium. Urethra (sheep). The surface layer of cells is cuboidal. H&E ×400.

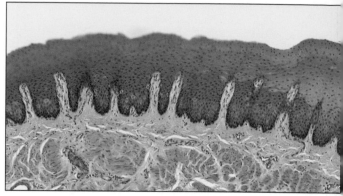

Figure 3.13 Stratified squamous non keratinised epithelium. Oesophagus (sheep). The polyhedral cells of the basal layer divide and the daughter cells are pushed towards the surface where the outer layers (dead squames) are shed. H&E ×150.

lining the mouth, oropharynx, oesophagus, and vagina (**Figure 3.13**). The basal cell layer, cuboidal in shape, is mitotically active, and the new cells are continually formed and pushed towards the surface, moving away from the nourishing capillary bed beneath the epithelium. These cells are dead or dying by the time the surface is reached. There they lose their nuclei and become detached (desquamate); only on the surface layers are the cells squamous. In some sites, such as the epidermis and the tongue, the cells become keratinised and form a protective waterproof layer on the surface (**Figures 3.14–3.16**).

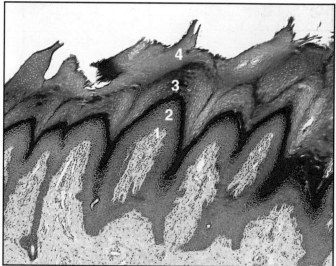

Figure 3.14 Stratified squamous keratinised epithelium. Footpad (dog). (1) Stratum germinativum; the basal layer of simple columnar cells. (2) Stratum spinosum; several layers of pear-shaped cells. (3) Stratum granulosum; layer of cells deeply stained containing keratohyalin granules. (4) Stratum corneum; multilayered zone of anucleate squames. H&E ×100.

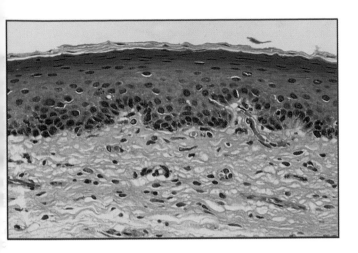

Figure 3.15 Stratified squamous keratinised epithelium. Skin (cow). Relatively fewer layers of cells; the surface is covered with keratin. H&E ×100.

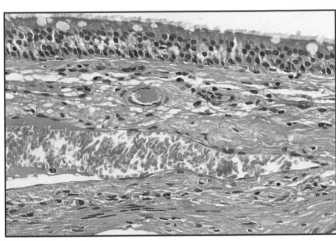

Figure 3.17 Pseudostratified ciliated columnar epithelium with goblet cells; respiratory epithelium lining the nares (horse). The cilia appear as a fringe on the free surface, the goblet cells, as clear rounded spaces. There are several layers of nuclei, but all the cells rest on the basement membrane. H&E ×100.

Figure 3.16 Stratified squamous epithelium. Teat (cow). The stratum lucidum is present in thick skin as a clear translucent layer. Phosphotungstic acid haematoxylin ×100.

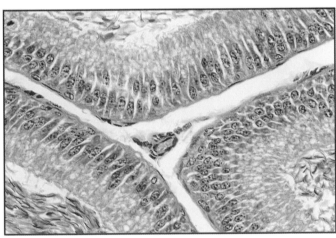

Figure 3.18 Pseudostratified epithelium. Epididymis (bull). Several rows of nuclei create the appearance of a stratified epithelium, but all the cells rest on the basement membrane. The stereocilia give a fringe effect to the luminal surface. H&E ×200.

Special Types of Epithelium

Pseudostratified Columnar Epithelium

Pseudostratified columnar epithelium appears to consist of more than one layer of cells. All the cells are in contact with the basement membrane, but not every cell reaches the luminal surface. The nuclei lie at different levels, causing the stratified appearance. This type of membrane is seen in the respiratory tract (**Figure 3.17**) and in the genital tract (**Figure 3.18**), where cells may be secretory or ciliated.

Transitional Epithelium (Urothelium)

Urothelium lines most of the renal pelvis, the ureters, and the urinary bladder and is designed to allow stretching of the membrane without rupture. It can be classified as pseudostratified because in the relaxed, unstretched state a number of layers of cells are present. The basal cells are cuboidal, and above them, several layers of polyhedral cells can be found. The surface cells are dome-shaped and occasionally binucleated, but when the epithelium is distended become flattened (**Figure 3.19**). In the stretched state, the appearance is that of stratified squamous epithelium. The cells have the ability to stretch and distort, without pulling apart, and are ideally suited to the demands of the bladder and ureters. The surface is thickened and gives a waterproof coating.

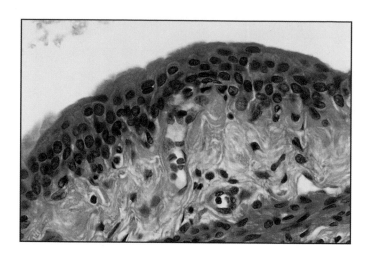

Figure 3.19 Urothelium. Urinary bladder (dog). The surface cells have a rounded appearance, the cells of the middle layer are pear shaped, and the basal layer is columnar. H&E ×400.

Glands

All glands are derived from either ectoderm and mesoderm or endoderm and mesoderm. The ectoderm and endoderm form the epithelial secretory cells, the parenchyma. The mesoderm forms the supporting connective tissue framework, the stroma. Where the demand for secretion is low, a single secretory cell is sufficient. The mucus-secreting goblet cell of the small intestine, for example, is adequate. At the other extreme is the liver, the largest gland in the body, which is required to cope with the food absorbed by the intestines.

Exocrine Glands

Exocrine glands secrete onto a body surface directly or by way of a duct. They are classified according to the nature of their secretion (mucous, serous, or mixed), their mode of secretion (merocrine or holocrine) and the number of cells (unicellular or multicellular).

In the merocrine gland, the vesicles containing the secretion in the cytoplasm fuse with the cell membrane and release the contents onto the cell surface (e.g. parotid salivary gland). In the holocrine gland, the cell builds up the secretion in the cytoplasm, migrates away from the basement membrane and the source of nutrient, and dies. The cell debris itself becomes the secretion (e.g. sebaceous glands (**Figure 3.20**)).

Multicellular glands can be classified as simple (cells secreting into a duct opening into the lumen) or compound (branched duct system draining several secretory units). The connective tissue capsule extends into the gland, carrying blood vessels and nerves, and divides it into lobes and lobules. The secretory units may be tubular (elongated), acinar (spherical or flask-shaped), or tubuloacinar (a secretory tubule and a cluster of acini that open into the tubule) (**Figures 3.21–3.24**).

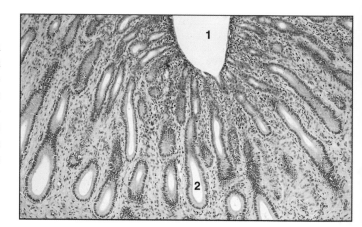

Figure 3.20 Sebaceous gland. Skin (dog). The gland consists of pale staining cells filled with sebum, fatty substance. This forms the secretion, an example of a holocrine gland. Gomori's trichrome ×125.

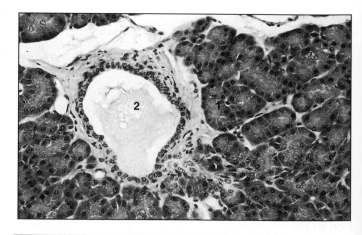

Figure 3.21 Simple tubular glands. Uterus (cat). (1) Lumen of the uterus. (2) Simple tubular glands in the endometrium of the uterus. H&E ×20.

Figure 3.22 Compound acinar gland. Pancreas (dog). (1) The acinus is lined by secretory epithelial cells with a basal nucleus. (2) The excretory duct is lined by a stratified cuboidal epithelium. H&E ×250.

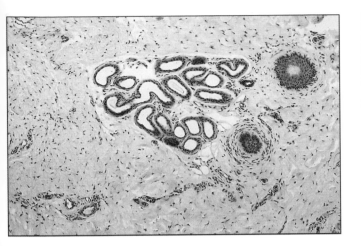

Figure 3.23 Alveolar gland. Carpal skin (pig). The secretory alveoli are cut in cross-section; the diameter of the lumen exceeds the height of the lining secretory cells. H&E ×100.

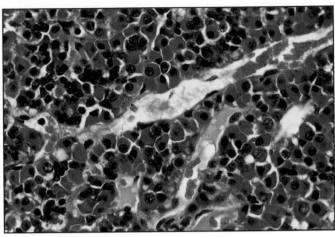

Figure 3.25 Pars distalis of the adenohypophysis (pituitary gland; cat). The secretory cells are closely associated with a rich network of blood vessels. H&E ×400.

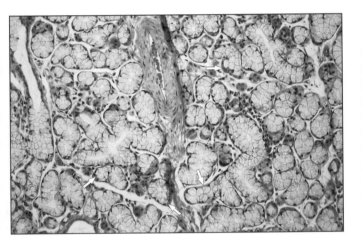

Figure 3.24 Compound tubuloacinar gland. Mixed seromucous salivary gland. The pale staining mucus-secreting cells are filled with secretion and almost obliterate the lumen. The serous cells form a darkly stained demilune around the mucous cells (arrowed). A thick strand of connective tissue with a blood vessel represents the supporting framework of the gland. H&E ×125.

Myoepithelium

Myoepithelial cells, derived from ectoderm and endoderm, are found in sweat and mammary glands and lie between the secretory epithelial cell and the basement membrane. Myoepithelial cells are also found in the modified salivary (venom) glands and ducts of venomous snakes. These epithelial cells have some characteristics of smooth muscle cells; their cytoplasm contains myofilaments and are capable of contraction, thus assisting the expulsion of the secretions from these structures (**Figures 3.26–3.28**).

Endocrine Glands

Endocrine glands have no ducts. Small groups of cells secrete hormones, chemical messengers, into the capillary network, lymphatic vessels, or tissue fluid for transmission to the target cell or organ at a variable distance away (**Figure 3.25**). Secretory cells can be organised in cords or follicles. A detailed discussion of these glands can be found in Chapter 11.

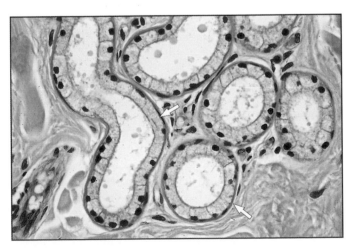

Figure 3.26 Sweat gland. Skin (horse). Myoepithelial cells lie between the simple columnar secretory epithelium and the basement membrane and appear as a deep pink line (arrowed). H&E ×300.

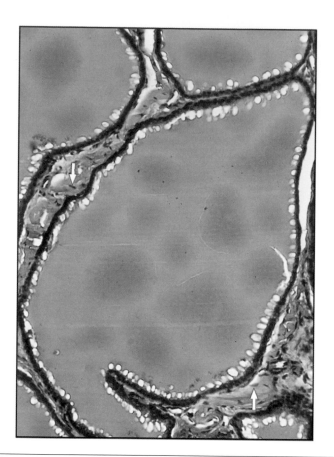

Figure 3.27 The paired venom glands of many venomous snakes [and the four pairs in the single genus of venomous lizards (*Heloderma*)] are composed of thin-walled, follicle-like structures lined by a single layer of non keratinised squamous-to-plump cuboidal epithelial secretory cells. These dilated follicles store the pink staining, protein-rich venom until it is delivered via the duct system and fangs. Contractile myoepithelial cells (arrowed) and skeletal muscle aid in the expression and delivery of venom through the coiled ducts and into the hollow fangs of snakes (and to external grooves in the solid teeth of Helodermatid lizards). Illustrated is a section of the venom gland of a small Mexican rattlesnake (*Crotalus enyo*). H&E × 85.

Figure 3.28 The much-coiled venom duct of this rattlesnake is thin-walled and is lined by low cuboidal cells with dark staining basal nuclei. The coiled sections of the duct are separated from each other by connective tissue in which many myoepithelial cells (arrowed) are embedded. The entire duct is surrounded by the temporal and masseter skeletal muscles, which, when they contract, augment the myoepithelial cells in forcing venom to and through the fangs. H&E ×400.

CLINICAL CORRELATES

A large variety of benign and malignant epithelial tumours are recognised in domestic animals. A squamous cell carcinoma, a malignant tumour of squamous epithelium taken from the eyelid of a cow, is shown in **Figure 3.29**. The neoplastic cells are large with abundant eosinophilic cytoplasm and form nests where cells differentiate and keratinise towards the centres. There is variation in nuclear and cellular size and mitotic figures can be seen.

Bovine ocular squamous carcinoma is relatively common. Its occurrence is related to ultraviolet light exposure, particularly in white-faced breeds such as Herefords. The tumour develops through premalignant stages before progressing to carcinoma *in situ* and finally to invasive carcinoma.

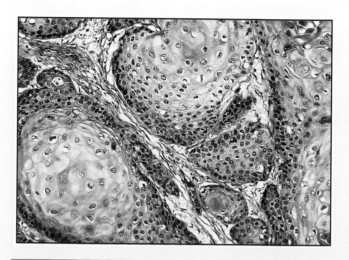

Figure 3.29 Squamous cell carcinoma from the eyelid of a cow. H&E ×200.

An ovine pulmonary adenocarcinoma, a contagious lung malignant tumour of epithelial origin caused by retrovirus infection (jaagsiekte sheep retrovirus), is shown in **Figure 3.30**. Cuboidal to columnar tumoural cells grow forming acinar to papillary structures. The alveoli adjacent to neoplastic lesions appear full of alveolar macrophages.

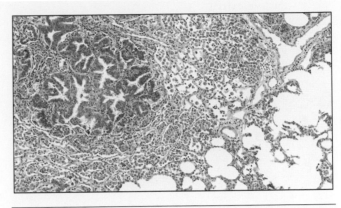

Figure 3.30 Ovine pulmonary adenocarcinoma in a sheep. H&E ×100.

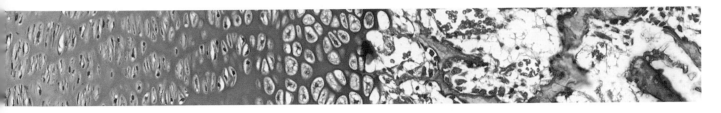

4

CONNECTIVE TISSUE

Connective tissues support and connect other tissues providing a framework for support of the entire body.

Connective tissue can be classified as follows:

- Embryonic connective tissues
  - Mesenchymal and mucous connective tissues
- Connective tissue proper
  - Loose (areolar) connective tissue
  - Dense (regular and irregular) connective tissues
  - Reticular tissue
  - Adipose tissue
- Specialised connective tissues
  - Cartilage and bone
  - Blood cells and blood-forming tissues (*see* Chapter 5)

Embryonic Connective Tissue (Mesenchymal and Mucous Connective Tissues)

Connective tissue is derived from mesenchyme, the loose embryonic packing tissue of mesodermal origin. Mesenchymal cells have long slender processes and are embedded in an amorphous gelatinous substance, the extracellular matrix (**Figure 4.1**). Both types are found in the embryo. Mucoid connective tissue is found in the umbilical cord and subdermal connective tissue of embryos and also occurs in limited regions in adult animals, the comb and wattle of the chicken, and around a healing wound. There are few cells, which are usually stellate undifferentiated fibroblasts; the ground substance is abundant and gelatinous with very few fibres (this tissue is also known as Wharton jelly). This type of tissue stains poorly with haematoxylin and eosin (H&E) (**Figure 4.2**) but stains well with mucin dyes.

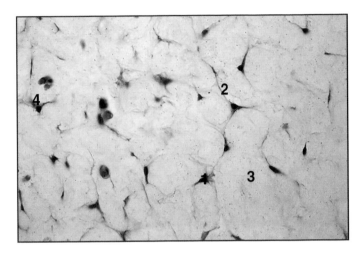

Figure 4.1 Umbilical cord (foal). (1) Nucleus of the stellate mesenchymal cell. (2) Long cell processes. (3) Extracellular matrix. (4) Blood vessels. H&E ×125.

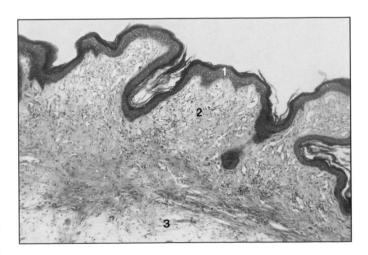

Figure 4.2 Mucoid connective tissue. Comb (chicken). (1) Stratified squamous epithelium. (2) Lamina propria. (3) Mucoid connective tissue. H&E ×50.

DOI: 10.1201/9781003333807-4

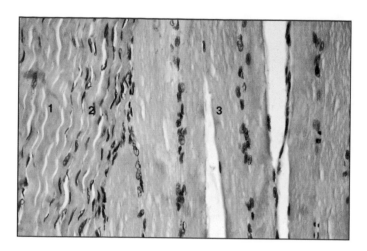

Figure 4.3 Tendon/muscle insertion (dog). (1) Collagen fibres; note wavy appearance. (2) Fibrocytes. (3) Striated muscle fibres. H&E ×200.

Connective Tissue Proper

Connective tissue proper fills the interstices of tissues and organs and forms a continuous structure that carries blood vessels and nerves throughout the body. The relative proportions of the basic components – fibres, cells, and extracellular matrix – determine the functional characteristics of the tissue, giving it tensile strength in ligaments and tendons and mechanical stability in cartilage and bone, and acting as a fluid-transport medium: blood.

Collagen fibres are thick (2–10 μm in diameter) and unbranched; when fresh, they are white when unstained and wavy in section. They stain pink with eosin and green with Masson's trichrome, which distinguishes them from muscle fibres (**Figures 4.3** and **4.4**). They are composed of tropocollagen subunits and ultrastructurally display cross banding at regular intervals. Elastic fibres are relatively thin (about 1 μm in diameter), less wavy than collagen fibres, unbranched, and yellow in unstained material. Because they stain poorly with H&E, selective staining is often used (**Figures 4.5–4.7**). They are composed of elastin, fibrilin-1, fibulin-5, and type VIII collagen. Reticular fibres, narrow bundles of type III collagen fibrils, are fine and delicate,

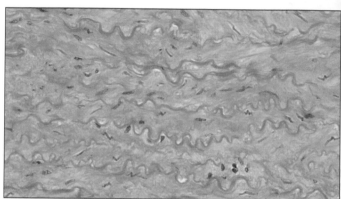

Figure 4.5 Aorta (horse). The elastic fibres are stained dark pink and the collagen fibres light pink. H&E ×400.

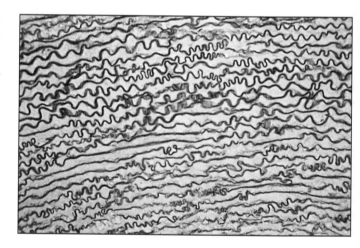

Figure 4.6 Elastic artery (horse). The elastic fibres are selectively stained. Weigert's elastin ×62.5.

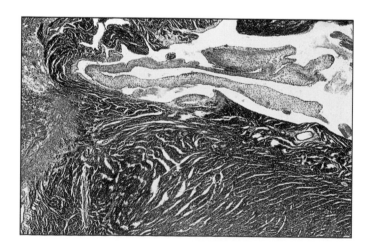

Figure 4.4 Heart (kitten). The collagen fibres and valve cusps are stained green, and the heart muscle is stained red. Masson's trichrome ×50.

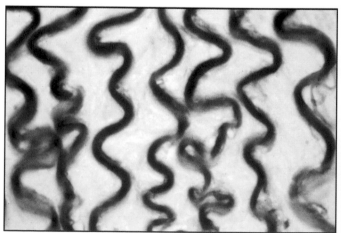

Figure 4.7 Elastic artery (horse). The elastic fibres are selectively stained. Weigert's elastin ×500.

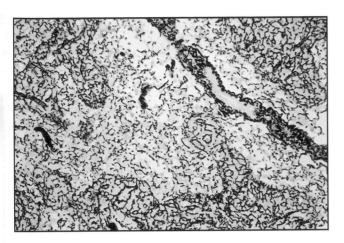

Figure 4.8 Lymph node (dog). The reticular fibres are silver plated by this method and appear as black strands. The delicate network of these fibres supports the lymphatic tissue. Gordon and Sweet ×125.

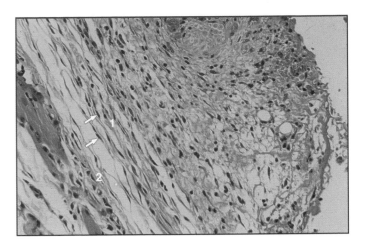

Figure 4.10 Fibroblasts in a healing wound (dog). Nucleus (arrowed) of the fibroblast. (1) Cytoplasm of the fibroblast. (2) Collagen fibres. H&E ×125.

branching extensively to form a supporting network. These also do not stain with standard methods, so a selective process such as silver impregnation is used (**Figure 4.8**).

Cell Types

Fibroblasts/fibrocytes are the commonest cell type, synthesising collagen, elastin and reticular fibres, and the extracellular matrix. The fibroblast represents the active form, is elongated and spindle-shaped, and has abundant cytoplasm and an oval- or cigar-shaped nucleus. It is found in sites of active repair or growth. The fibrocyte represents the less active stage of the cell, acting in a maintenance capacity. The cytoplasm is reduced in volume and is less reactive to stains. The nucleus is flattened and the chromatin is condensed (**Figures 4.9** and **4.10**).

Macrophages, also referred to as histiocytes, are part of the mononuclear phagocyte system. They are large, free, mobile phagocytic cells with a basophilic cytoplasm

that contains many small vacuoles and small dense granules and with an eccentric ovoid and usually indented on one side (horseshoe shaped) nucleus. They are part of a large population of scavengers capable of phagocyting cell debris, taking an active part in the protection of the body by eliminating some micro-organisms. Identification may be achieved by using the ability of the cell to engulf particulate matter, such as carbon particles injected *in vitro* (**Figures 4.11** and **4.12**).

Plasma cells have a basophilic cytoplasm as a result of a well-developed rough endoplasmic reticulum (rER), and the nucleus is eccentric with densely clumped chromatin distributed beneath the nuclear membrane to give a characteristic clock face appearance. Plasma cells are derived from B lymphocytes and are involved in the body's immune response representing the cellular source of circulating immunoglobulin (antibody). They are common both to loose connective tissue and to the lymphatic system (**Figures 4.13** and **4.14**).

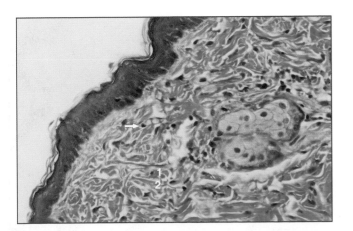

Figure 4.9 Dense irregular connective tissue (dog). Nucleus of the fibrocyte (arrowed). (1) Collagen fibres cut in transverse section. (2) Collagen fibres cut in longitudinal section. Masson's trichrome ×125.

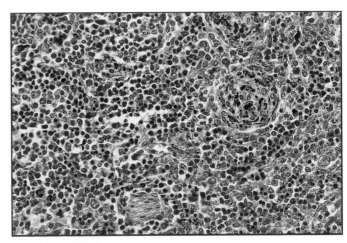

Figure 4.11 Spleen (dog). The macrophages have phagocytosed the carbon particles. Carbon injected, with H&E counterstain ×200.

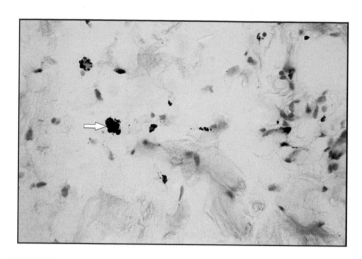

Figure 4.12 Loose connective tissue (dog). A tissue macrophage is arrowed. Carbon injected, with H&E counterstain ×200.

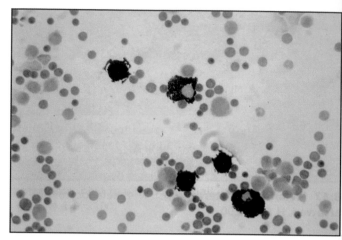

Figure 4.15 Mast cell (sheep). The granules stain purple with the blue dye, toluidine blue. Metachromasia. Toluidine blue ×500.

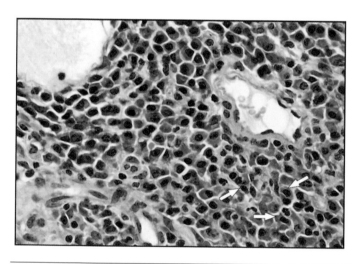

Figure 4.13 Plasma cell (dog). Lymph node. Plasma cells are arrowed. H&E ×200.

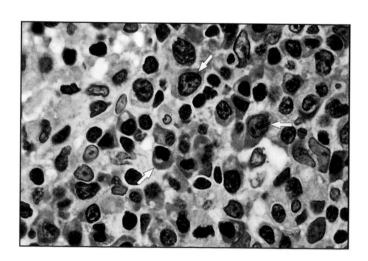

Figure 4.14 Plasma cell (dog). Lymph node. Plasma cells are arrowed. H&E ×500.

Mast cells are round or ovoid, and the nucleus is spherical, centrally placed and often obscured by the cytoplasmic granules. They are metachromatic (stained purple with a blue dye such as toluidine blue or Giemsa) and contain heparin and histamine and other mediators of inflammation. Mast cell degranulation causes a local irritant effect from the release of histamine (**Figure 4.15**).

Fat cells (adipocytes) arise from pericapillary mesenchymal cells. They accumulate fat in the cytoplasm as lipid droplets that coalesce until the cell is filled with one large droplet. The cytoplasm forms a peripheral rim and the nucleus is displaced to lie immediately beneath the cell membrane. Fat cells may occur singly or in groups in loose connective tissue to become adipose tissue. Fat may be white, in which each cell has a single large droplet (unilocular fat cells), or brown, in which small individual droplets are scattered throughout the cytoplasm (multilocular fat cells). Tissue processing and paraffin embedding dissolves the fat and leaves a network of lacy, empty spaces in H&E sections. Freeze fixation allows fats to be revealed by specific stains (e.g. Oil Red O). The unilocular adipocytes are usually described as appearing like a signet ring with the nucleus constituting the signet (**Figure 4.16**).

Pigment cells are derived from neural crest ectoderm. However, cells carrying pigment granules (melanin, erythrin, xanthin) are often found in connective tissue and are called chromatophores (literally, pigment carriers) and can be macrophages (melanophages, erythrophages, xanthophages; **Figures 4.17–4.20**).

Neutrophils (polymorphonuclear leukocytes in mammals; heterophils, see p. 62, in some lower vertebrates), eosinophils, lymphocytes, and monocytes are blood cells commonly found in loose connective tissue. They are migratory and move freely between the blood vessels and

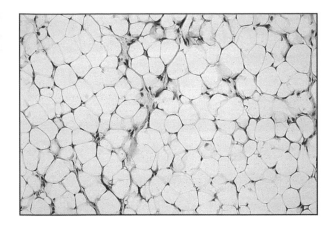

Figure 4.16 Adipocytes. Foot pad (dog). The fat has been lost during processing and the lacy network is formed by cytoplasm and cell membranes. H&E ×50.

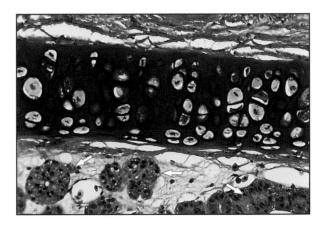

Figure 4.18 Nasal septum of green iguana. A supporting sheet of hyaline cartilage is seen in the centre. Immediately beneath are saline-secreting nasal salt glands (arrowed). H&E ×100.

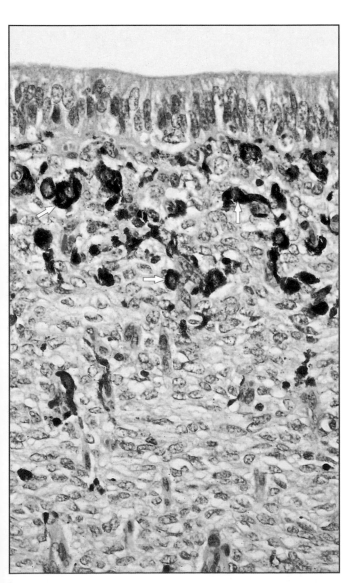

Figure 4.17 Melanocytes. Uterus (ewe). The melanocytes are arrowed. H&E ×200.

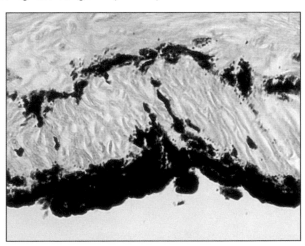

Figure 4.19 Serosal surface of chameleon small intestine. Note heavily pigmented coelomic surface and zone between longitudinal and circular muscular layers. H&E ×50.

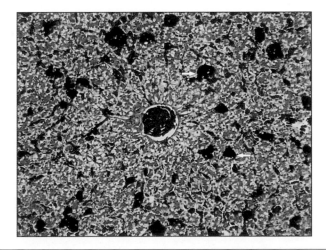

Figure 4.20 The hepatic parenchyma of many amphibians and reptiles is characterised by abundant aggregates of melanin pigment contained in melanophages (arrowed). Illustrated is a section of liver from an African clawed frog (*Xenopus laevis*). The hepatocytes are arranged in ray-like cords extending outward from a thin-walled central vein, next to which are small tributaries of the hepatic artery and bile duct. H&E ×100.

the surrounding tissue, especially during inflammation (**Figures 4.21–4.23**).

Endothelial cells and pericytes form a special cell population in connective tissue, retaining the capacity to divide and to synthesise collagen and the extracellular ground substance. The endothelium is often fenestrated in the capillary bed, and it controls the tissue fluid content locally. Pericytes are pale staining, connective tissue cells lying adjacent to endothelial cells of capillaries and small venules. They are comparatively undifferentiated and can give rise both to fibroblasts and to smooth muscle cells in areas of tissue repair, as well as assist in the revascularisation and repair of damaged blood vessels (**Figure 4.24**).

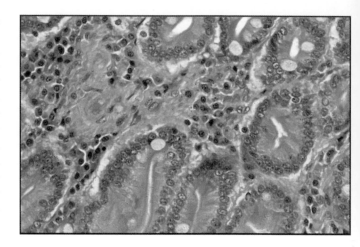

Figure 4.23 Lymphocytes. Duodenum (ox). Lymphocytes are present in the lamina propria of the duodenum. H&E ×125.

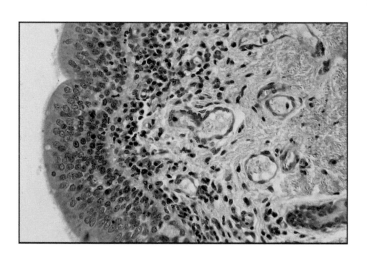

Figure 4.21 Polymorphonuclear leucocytes. Bronchus (ox). The polymorphonuclear leucocytes have invaded the connective tissue of the lamina propria in response to infection. H&E ×125.

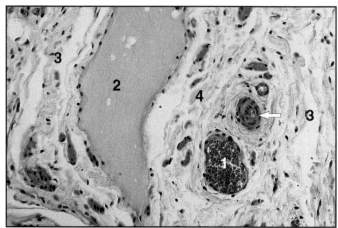

Figure 4.24 Pericyte. Loose connective tissue (cat). The pericyte is arrowed in the wall of the arteriole. (1) Vein. (2) Lymphatic vessel. (3) Fibrocyte. (4) Extracellular matrix. H&E ×100.

Loose Areolar Connective Tissue

Loose areolar connective tissue is found as a packing material throughout the body and carries the blood vessels and nerves. It contains many scattered cells of various types (fibroblasts, adipocytes, macrophages, mast cells, and some undifferentiated cells), blood and lymphatic vessels, and a loose network of fine collagenous, reticular, and elastic fibres. It is widespread throughout the body, surrounding vessels, and nerves and is found in serous membranes, the lamina propria of mucous membranes, subcutaneous tissue, and the superficial layer of the dermis (**Figure 4.25**). Amorphous ground substance is particularly abundant in loose connective tissue. It is composed of a group of carbohydrates, the glycosaminoglycans, which may be complexed with a protein to form proteoglycans, and also by glycoproteins. These substances stain poorly with H&E.

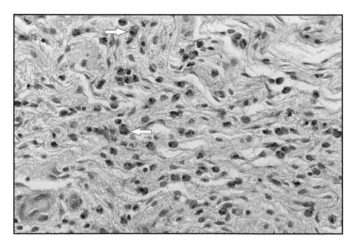

Figure 4.22 Eosinophils. Colon (horse). The eosinophils are arrowed. H&E ×125.

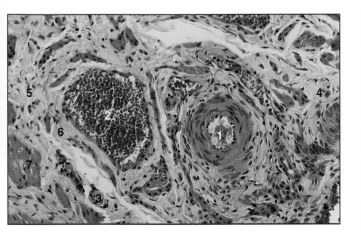

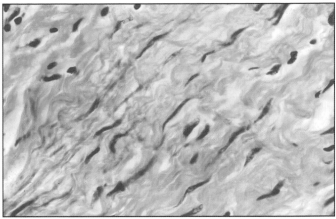

Figure 4.25 Loose connective tissue (cat). Blood vessels: (1) artery and (2) vein. (3) Nerves. (4) Extracellular matrix. (5) Fibrocytes. (6) Collagen fibres. (7) Smooth muscle of the uterine wall. H&E ×100.

Figure 4.27 Dense connective tissue. Vagina (sheep). The collagen fibres are stained green. Masson's trichrome ×160.

Dense Connective Tissue

Composed principally of thick collagen fibres, dense connective tissue contains few cells. Fibrous elements predominate and the commonest cell is the fibrocyte. Dense regular connective tissue is characterised because the fibres are arranged in parallel or in an organised fashion. In dense regular connective tissue, the collagen fibres may be arranged in rows to provide tensile strength in tendons and ligaments, and as sheets in aponeuroses (*see* **Figure 4.3**). In dense regular elastic connective tissue predominate elastic fibres with only a few collagen fibres and they are arranged in parallel, as in the ligamentum nuchae of the horse. Dense irregular connective tissue is characterised by randomly arranged fibre bundles. In dense irregular connective tissue, the fibres are irregularly arranged in different planes to allow stretching without tearing the surface membrane, as in the dermis and the vagina (**Figures 4.26** and **4.27**).

Special Types of Connective Tissue

Reticular tissue is composed of numerous reticular fibres and stellate reticular cells, forming a supportive network for structures such as the spleen, lymph node, kidney, liver sinusoids, and bone marrow. Adipose tissue consists of groups of adipocytes (see above).

Cartilage and Bone

Cartilage

Cartilage is a specialised form of connective tissue combining a degree of rigidity with flexibility and strength. There are three types of cartilage: hyaline, elastic, and fibrocartilage. They differ only in the distribution of the main components: the cells, fibres, and matrix.

Hyaline Cartilage

This type of cartilage is bluish/white and semitranslucent in the fresh state and is the most prevalent form. In the embryo, the precursors of the long bones begin as cartilage models (**Figure 4.28**). As the neonate grows, the cartilaginous template undergoes progressive mineralisation. In postnatal life, hyaline cartilage is present in tracheal rings and bronchi and in plates in the larynx and nose. With ageing and under certain conditions of hypervitaminosis-D_3 and hypercalcaemia, cartilage may become pathologically mineralised. Cartilage also caps the ends of bones in articulating joints (**Figure 4.29**).

At predetermined sites in the embryo, mesenchymal cells round off and differentiate into chondroblasts (cartilage-forming cells) and secrete a matrix consisting of

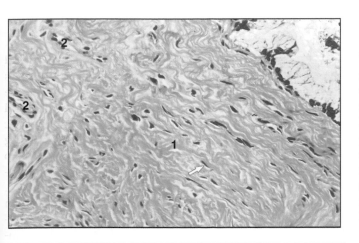

Figure 4.26 Dense connective tissue (dog). Fibrocytes (arrowed). (1) Collagen fibres. (2) Blood vessels. Masson's trichrome ×100.

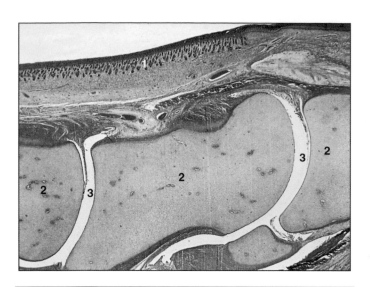

Figure 4.28 Developing hoof (foal). (1) Skin. (2) Hyaline cartilage models of the digits. (3) Joint cavity. H&E ×25.

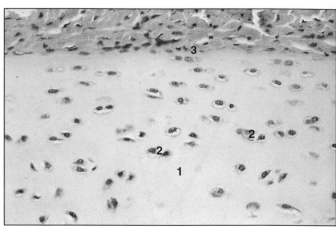

Figure 4.30 Hyaline cartilage. Costal cartilage (dog). (1) Hyaline cartilage matrix. (2) Chondrocytes. (3) Perichondrium. H&E ×100.

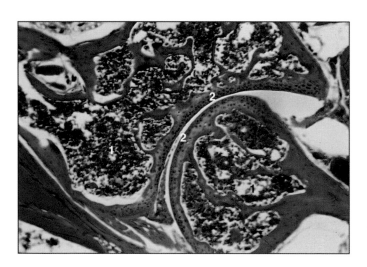

Figure 4.29 Scapulohumeral articulation of a small skink, *Scincella lateralis*. (1) Both the humeral head (lower right) and the scapula (upper left) contain cancellous spaces filled with bone marrow. (2) Articular cartilage. H&E ×250.

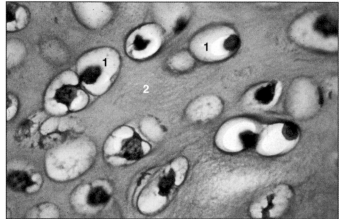

Figure 4.31 Hyaline cartilage. Costal cartilage (dog). (1) Chondrocyte in a lacuna. (2) Extracellular matrix. H&E ×400.

glycosaminoglycans, proteoglycans, and collagen fibrils. The space occupied by each cell is a lacuna and once the matrix is laid down, the cells are called chondrocytes (cartilage cells; **Figure 4.30**). Chondrocytes are capable of dividing, and several cells may come to occupy a lacuna; then, they are known as an isogenous group or cell nest (**Figure 4.31**). Compared with the bulk of the matrix, which stains poorly with H&E, the matrix in the immediate vicinity of the cells stains intensely with metachromatic dyes because of the presence of glycosaminoglycans. Mesenchymal tissue surrounds the developing cartilage and forms a fibrous covering, the perichondrium,

responsible for the growth and maintenance of the cartilage. The outer layer of the perichondrium is composed of fibrous connective tissue and the inner layer is capable of generating new chondroblasts. Cartilage is thus able to grow from the perichondrium by appositional growth and by interstitial growth from within by chondrocyte division and deposition of new matrix. It is avascular and lacks nerves or lymphatic vessels; the cells are nourished by diffusion from blood vessels of surrounding connective tissue.

Elastic Cartilage

Elastic cartilage is specially adapted to give resilience and withstand repeated bending. The matrix contains elastic fibres, so in fresh state is yellowish and more opaque than

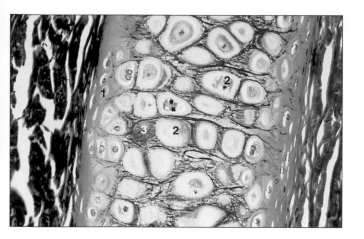

Figure 4.32 Elastic cartilage. Pinna (dog). (1) Perichondrium. (2) Chondrocyte in lacuna. (3) Extracellular matrix with red elastic fibres. Masson's trichrome ×160.

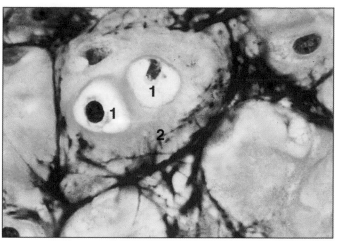

Figure 4.34 Elastic cartilage. Pinna (dog). (1) Chondrocyte in lacuna. (2) Extracellular matrix with red elastic fibres. Gomori's trichrome ×400.

hyaline cartilage. Chondrocytes of elastic cartilage are larger and more abundant than in hyaline cartilage. Examples of elastic cartilage can be found in the epiglottis and the pinna of the ear (**Figures 4.32–4.34**).

Fibrocartilage

Fibrocartilage occurs at the site of tendon insertions, pubic symphysis, and in intervertebral discs, where firm support and tensile strength are necessary. The chondroblasts lie in rows between parallel bundles of collagen fibres and secrete cartilage matrix (**Figures 4.35–4.37**). This type of cartilage lacks perichondrium.

Bone

Bone is a rigid form of connective tissue composed of cells embedded in an intercellular matrix of collagen fibres, glycosaminoglycans, and calcium phosphate deposited as hydroxyapatite crystals being one of the hardest

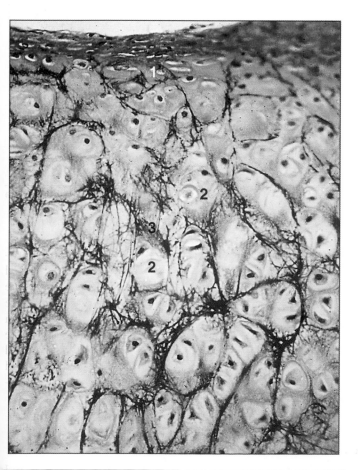

Figure 4.33 Elastic cartilage. Pinna (dog). (1) Perichondrium. (2) Chondrocyte in lacuna. (3) Extracellular matrix with red elastic fibres. Gomori's trichrome ×160.

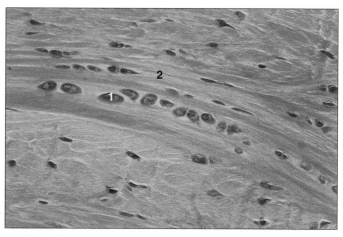

Figure 4.35 Fibrocartilage (ox). (1) Chondrocyte. (2) Collagen fibres arranged as dense regular connective tissue. H&E ×125.

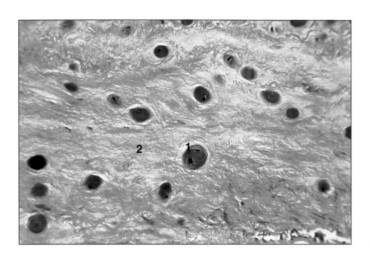

Figure 4.36 Fibrocartilage (ox). (1) Chondrocyte. (2) Collagen fibres arranged as dense regular connective tissue. Gomori's trichrome ×160.

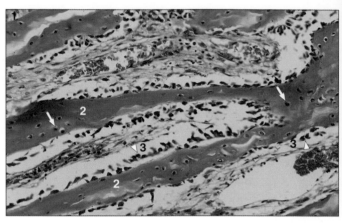

Figure 4.38 Spiculated bone (dog). (1) Periosteum. (2) Bone spicules with osteocytes in the lacunae (arrowed). (3) Bone marrow-filled spaces lined by osteoblasts (arrowhead). H&E ×100.

substances of the body. Bone provides the framework of the body and serves as a lever for muscle action, as protection for viscera, as a haemopoietic organ, and as a reservoir of body stores of calcium and phosphorus. Small bones are also found in soft tissues to provide extra rigidity, e.g. the *os penis* in the dog and *ossa cordis* in the ox. Bone is a living tissue that is supplied with blood vessels and nerves and is constantly changing in response to body stresses and circumstances.

There are two forms of bone tissue: cancellous (spongy, medullary) and compact (cortical, dense). All bones have both cancellous and compact forms of bone deposition. Cancellous bone consists of irregular interconnecting bars, the trabeculae, and spicules, forming a three-dimensional network of lined spaces filled with bone marrow (**Figures 4.38** and **4.39**). Compact bone is

a solid continuous mass in which the spaces are only visible with the aid of a microscope (**Figure 4.40**).

Bones are covered with a specialised layer of connective tissue, the periosteum, that have an outer layer of fibrous connective tissue and an inner layer which is osteogenic (capable of laying down new bone; **Figure 4.40**). Spaces in bone, like the marrow cavity (**Figures 4.38–4.40**) and the canal system (**Figures 4.40–4.42**), are lined with a single layer of osteogenic cells, the endosteum. The characteristic feature of all bone tissue is the arrangement of the mineralised bone matrix in layers (parallel or concentric), the lamellae. Small lacunae present in the lamellae are occupied by a single bone cell, the osteocyte. Tubular passages, the canaliculi, radiate from each lacuna and link up with canaliculi from adjacent lacunae to create an extensive system of interconnecting canals. This arrangement is clearly defined in the compact bone where the lamellae are arranged concentrically around a longitudinal canal to form an osteon (Haversian System; **Figures 4.41–4.43**).

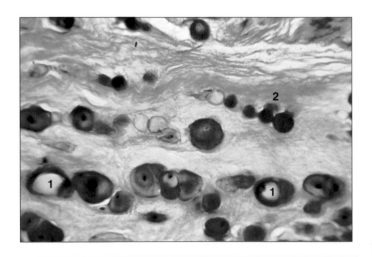

Figure 4.37 Fibrocartilage (ox). (1) Chondrocyte. (2) Collagen fibres. Gomori's trichrome ×250.

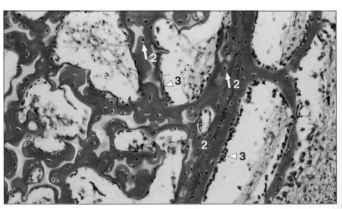

Figure 4.39 Spiculated bone (dog). (1) Periosteum. (2) Bone spicules with osteocytes (arrowed). (3) Osteoblasts (arrowhead) on the free surface of the bone. H&E ×160.

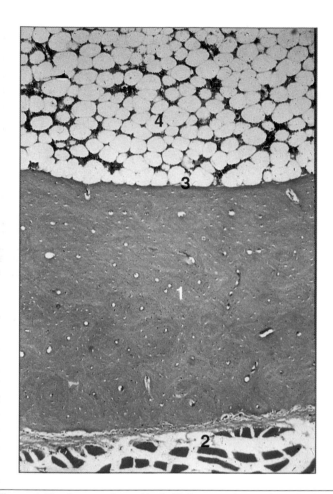

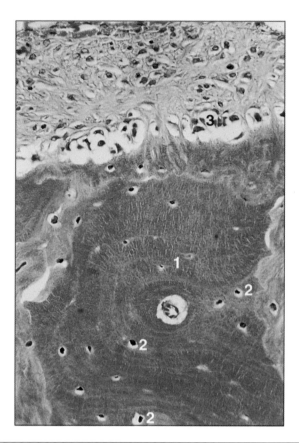

Figure 4.40 Compact bone (dog). (1) Compact bone of the diaphysis of the femur. (2) Periosteum. (3) Endosteum. (4) Marrow with a high proportion of fat cells. H&E ×20.

The central canal of the osteon carries blood vessels and branches of the perpendicular canals (of Volkmann). These are part of the main blood supply to the bone and link the endosteal and periosteal surfaces. The lamellae may be regular circular rings as in the osteon, surround the shaft of the bone as circumferential lamellae, or fill in

Figure 4.42 Compact bone (dog). (1) The osteon fills the field. The central canal is lined by endosteum and carries blood vessels and nerves. (2) Osteocytes in lacunae in the circumferential lamellae. (3) Periosteum. H&E ×100.

the angular spaces between lamellae as interstitial lamellae. They are often the result of bone remodelling. Bone contains a vast continuous network of canals that are essential for the nutrition of the bone cells in the lacunas, the osteocytes, and allows the flow of substances as hormones, ions, and waste products to and from them

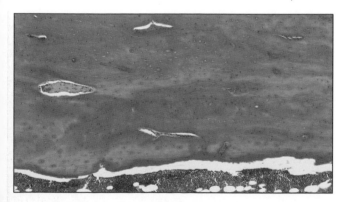

Figure 4.41 Compact bone (dog). Compact bone with osteocytes in lacunae arranged in lamellae with a central canal carrying blood vessels and nerves. The canal is lined by endosteum. Bone marrow observed at the bottom of the image. H&E ×100.

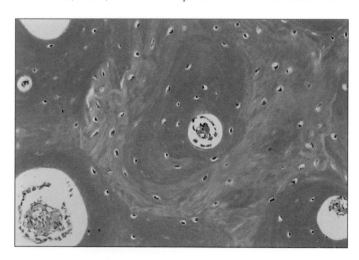

Figure 4.43 Compact bone (dog). This is a high-power view of **Figure 4.42**. H&E ×250.

Figure 4.44 Developing periosteal bone (foal). (1) Periosteum. (2) Bone spicules. (3) Osteogenic tissue fills the spaces between the spicules. (4) Endosteum. (5) Bone marrow. H&E ×62.5.

(**Figures 4.41–4.45**). The cell body lies in the lacuna and extends in long processes into the canaliculi to contact similar processes from adjacent osteocytes.

Bone is a living tissue and is constantly being remodelled; osteoclasts are multinucleate nondividing cells with acidophilic cytoplasm that are found in resorption bays (Howship's lacunae) at the site of bone remodelling (**Figure 4.46**). Osteoclasts are derived from a progenitor cell in bone marrow. The cell migrates to the developing tissue. The ruffled (brush) border is the undulating mobile cell membrane and is the part of the cell directly involved in resorption of bone.

Osteoblasts are present on the inner surface of the periosteum and endosteum and cover the surface of bone spicules in an active osteogenic area (**Figures 4.38**, **4.39**, and **4.46**). The cell has a central or slightly eccentric nucleus with chromatin granules, and the cytoplasm stains deeply with haematoxylin because of the concentration of organelles. Long cytoplasmic processes extend out from the osteoblast contacting with those of neighbouring cells and the matrix is deposited around them. This provides a fine canalicular network and allows nutrients

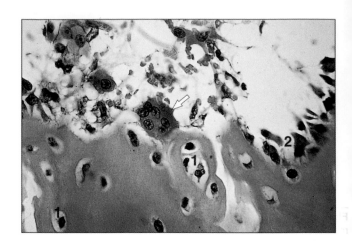

Figure 4.46 Compact bone (cat). Osteoclast (arrowed). (1) Osteocytes in lacunae. (2) Osteoblasts on the free surface of the bone. H&E ×250.

to pass to the bone cells. Unlike cartilage, there is no diffusion in bone. Once the bone has been deposited, the trapped cells (osteocytes) function is to maintain the bone and retain contact with the blood vessels through the canalicular system. Osteocytes conform to the shape of the lacunae and have a flattened nucleus and a cytoplasm poor in organelles.

Endochondral Ossification

In the embryo, mesenchymal cells at predetermined sites differentiate into chondroblasts and lay down cartilage models of the long bones (**Figure 4.47**). Some irregular bones (with complex shape) like vertebrae also begin endochondral ossification during the embryonic period (**Figure 4.48**). Later in gestation, the mesenchyme surrounding the cartilage model becomes very vascular, chondrocytes in the centre of the cartilage model hypertrophy, and the cartilage matrix is calcified. Nutrients are unable to diffuse through this calcified cartilage and the cells die. Cell

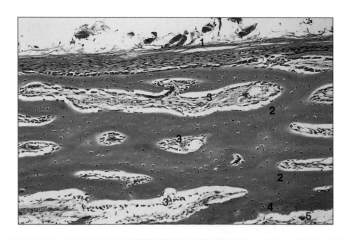

Figure 4.45 Developing periosteal bone (foal). (1) Periosteum. (2) Bone spicules. (3) Osteogenic tissue fills the spaces between the spicules. (4) Endosteum. (5) Bone marrow. H&E ×125.

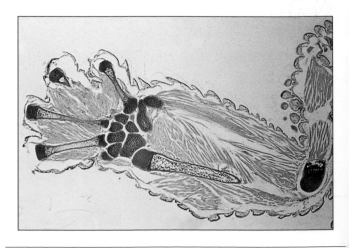

Figure 4.47 Whole mount section of the left forelimb of a small viviparous yucca night lizard (*Xantusia vigilis*). Note the transition between the blue staining cartilaginous ends and the mineralised diaphyseal compact bone. H&E ×5.

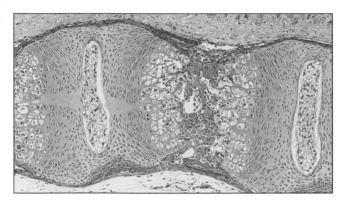

Figure 4.48 Endochondral ossification of two vertebrae within a mouse embryo. H&E ×100

death is followed by a breakdown of the matrix. Vascular mesenchymal tissue moves in, differentiates into osteogenic tissue, and deposits bone on the remains of the calcified cartilage (**Figures 4.49** and **4.50**). This is an ossification centre. The cartilage is replaced by bone beginning at the centre of the diaphysis (primary centre of ossification) and

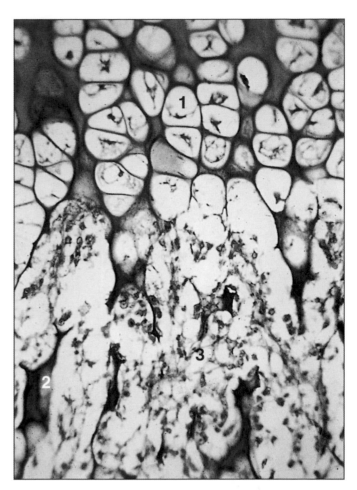

Figure 4.50 Endochondral ossification (cat). (1) Zone of hypertrophied chondrocytes. (2) Calcified cartilage with bone deposits. (3) Osteogenic tissue. H&E ×400.

extending to the epiphysis (secondary centre of ossification). The epiphysis has a separate centre of ossification and a plate of cartilage persists, the epiphyseal plate; this separates the epiphysis from the diaphysis (**Figure 4.51**). The perichondrium becomes the periosteum.

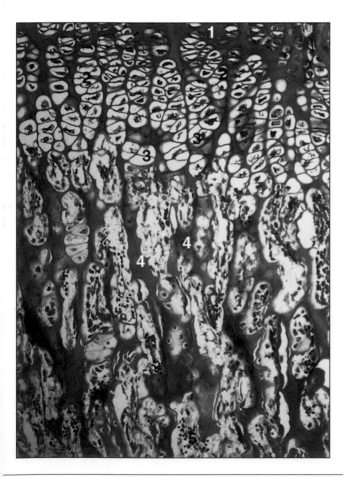

Figure 4.49 Endochondral ossification (cat). (1) Zone of resting cartilage. (2) Zone of proliferating cartilage. (3) Zone of hypertrophied chondrocytes. (4) Basophilic calcified cartilage with freshly deposited eosinophilic bone. (5) Osteogenic tissue. H&E ×400.

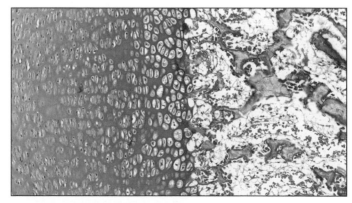

Figure 4.51 Epiphyseal plate (pig). Growth area of endochondral ossification (left) adjacent to spiculated bone (right). H&E ×200.

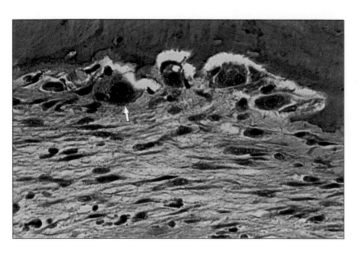

Figure 4.52 Section of femur from a green iguana (*Iguana iguana*) with metabolic bone disease. Note the multinucleated osteoclasts (arrowed) that have been removing ossified tissue and fibrous connective tissue that has replaced the cancellous bone. H&E ×250.

Multinucleated osteoclasts remove mineralised osseous matrix (**Figure 4.52**). In metabolic bone disease (fibrous osteodystrophy, secondary nutritional hyperparathyroidism, 'rubber jaw', 'renal rickets'), which may be caused by an improper dietary calcium: phosphorus ratio, or in some cases of severe chronic renal disease, the osteopenic bone is replaced by fibrous connective tissue (**Figure 4.52**).

Growth of a Long Bone

Bones are able to grow in width and length and do so until the adult size is reached. Thereafter, bone continues to remodel as circumstances demand. Growth in width is appositional from the osteogenic inner layer of the periosteum, and new bone is formed locally, the periosteal collar (**Figure 4.44**). This ability to deposit bone is utilised in bone grafts. Growth in length is accomplished at the diaphyseal/epiphyseal junction at the persistent layer of the epiphysis. Cartilage grows interstitially on the epiphyseal side and the replacement by bone takes place at the diaphyseal side of the plate. Osteogenic tissue invades and deposits bone, the epiphyses are pushed apart, and the bone grows in length (**Figure 4.51**).

Intramembranous Ossification

Small bones and flat bones develop directly from mesenchymal cells; these differentiate into osteoblasts. Bone is laid down as a network of bony spicules that are gradually enlarged by the deposition of new bone lamellae. The spaces are filled in and compact bone is formed in the outer plates which are continuous with the more central cancellous bone. The surrounding mesenchyme condenses to become the periosteum; the inner layer is osteogenic (*see* **Figure 4.44**). The osteoblasts on the inner surface lining the spicules of cancellous bone also retain their osteogenic ability.

Some special features of avian, reptilian, amphibian and fish bone, and bone marrow are described in Chapter 5.

CLINICAL CORRELATES

Although a variety of benign and malignant tumours of various origins arise in the skeleton, osteosarcomas, a malignant mesenchymal tumour in which the neoplastic cells produce tumour osteoid or bone, account for approximately 85% of canine and 70% of feline malignant bone tumours. Osteosarcomas can be subdivided according to their location within bone or histological appearance. An osteogenic, or bone producing, osteosarcoma that has multifocal formation of well mineralised new bone is shown in **Figure 4.53**. This was taken from an 8-year-old St Bernard dog. An example of a giant cell osteosarcoma (**Figure 4.54**) from a German Shepherd dog shows numerous multinucleate syncytia formed by fusion of the neoplastic cells.

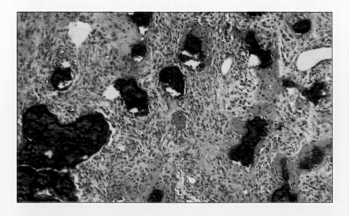

Figure 4.53 Osteogenic osteosarcoma (dog) showing tumour bone formation. H&E ×200.

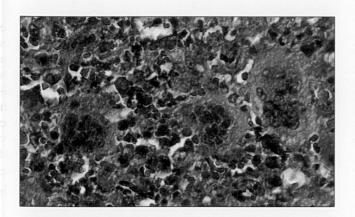

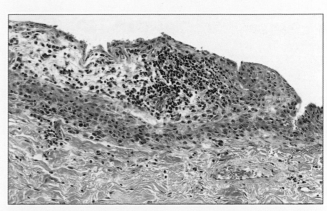

Figure 4.54 Giant cell osteosarcoma (dog). Note the numerous multinucleate cells. H&E ×250.

Figure 4.55 Plasmacytic arthritis in the dog. The synovial membrane is infiltrated by abundant inflammatory cells, predominantly plasma cells. H&E ×125.

Arthritis is a general term for inflammatory disease of the joints. There are many causes and many types of arthritis but often a similar pattern of reaction and joint damage may be observed. A joint biopsy from an adult German Shepherd dog in which the thickened synovial membrane is invaded by large numbers of inflammatory cells, predominantly plasma cells, is shown in **Figure 4.55**. The predominance of plasma cells suggest an immunological basis for the disease in this case.

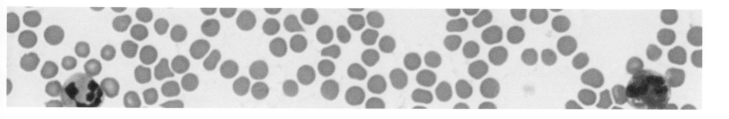

5

BLOOD AND BONE MARROW

Blood is a special type of connective tissue that consists of specialised cells derived from the bone marrow suspended in plasma, a liquid containing organic and inorganic substances. The three cell types we can find in the blood are erythrocytes (red blood cells), leukocytes (white blood cells), and thrombocytes or platelets. Both red and white blood cells are derived from the same primitive cell, the haemocytoblast. The red blood cells are contained within the blood vessels and carry oxygen and carbon dioxide. The white blood cells are part of the body's defence mechanism and use circulation as a means of transport to particular sites where they leave the blood vessel and enter the tissues. The cellular elements account for approximately 45% of whole blood and the plasma for 55%.

The preparation of blood for microscopic examination is remarkably simple (**Figure 5.1**). A drop of blood is spread on a glass slide, fixed immediately in air and by immersion in absolute methanol, and stained with one of the Romanowsky-type stains (including Giemsa, Jenner, Wright, Leishman, or May-Grunwald), containing mixtures of methylene blue and the derivatives azure blue and eosin. Cell nuclei stain purple; the haemoglobin-containing erythrocytic cytoplasm stains pink or tan; and the cytoplasm of leukocytes assumes a blue to blue–grey hue. Cytoplasmic granules react either with eosin and are eosinophilic or with the blue stains and are basophilic. They often display some degree of refractility. Some granules do not stain with either and are regarded as neutral. This method allows for the identification of all the cell types in a blood smear, and differential cell counts are used to assess the blood for changes in the elative numbers of blood cells or changes in the shape and size of blood cells in pathological conditions (**Figure 5.2**) (*see* **Appendix Table 2** for species variation).

CLINICAL CORRELATE

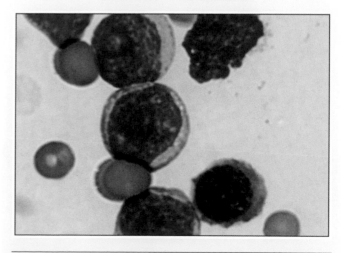

Figure 5.2 Feline prolymphocytic leukaemia. The nuclei from lymphocytes are larger than normal, and the nucleoli are retained. Wright's stain ×600.

Figure 5.1　Blood (ox). (1) Erythrocytes. (2) Large lymphocyte. (3) Small lymphocyte. Leishman ×200.

DOI: 10.1201/9781003333807-5

Erythrocytes

The mature mammalian erythrocyte is a highly differentiated cell lacking a nucleus, ribosomes, and mitochondria. It is a biconcave disc with the cell membrane enclosing the cytoplasm and is filled with haemoglobin, the protein carrier of oxygen, and carbon dioxide (**Figure 5.1**). The diameter varies in size according to species. Mammalian erythrocytes are round, and they measure from 4.1 μm in the goat to 7 μm in the dog. Interestingly, camelids have ellipsoid and not round erythrocytes. The erythrocytes of birds, fish, amphibians, and reptiles are elongated and nucleated, and they can measure nearly 20×100 μm in some amphibians and reptiles. Erythrocytes from one animal are the same size, except in the cow and sheep, where variation in size (anisocytosis) is not unusual. The erythrocyte is flexible enough to pass through the smallest capillary. The average life span is 3 months in mammals, but in some reptiles, it is as long as 3 years. Immature erythrocytes may appear in the circulation in response to urgent need; the commonest of these is the reticulocyte (**Figure 5.3**). Pronormoblasts, basophilic normoblasts, and polychromatic or orthochromatic normoblasts are also seen occasionally (**Figure 5.4**). The nuclei of avian, amphibian, and reptilian erythrocytes are elongated and retained into maturity. Nucleus-free, but otherwise intact, erythrocytes (erythroplastids) are occasionally found in reptilian blood. Similarly, nuclei with a tiny amount of haemoglobin-containing cytoplasm may be seen. These cell-like objects, called haematogones, appear to be intact erythrocytic nuclei or nuclear remnants that have been extruded from other red blood cells. Because they are nucleated, the cells of birds, fish, amphibians, and reptiles are capable of mitosis and amitotic division even

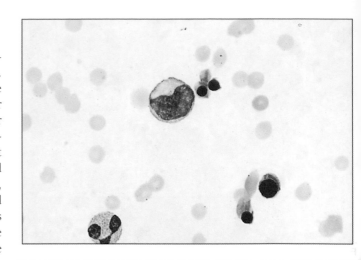

Figure 5.4 Blood (dog). Normoblasts are nucleated early-stage erythrocytes seen on rare occasions in circulating blood. Leishman ×200.

after they have matured. Although the erythrocytes of the lower vertebrates retain their nuclei, the presence of their immature phase, the reticulocyte, is easily demonstrated by staining unfixed blood films with the supravital dye new methylene blue.

During metamorphosis in amphibians, the erythrocytes change from the larger larval cells to the smaller adult cells. The nucleus is retained in both, but the amount of the endoplasmic reticulum is greater in the adult cells. The haemoglobin also changes during the development of larval amphibians into adults. Larval haemoglobin possesses a greater affinity for oxygen; this is consonant with the aquatic habitat. Larval erythrocytes are more able to incorporate the amino acids uridine and thymidine than are the adult cells. The site of haemopoiesis shifts during larval and metamorphic development. As an embryo, the erythrocytes are formed in the ventral blood islands. Later, the pronephric and nephric kidneys are major sites for red blood cell production. In late metamorphosis and into adulthood, the liver and then the spleen and bone marrow predominate as the sites where erythropoiesis occurs.

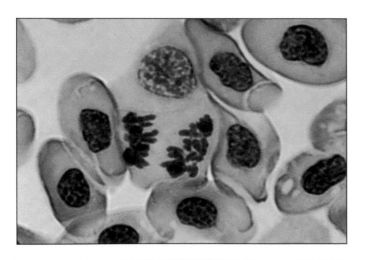

Figure 5.3 Mitotic division in a previously mature, haemoglobin-containing erythrocyte of a western diamondback rattlesnake (*Crotalus atrox*) demonstrates the ability of these cells, under certain conditions of anaemia, to revert to a blastic phase and undergo mitosis. Benzidine peroxidase ×600.

Leukocytes

There are two main types of leukocytes, granulocytes and agranulocytes, both of which can leave and enter the circulating blood by crossing the capillary wall between the endothelial cells. They are part of the body's defence system and are involved in the immune response. The most numerous of the white cells in domestic animals, with the exception of many ruminants and other animals, are the polymorphonuclear leukocytes (PMLs), large granulocytes (10–12 μm in diameter) with lobed nuclei, usually five lobes joined by thin strands of chromatin. They are described as PMLs

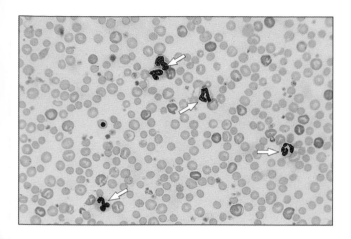

Figure 5.5 Blood (dog). Polymorphonuclear leukocytes are arrowed. Leishman ×160.

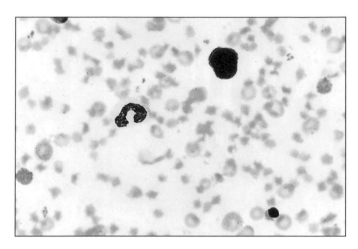

Figure 5.7 Basophil. Blood (dog). The dense basophilic granules obscure the cell. Leishman ×200.

(**Figure 5.5**). The staining of the granules is very variable, and the individual granules are small and difficult to identify. The eosinophilic leukocytes (eosinophils, acidophils) have large red granules in the cytoplasm (particularly large in the horse), bilobed nuclei, and form about 5% of the white cell population (**Figure 5.6**). They are typically associated with response to parasitic infections and allergic reactions.

The heterophil is another granulocyte that is present in the lower vertebrates. It is the functional and numerical equivalent of the PML in mammals. Heterophils display the greatest diversity in size and granule shape from species to species of any of the leukocytes. In most amphibians, lizards, snakes, and crocodilians, the heterophil granules are small, round, and orange. In chelonians (turtles, tortoises, and terrapins), they are elongated, needle-shaped, and often a muddy brown colour. Like their mammalian PML counterpart, the heterophils are lysosomally active, recruited to sites of bacterial infection and can act as phagocytes. Because of their reddish-orange coloured granules, heterophils are often mistaken for eosinophils.

The basophilic leukocytes (basophils) are the rarest of the granulocytes in mammals, forming only 0.5% of the white cell population. However, as one descends the phylogenetic scale, the number of basophils enumerated in a differential blood cell count increases markedly to as many as 10% in some species. Characteristically, large basophilic granules, containing histamine and heparin, obscure the bilobed, multilobed, or unlobed nuclei, and the cells are essentially similar to mast cells (**Figure 5.7**).

The most common of the agranulocytes is the lymphocyte, which accounts for 20–40% of all circulating leukocytes in dogs, cats, and horses but may be 50–60% in ruminants, mice, and pigs. Small lymphocytes are 7–8 µm in diameter, with a dark blue nucleus and a thin rim of cytoplasm (**Figures 5.1**, **5.8**, and **5.9**). Medium and large lymphocytes are up to 10 µm in diameter with a dark blue nucleus and pale

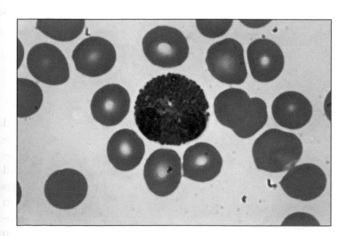

Figure 5.6 Eosinophil. Blood (dog). Large granules are present in the cytoplasm. Leishman ×600.

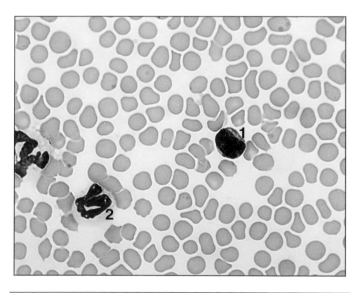

Figure 5.8 Blood (dog). (1) Small lymphocyte. (2) Polymorphonuclear leukocytes. Leishman ×200.

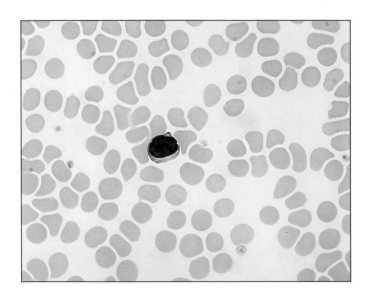

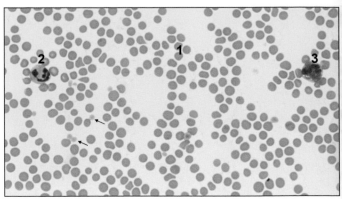

Figure 5.11 Blood (cat). Abundant erythrocytes (1), one neutrophil (2), and one eosinophil (3) with some interspersed platelets (arrows, small but of different size and shapes, typical of this species) ×100.

Figure 5.9 Lymphocyte. Blood (dog). A small lymphocyte lies in the middle of the field. Leishman ×200.

blue cytoplasm (**Figures 5.9** and **5.10**). The largest cell of the agranulocytic series is the monocyte, which is 12–20 μm in diameter, with a horseshoe-shaped nucleus lying in abundant cytoplasm (**Figure 5.12**). It leaves the blood and transforms into a variety of macrophages at various sites in the body as part of the mononuclear phagocyte system.

Platelets

Platelets or thrombocytes are elongated and very small in size: 1–4 μm in width and 5–7 μm in length, depending on the species (**Figure 5.11**). Platelets derive from large

multinucleated cells in the bone marrow called megakaryocytes. These cells function with fibrinogen to repair damaged blood vessels and stop blood leaking from the vessels by the coagulation (clotting) process (**Figure 5.13**). In birds, fish, amphibians, and reptiles, thrombocytes are nucleated, elongated, and usually somewhat smaller than an erythrocyte. It possesses a pale blue, agranular cytoplasm and is capable of active phagocytosis, amitotic division, and has the ability under conditions of severe acute and chronic blood loss to be transformed into a haemoglobin-rich, respirationally functional erythrocyte. In reptiles, thrombocytes bud off from large multinucleated cells with distinctly granular cytoplasm that are present in the bone marrow. Under conditions of severe anaemia, they can also be found in extramedullary sites such as the liver and

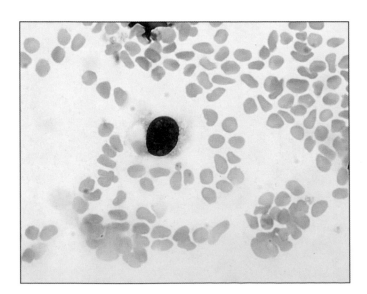

Figure 5.10 Lymphocyte. Blood (ox). A large lymphocyte lies near the middle of the field. Leishman ×200.

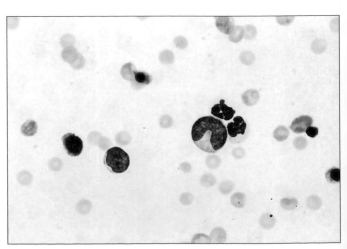

Figure 5.12 Monocyte. Blood (dog). A monocyte is the largest present cell with a horseshoe-shaped nucleus. Leishman ×200.

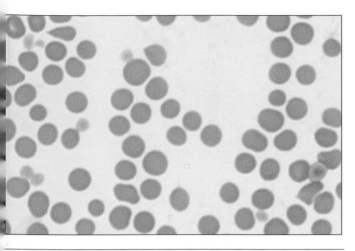

Figure 5.13 Platelets. Blood (cat). Several platelets of different sizes and shapes interspersed among erythrocytes. Giemsa ×400.

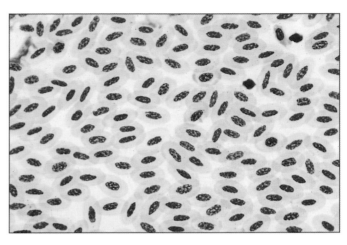

Figure 5.15 Blood (bird). The avian erythrocytes are nucleated. Leishman ×200.

spleen. Because of their delicacy, the megakaryocyte-like giant cells are best seen in bone marrow *touch* preparations (**Figure 5.14**).

Avian Blood Cells

The various corpuscles of mammalian blood are present in birds but exhibit distinctive features. These include retention of the nucleus by the mature erythrocyte and the presence of a true thrombocyte more complex than mammalian platelets. The erythrocyte is a nucleated oval cell with an eosinophilic cytoplasm, measuring approximately 9–12 μm

in diameter (**Figure 5.15**). The thrombocyte is analogous to the mammalian platelet, is oval or round in shape, and is smaller than the erythrocyte. The nucleus stains deeply basophilic, as does the cytoplasm. Prominent red granules are commonly present in vacuoles adjacent to the nucleus (**Figure 5.16**).

Granular leukocytes account for approximately 24% of the white cells and are classified as heterophils, eosinophils, and basophils. The heterophils are 8–10 μm in diameter and are so-called because of the variable staining of the cytoplasmic rods; the nucleus is bilobed or trilobed (**Figure 5.17**). The eosinophils are about 7 μm in diameter and rounded granules are present in the cytoplasm. They are less regularly rounded than are the heterophils and most have a bilobed nucleus. The specific granules are round or oval and smaller than the granules of the heterophils. The cytoplasm is eosinophilic (**Figure 5.19**). The basophil is more numerous than in the mammal; the

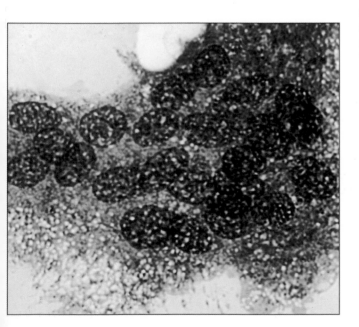

Figure 5.14 Megakaryocytic-like multinucleated cell from which reptilian thrombocytes originate. Jenner–Giemsa ×600.

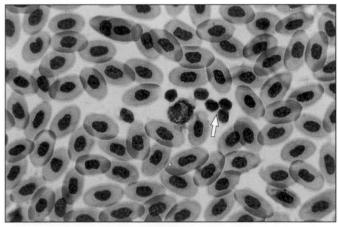

Figure 5.16 Blood (bird). A group of thrombocytes is arrowed. Leishman ×250.

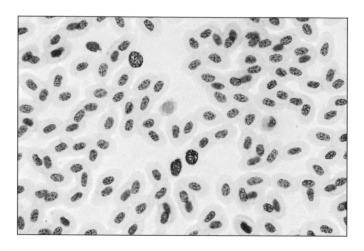

Figure 5.17 Blood (bird). The cytoplasm of the avian heterophil is filled with eosinophilic granules. Leishman ×200.

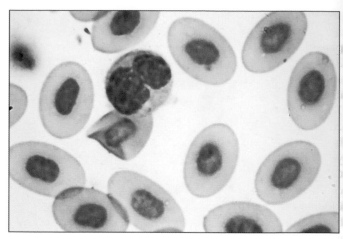

Figure 5.20 Blood (bird). Monocyte. Leishman ×600.

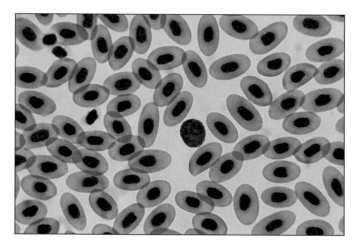

Figure 5.18 Blood (bird). A small lymphocyte. Leishman ×480.

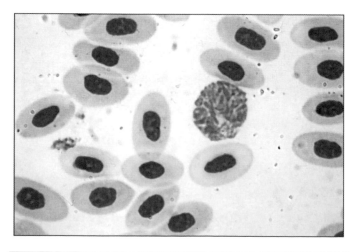

Figure 5.19 Blood (bird). Eosinophil. Leishman ×600.

granules are deep blue staining. Agranular leukocytes are identical to the lymphocytes and monocytes of the mammal and form 72% of the white cell population (**Figures 5.18** and **5.20**). The bone marrow haemopoietic tissue is basically similar to the mammalian, with the notable absence of megakaryocytes.

Reptilian, Amphibian, and Fish Blood Cells

The formed cellular components of amphibian and reptilian blood are similar to those present in avian blood. Erythrocytes are elongated and vary widely between species. The largest erythrocytes are found in the tuatara, *Sphenodon punctatus* (a quadruped reptile that superficially resembles a large lizard), and in some aquatic salamanders, particularly *Amphiuma* spp. In some instances, the erythrocytes of animals are approximately 100 μm long and 20 μm wide. Generally, lizards and snakes possess smaller erythrocytes and leukocytes than do chelonians and crocodilians. However, there are individual variations between species.

Although all of the blood cells of fish, amphibians, and reptiles retain their nuclei throughout their lifespans, true reticulocytes can be demonstrated by supravital staining. Because the blood cells retain their nuclei, it is not unusual to find both mitotic and amitotic division in stained films of peripheral blood from amphibians and reptiles. These cells in division usually reflect a response to blood loss and, thus, do not necessarily suggest neoplastic proliferation.

The leukocytes of amphibians and reptiles are also similar to those present in birds. The granulocytes include acidophils (amphophils) comprising heterophils and eosinophils; the intracytoplasmic granules of which stain reddish-orange or muddy brown with Romanowsky dyes are ovoid to distinctly spindle shaped and possess pale blue lobed or unlobed nuclei that tend to be eccentrically placed. The heterophil is one of the major leukocytes that function enzymatically in opsonisation and chemotaxis in amphibians and reptiles. It is also capable of phagocytosing particulate matter, such as bacteria and cellular detritus. The percentage of heterophils in the differential leukocyte counts of reptiles and amphibians normally ranges widely from 2 to 65%, depending on the species, season of the year, and sex of the animal from which they were obtained. Clinically, they usually range from 20 to 55%. The granules of eosinophils are spherical and stain deeply red with the identical dyes used to stain blood films of mammals and birds. In some instances, one or more giant red granules are found rather than dozens of smaller ones. The nuclei of eosinophils tend to be more centrally located and may be lobed, unlobed, or concentric. Eosinophils may account for 0–10% (or more) of the differential leukocyte count of most reptiles. They increase during parasitism and other antigenic challenges. The basophil granulocytes are probably the most readily identified because of their uniformly dense and dark blue or purple staining. Their nuclei may be lobed or unlobed and centrally or eccentrically located against the inner surface of the cell membrane. Basophils account for 0–10% of the leukocytes of most normal amphibians and reptiles.

Another granulocyte observed in amphibians and reptiles is the azurophil. This cell is characterised by its large, usually unlobed, nucleus composed of densely clumped chromatin and finely granular cytoplasm, which usually contains tiny azurophilic granules. Amphibian and reptilian azurophils are often called PMLs. In instances of bacteraemia, azurophils may engulf particulate matter. Azurophils account for approximately 2–10% of the differential cell count.

The mononuclear leukocytes of amphibians and reptiles include small and large lymphocytes, plasma cells, and monocytes. Each of these cell types resembles its counterpart in mammals and birds.

Unlike mammals, for which the cellular component of the blood clotting mechanism involves nucleus-free platelets, fish, amphibians, and reptiles possess nucleated thrombocytes. These elongated cells characteristically contain a pale blue, usually nongranular, cytoplasm and a single elongated central nucleus. Superficially, they resemble erythrocytes, but they lack haemoglobin. Under certain conditions involving acute and chronic blood loss, thrombocytes display pluripotentiality: they can be transformed into erythrocytes. In doing so, they acquire increasing amounts of haemoglobin and may undergo mitotic or amitotic division. Like their leukocytic counterparts, the amphibian and reptilian thrombocytes can also serve as phagocytes and engulf bacteria and cellular detritus, including senescent erythrocytes.

Because they are nucleated, all of the cellular elements of amphibian and reptilian blood can be involved in

Figure 5.21　Baso-eosinophilic myelogenous leukaemia in a boa constrictor *(Boa C. constrictor)*. Note the dual population of eosinophilic and basophilic granules that characterise a clonal population of neoplastic granulocytic leukocytes. Wright's ×600.

haemopoietic neoplastic disorders (**Figure 5.21**), in addition to the many lymphoreticular and myelopoietic disorders that occur in mammals and birds.

Bone Marrow

Haemopoietic tissue located in the adult bone marrow is responsible for the production of red blood cells, granular and agranular white blood cells, and platelets. The precursors of these cells are present in bone marrow (all of which derive originally from the multipotential stem cell, the haemocytoblast), but as they develop, the individual characteristics of each cell type predominate and the various stages may be identified (**Figures 5.22–5.26**).

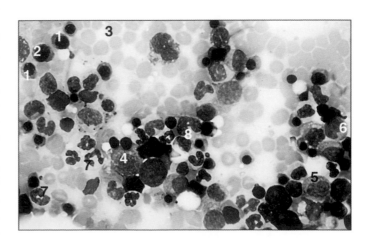

Figure 5.22　Bone marrow (dog). (1) Basophilic erythroblast. (2) Polychromatophil erythroblast (3) Erythrocytes. (4) Myeloblast. (5) Promyelocyte. (6) Myelocyte. (7) Polymorphonuclear leukocytes. (8) Monocyte. Giemsa ×250.

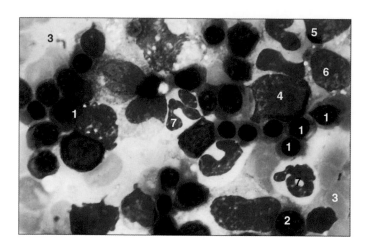

Figure 5.23 Bone marrow (dog). (1) Basophilic erythroblast. (2) Polychromatophil erythroblast. (3) Erythrocytes. (4) Myeloblast. (5) Promyelocyte. (6) Myelocyte. (7) Polymorphonuclear leukocytes. Giemsa ×500.

The erythrocyte series begins with the proerythroblast, a large cell with a large clear nucleus with two nucleoli in a basophilic cytoplasm. The basophilic erythroblast is smaller, with a dense nucleus and diffuse basophilic cytoplasm. The polychromatophilic erythroblast has a dense nucleus in grey-pink cytoplasm, which is caused by the synthesis of haemoglobin. There are no further cell divisions; the nucleus condenses, the haemoglobin content of the cytoplasm increases, and the cell is called a normoblast. The nucleus then becomes extruded and the cell becomes a reticulocyte (an immature form of erythrocyte), small numbers of which are present in circulating blood.

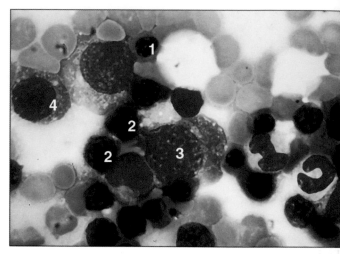

Figure 5.25 Bone marrow (dog). (1) Basophilic erythroblast. (2) Polychromatophil erythroblast. (3) Promyelocyte. (4) Myelocyte. Giemsa ×480.

A similar series forms the granulocytes, beginning with the myeloblast, a large cell with a clear nucleus and a pale rim of agranular cytoplasm. This divides and forms the promyelocyte, a very large cell with a large clear nucleus and some granules in the basophilic cytoplasm. The myelocyte is a markedly smaller cell. The deeply indented nucleus is eccentrically disposed in the cytoplasm, and specific granules indicate the type of leukocyte. Late metamyelocytes have a band nucleus; again, these are found in the circulating blood. The mature leukocyte has a lobed nucleus.

Where haemopoietic tissue predominates, with small amounts of adipose tissue, the active marrow is red in appearance (**Figures 5.22**, **5.23**, and **5.26**). Where

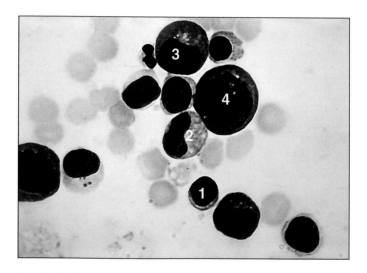

Figure 5.24 Bone marrow (dog). (1) Basophilic erythroblast. (2) Polychromatophil erythroblast. (3) Promyelocyte. (4) Myelocyte. Giemsa ×480.

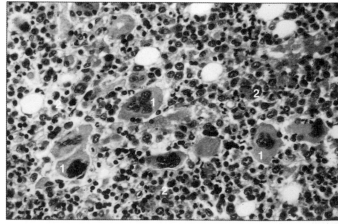

Figure 5.26 Red bone marrow (dog). (1) Megakaryocytes are separated by haemopoietic tissue. (2) Blood vessel. Giemsa ×250.

adipose tissue predominates, the inactive marrow is yellow (**Figure 5.27**). In some conditions, yellow bone marrow can become red bone marrow if the body needs to increase the supply of blood cells. Blood sinusoids lined by reticuloendothelial cells are present in the marrow (**Figure 5.27**).

Lymphopoiesis is complicated by the fact that although lymphocytes arise from stem cells in the bone marrow, they become two separate and functionally different cell populations: T lymphocytes and B lymphocytes. T lymphocytes from the thymus are segregated from the blood by the thymic barrier and are responsible for cell-mediated immunity. When they leave the thymus, T lymphocytes are found in diffuse sites in secondary lymphatic tissue, such as the paracortex of a lymph node. The B lymphocytes are present in the bone marrow of mammals. As a result of antigen stimulus (the humoral response), they become plasma cells and produce immunoglobulin.

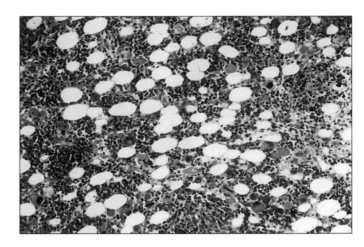

Figure 5.27 Yellow bone marrow (dog). The empty spaces indicate the presence of adipose tissue. Giemsa ×250.

CLINICAL CORRELATES

Disorders of the blood can be attributed to abnormalities in the production of blood cells from the bone marrow or other central blood-forming organs or to abnormal loss or consumption of these cells. Anaemia, the deficiency of oxygen-carrying erythrocytes, can be regenerative (i.e. the bone marrow is able to respond by releasing new cells, including immature cells into the circulation) or nonregenerative (in which this response does not occur). A blood film that illustrates haemolytic anaemia in a dog is shown in **Figure 5.28**. In this regenerative anaemia, there is variation in cell size (anisocytosis) with red cell precursors, such as the large, stippled reticulocyte (1) and nucleated normoblast (2), visible. Some of the red cells have small, dense dots (Howell–Jolly bodies) (3), which represent remnants of nuclear chromatin and are characteristic of regenerative anaemias. Spherocytes, red cells that have lost their biconcave shape and become globular due to immunological attack, are present. There is an increased neutrophil count with some immature neutrophils (reactive neutrophilia with a left shift), which typically accompanies haemolysis. These findings are diagnostic of autoimmune haemolytic anaemia, the most common type of anaemia in the dog.

In immune-mediated anaemia, erythrocyte destruction follows the attachment of antibody to the cell membrane. In most cases, the aetiology is unknown.

The blood film in **Figure 5.29** was prepared from a dog with acute lymphoblastic leukaemia. Very high numbers of malignant lymphoblasts, which are larger than normal lymphocytes and have bluer cytoplasm, are seen. In some cases of leukaemia, although the

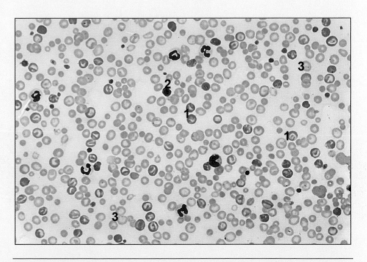

Figure 5.28 Haemolytic (regenerative) anaemia in a dog. Note the variation in the appearance of the red cells. Stippled Reticulocytes (1), nucleated normoblasts (2), and erythrocytes with Howell-Jolly bodies (3) are present. Giemsa ×200.

bone marrow is invaded by neoplastic haemopoietic cells, no abnormal cells are released into the peripheral blood. Such cases may be termed aleukaemic leukaemias. A nonregenerative anaemia is a common finding with many leukaemias.

The liver of a cat in which the sinusoids (vascular channels) are heavily infiltrated by large neoplastic megakaryocytes with multilobed nuclei is shown in **Figure 5.30**. This is megakaryocytic myelosis, a rare, chronic disease of dogs and cats characterised clinically by bleeding and thrombosis of the ear and tail tips.

The presence of numerous megakaryocytes in the splenic red pulp is a common finding in the healthy house mouse, *Mus musculus,* whereas extramedullary haemopoiesis is usually considered an abnormal finding in most other animals at adult stages (**Figure 5.31**).

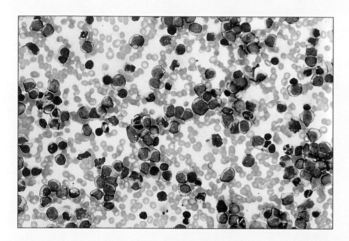

Figure 5.29 Acute lymphoblastic leukaemia (dog) showing large numbers of malignant lymphoblasts. Giemsa ×250.

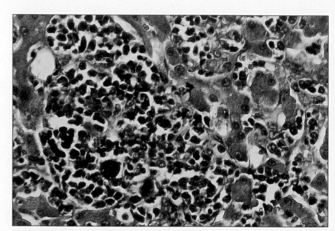

Figure 5.30 Megakaryocytic myelosis in the liver of a cat. Note the megakaryocyte with the multilobed nuclei in the cellular infiltrate. H&E ×250.

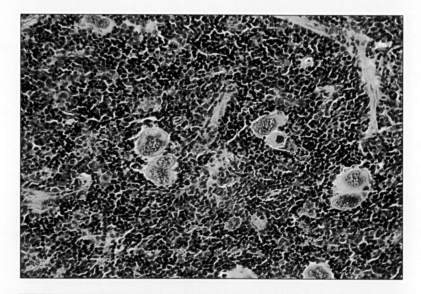

Figure 5.31 Extramedullary haemopoiesis in the spleen of a domestic mouse (*Mus musculus*). Note the multinucleated megakaryocytes which are normal in murine splenic tissue. H&E ×200.

Avian, Reptilian, Amphibian, and Fish Bone and Bone Marrow

A unique feature of avian bone in egg-producing birds is the accumulation of spiculated bone in the medullary cavity, under the combined influence of oestrogens and androgens. This medullary bone is particularly labile, and the stored calcium is utilised in the formation of the calcareous egg shell. The medullary bone is basophilic and decreases during shell deposition and increases at other times. The basophilia is caused by a change in the density of the matrix and in the glycosaminoglycans that are present. During resorption, osteoclasts are active; during deposition, osteoblasts are prominent on the surface of the bone spicules (**Figures 5.32** and **5.33**).

Reptiles do not possess pneumatised bone such as that found in birds capable of flight. The box-like carapace and plastron of chelonians are composed of specialised bone that must be both strong and relatively lightweight in order for this bone to protect efficiently the delicate internal structures. Compressional stresses applied to the dorsal and ventral surfaces are distributed widely and are then borne

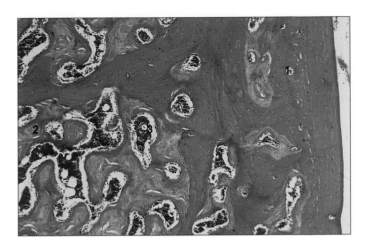

Figure 5.33   Head of the humerus (domestic hen). (1) The osteocytes in the eosinophilic cortical bone form an osteon. (2) Osteocytes in the open lacunae of the basophilic medullary bone. (3) Marrow cavity lined by osteoblasts. H&E ×125.

upon buttress-like vertical supporting pillars of bone that are at each end of the 'bridge' that joins the carapace to the plastron on each side. The strength of their shells is further enhanced by the curved shape and by additional internal struts that distribute compressive forces so that they are not concentrated onto a single focus. The shell is composed of parallel layers of inner and outer tables of compact bone with an intervening layer of spongy cancellous bone characterised by spaces filled with bone marrow. In form and function, the bony shell resembles the calvaria that covers the brain case of a mammalian skull.

The bone of amphibians and reptiles contains numerous sites of haemopoiesis; long bones (*see* **Figure 4.29**), ribs, skull, and mandibles are locations in which active blood cell formation normally occurs. During severe blood loss and a few other conditions, sites other than bone marrow are recruited for extramedullary haemopoiesis; liver, spleen, and kidney are then most often involved.

In fish, the major organ of haemopoietic activity is the cranial pole of each kidney (*see* Chapter 10), with lesser amounts occurring in other extramedullary sites during times of severe anaemia, infection, or stress.

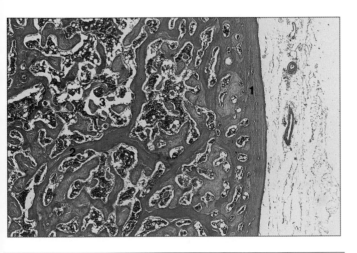

Figure 5.32   Head of the humerus (domestic hen). (1) Outer layer of eosinophilic cortical bone. (2) The spiculated medullary bone is basophilic. (3) The marrow cavity is lined by osteoblasts. H&E ×62.

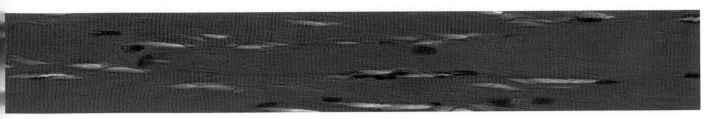

6

MUSCLE

Contractility, a fundamental property of many cells in the body with limited movement, is developed to a highly specialised degree in muscle tissue. Muscle tissue is present in three main areas of the body: the walls of hollow organs, the skeletal muscles, and the heart. The elongated muscle cell is commonly referred to as a myocyte, myofibre, or muscle fibre, the plasma membrane as the sarcolemma, and the cytoplasm as the sarcoplasm. There are basically two types of muscle: smooth, visceral, or involuntary muscle and striated muscle, which is further subdivided into skeletal voluntary muscle and cardiac involuntary muscle.

Muscle contraction depends upon the proteins actin and myosin in the sarcoplasm. In skeletal and cardiac muscle, the longitudinal arrangement of these proteins is aligned in register to give cross-striations, which are absent from smooth muscle.

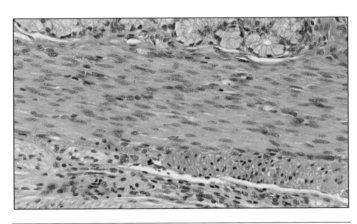

Figure 6.1   Smooth muscle. Duodenum (pig). H&E ×400.

Muscle Types

Smooth Muscle

Smooth muscle cells are elongated fusiform fibres (average length of about 20 μm with a diameter of 5–10 μm) with a single, centrally located oval nucleus with several nucleoli. These fibres may occur singly, as in the lamina propria of intestinal villi, but are more commonly arranged in sheets or layers in the walls of a wide range of hollow organs within the alimentary, urogenital, respiratory, and cardiovascular systems. The fibres are packed in a staggered fashion with the thickest nucleated portion of one fibre juxtaposed to the thin tapered end of an adjoining one. Individual smooth muscle fibres are supported by reticular fibres, with collagen and elastic fibres forming a supporting framework of connective tissue carrying blood vessels and nerves. Smooth muscle is under involuntary control, and it is innervated by the autonomic nervous system (**Figures 6.1–6.3**).

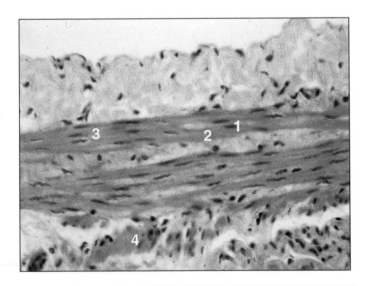

Figure 6.2   Smooth muscle. Uterus (cat). (1) Nucleus. (2) Sarcoplasm. (3) Longitudinal fibres. (4) Transverse fibres. H&E ×400.

DOI: 10.1201/9781003333807-6

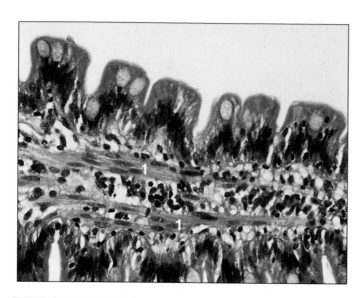

Figure 6.3 Smooth muscle. Duodenum (dog). (1) Fusiform smooth muscle fibre. H&E ×200.

Striated (Skeletal) Muscle

Skeletal muscle fibres are multinucleated cells that range in length from a few millimetres to several centimetres. The nuclei lie immediately beneath the sarcolemma, and the myofibrils give both a longitudinal and a cross-striation arrangement. The contractile myofilaments are arranged in alternating I bands (isotropic with polarised light) and A bands (anisotropic with polarised light). This imparts the cross-striated effect.

Skeletal muscle fibres are bound into large bundles by an outer dense irregular connective tissue investment, the epimysium. This dips into the muscle and surrounds bundles of muscle fibres (fascicles) in dense connective tissue, the perimysium. A network of reticular fibres, the endomysium, surrounds each individual muscle cell. The collagen fibres of the tendon extend into the epimysium and allow muscular contraction to affect movement (**Figures 6.4–6.9**). Every

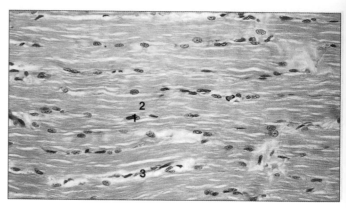

Figure 6.5 Striated muscle. Tongue (cat). (1) Nucleus. (2) Sarcoplasm. (3) Endomysium. H&E ×200.

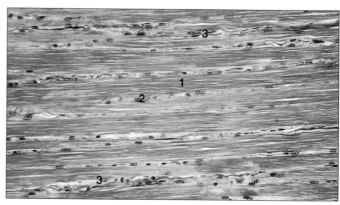

Figure 6.6 LS striated muscle. Tongue (cat). (1) Longitudinal striated muscle. (2) Nuclei. (3) Endomysium stained green. Masson's trichrome ×200.

skeletal muscle fibre receives an axon terminal from a motor neuron at the myoneural junction, the motor endplate (*see* Chapter 14). Each motor neuron and the muscle fibres controlled by it constitute a motor unit. Muscle spindles, which are attenuated skeletal muscle fibres, act as stretch receptors that are innervated both by motor and by sensory nerve

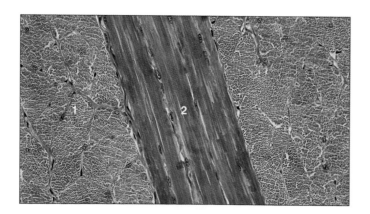

Figure 6.4 Striated muscle. Tongue (ox). (1) Transverse section (TS) muscle fibres. (2) Longitudinal section (LS) muscle fibres. Masson's trichrome ×125.

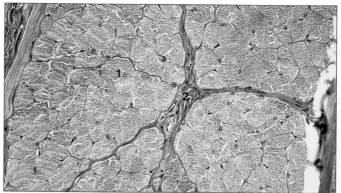

Figure 6.7 TS striated muscle. Tongue (cat). (1) Transverse section of muscle fibres. (2) Connective tissue perimysium. Masson's trichrome ×62.5.

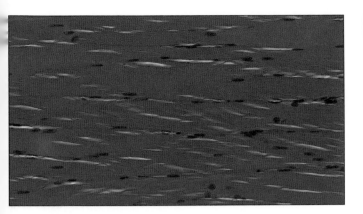

Figure 6.8 Striated muscle. Skeletal muscle (horse). Muscle fibres showing cross-striations. H&E ×400.

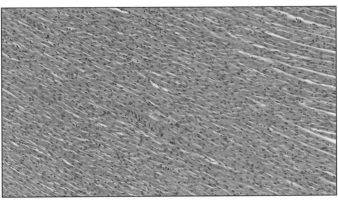

Figure 6.10 Cardiac muscle (pig). H&E ×150.

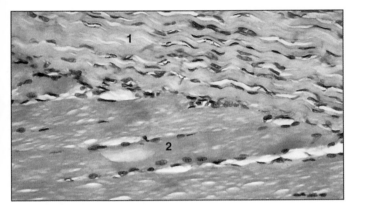

Figure 6.9 Tendon/muscle junction (dog). (1) Collagen fibres of the tendon. (2) Striated muscle fibres. H&E ×200.

terminals. Neurotendinous spindles (Golgi tendon organs) are located at the juncture of a muscle with its tendon and monitor the intensity of muscle contraction.

Cardiac Muscle (Myocardium)

Cardiac muscle is exclusive to the myocardium, the muscular wall of the heart. The fibres are smaller than skeletal muscle fibres (85–100 μm long with a diameter of about 15 μm) and branch repeatedly (**Figure 6.10**). The oval nucleus lies in the centre of the fibre (occasionally two nuclei are present), and the fibres have strong areas of attachment at the intercalated discs. These are visible as a dark cross-striation at the end of one fibre and the beginning of the next and confer structural integrity on the heart muscle to allow contraction to spread throughout the myocardium (**Figures 6.11** and **6.12**). These are represented at the ultrastructural level by fascia adherens and desmosomes in the transverse portion and gap junctions in the lateral portion. The cardiac conducting system is composed of several specialised muscle fibres in the sinoatrial and atrioventricular nodes. In the sinoatrial node, small cardiac muscle fibres, which are low in myofilaments, have an intrinsic ability to contract at a species-specific rate and act

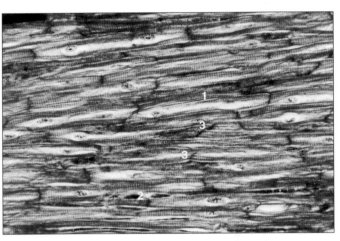

Figure 6.11 Cardiac muscle (horse). (1) Cardiac muscle fibre. (2) Vascular connective tissue of the endomysium. (3) Intercalated disc. Heidenhain's iron haematoxylin ×625.

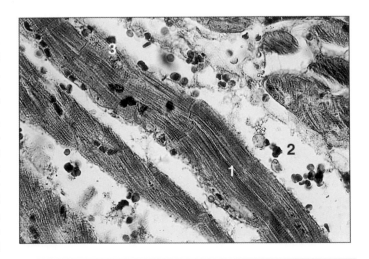

Figure 6.12 Cardiac muscle (horse). (1) Cardiac muscle fibre. (2) Vascular connective tissue of the endomysium. (3) Intercalated disc. Heidenhain's iron haematoxylin ×400.

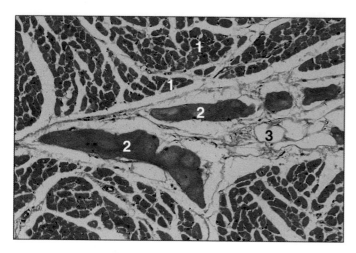

Figure 6.13 Cardiac muscle (horse). (1) Cardiac muscle. (2) Purkinje's fibres. (3) Loose vascular connective tissue. H&E ×125.

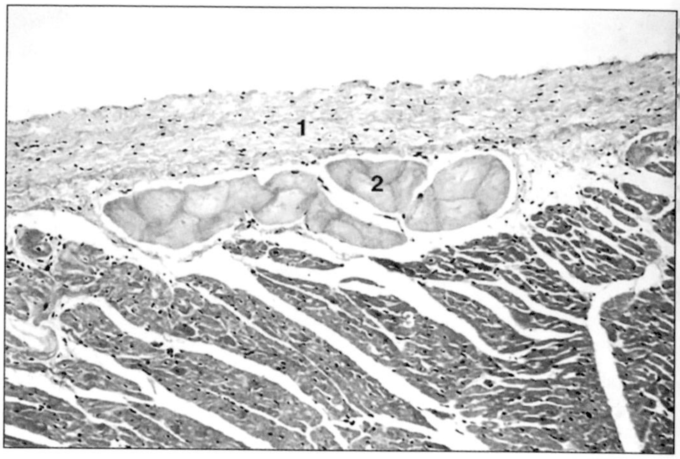

Figure 6.14 Cardiac muscle (horse). (1) Endocardium. (2) Impulse-conducting fibres (Purkinje's fibres). (3) Muscle fibre. H&E ×125.

as the pacemaker for cardiac muscle contraction. The atrial wave of depolarisation converges on the second node, the atrioventricular node, from where specialised large muscle fibres spread throughout the ventricular muscle and initiate contraction. These specialised conducting fibres (Purkinje fibres) are larger than other cardiac muscle cells and have a central nucleus and a pale area around it where glycogen is stored (**Figures 6.13** and **6.14**).

The various types of muscle tissue can be discerned, even in invertebrates low on the phylogenetic scale. For example, the striated muscle fibres of a spider are structurally similar to those found in mammals and the multiple heart-like pumping chambers that circulate the haemolymph and haemolymphocytes through the coelom of earthworms are formed from myocardium.

As all fish and amphibians and most reptiles (all except the crocodilians) possess a three-chambered heart with paired atria and a single ventricle, there are some structural differences between the hearts of these animals. For example, a ridge of myocardium helps direct the flow of blood through the ventricle so as to reduce the mixing of oxygenated and deoxygenated blood in noncrocodilian reptiles. Blood flow through twin aortic arches and atrioventricular valves is controlled by typical heart valve leaflets composed of myxoid connective tissue that are covered by a thin endothelial lining in all vertebrates.

CLINICAL CORRELATES

Myopathies, primary disorders of muscle structure, may be congenital, metabolic, or inflammatory. Viral, bacterial, fungal, parasitic, protozoal, and metazoal agents can be implicated in infective myopathies.

Where there is inflammation, the disease would be termed a myositis. In eosinophilic myositis, a condition that affects the masticatory muscles of dogs, in particular in the German Shepherd breed (**Figure 6.15**), the affected muscle is diffusely infiltrated by numerous eosinophils accompanied by lesser numbers of lymphocytes and other inflammatory cells. There is muscle degeneration and atrophy. The production of abnormal antibodies which attack these muscles is believed to initiate the process, which then becomes dominated by a cellular response. Progressive destruction of these muscles leads to fixation of the jaws. Granulomatous myositis can also be found in different skeletal muscles from pigs, with the presence of multinuclear giant cells between muscle

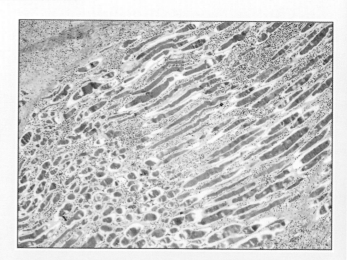

Figure 6.15 Eosinophilic myositis. Masticatory muscles from a dog. Phosphotungstic acid haematoxylin ×125.

fibres (**Figure 6.16**). This finding has been related to the infection by porcine circovirus type 2.

Other non-neoplastic conditions that affect the muscle may be loosely divided into neuropathies which result from disturbance of innervation and myasthenic conditions of the motor end plate. Muscle atrophy (reduction in cross-sectional area of the muscle fibres) will tend to occur.

Under certain circumstances, all types of muscles can be vulnerable to pathological deposition of mineral salts. This may be induced by conditions characterised by persistent hypercalcaemia (*see* **Figure 7.27**) or at sites of the previous injury.

Primary neoplasms of muscle are quite rare with malignant tumours (rhabdomyosarcoma, **Figure 6.17**) outnumbering benign ones (rhabdomyoma). Muscle can also be affected by neoplasms of associated connective tissue origin.

Sarcocystis spp. are protozoal parasites that can be found within the muscle of different animal species (**Figure 6.18**) as an incidental finding forming cysts without evidence of inflammation or clinical disease.

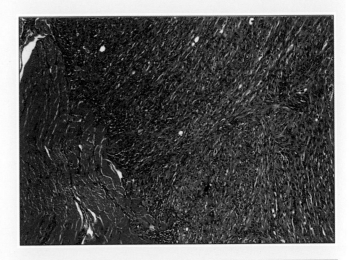

Figure 6.17 Rhabdomyosarcoma in a 10-year-old male cat. Large, rather pleomorphic, elongate to strap-like cells invade and replace skeletal muscle. These tumours, although uncommon, are highly malignant. H&E ×100.

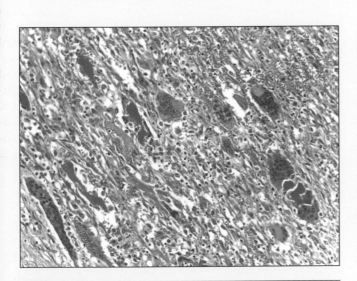

Figure 6.16 Granulomatous myositis (pig) with the presence of multiple inflammatory cells including multinucleated giant cells between the skeletal muscle fibres. H&E ×200.

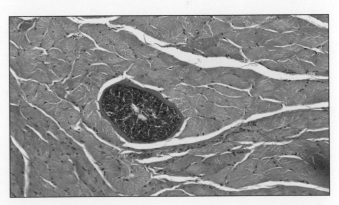

Figure 6.18 Oval cyst in muscle from a sheep produced by *Sarcocystis* spp. H&E ×200.

7

CARDIOVASCULAR SYSTEM

The mammalian circulatory system is a closed system of tubes with an endothelial lining in which blood flows from the heart through the arteries into the capillaries and back through the veins. The lymph passes through the lymph vascular system, which forms drainage channels joining major veins.

Arteries

Arteries are the conducting channels conveying blood from the heart to the capillary bed. There are three types of arteries: elastic (conducting), muscular (distributing), and arterioles. The arterial wall has a common structure: tunica intima (inner lining layer), tunica media (middle layer), and tunica adventitia (outer layer; **Figures 7.1** and **7.2**). The tunica intima consists of elongated, flattened endothelial cells resting on loose areolar connective tissue and an internal elastic lamina.

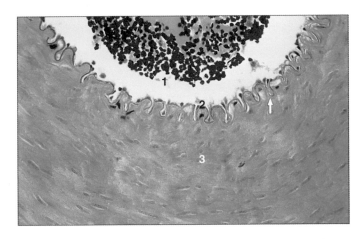

Figure 7.2 Artery (dog). (1) Lumen. (2) Tunica intima with the internal elastic lamina (arrowed). (3) Tunica media. Masson's trichrome ×125.

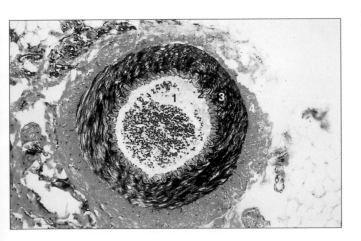

Figure 7.1 Artery (dog). (1) Lumen. (2) Tunica intima. (3) Tunica media. (4) Tunica adventitia. Masson's trichrome ×62.5.

The tunica media is the thickest layer of the vessel. In elastic arteries, this layer has a high proportion of concentric lamellae of fenestrated elastic fibres interspersed with smooth muscle fibres (**Figure 7.3**). These allow the vessels to dilate. The recoil sends the blood onwards, creating the pulse in the major elastic artery, the aorta.

In muscular (distributing) arteries, the elastic content is reduced and the smooth muscle increased. Larger muscular arteries have an external elastic lamina (**Figures 7.2** and **7.4–7.6**).

The arterioles (**Figure 7.7**) reduce the pressure of the blood and supply the capillary bed. The tunica media may consist of only one layer of smooth muscle cells, and the luminal diameter is less than the thickness of the wall (**Figure 7.8**). The tunica adventitia is composed of collagen and elastic fibres, contains the vasa vasorum, the small nutrient arteries and veins in the walls of the larger blood vessels, and blends into the surrounding connective tissue.

DOI: 10.1201/9781003333807-7

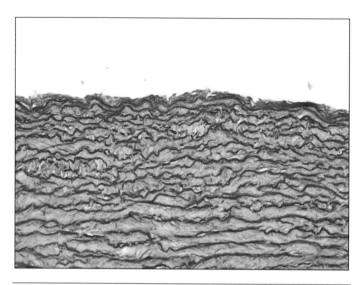

Figure 7.3 Aorta (pig). Elastic artery where the wavy elastic fibres can be identified. Orcein stain ×200.

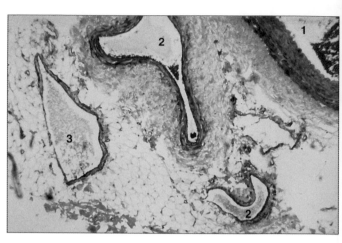

Figure 7.6 Artery and vein. Stomach (dog). (1) Small artery. (2) Small vein. (3) Lymphatic vessel. H&E ×62.5.

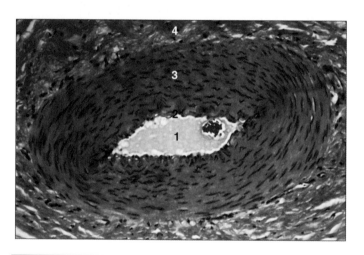

Figure 7.4 Muscular artery (sheep). (1) Lumen. (2) Tunica intima with the internal elastic lamina. (3) Tunica media. (4) Tunica adventitia. H&E ×125.

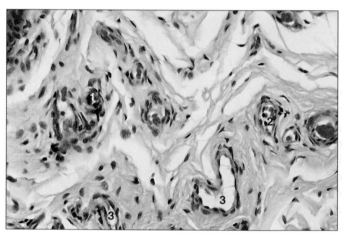

Figure 7.7 Arteriole, vein, and a lymphatic vessel in connective tissue. Tongue (ox). (1) Arteriole. (2) Vein. (3) Lymphatic vessel. H&E ×200.

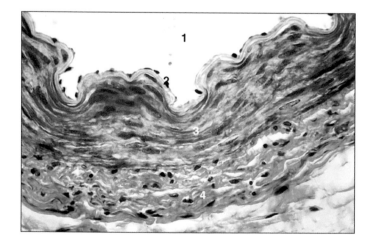

Figure 7.5 Muscular artery (sheep). (1) Lumen. (2) Tunica intima. (3) Tunica media. (4) Tunica adventitia. Masson's trichrome ×250.

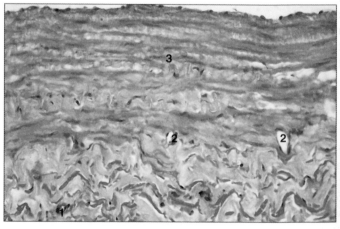

Figure 7.8 Arterioles in the wall of a larger artery (horse). (1) Arteriole. (2) Veins. (3) Artery. H&E ×160.

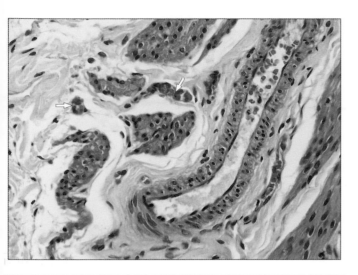

Figure 7.9 Capillaries (sheep) in the connective tissue of the cervix (arrowed). Masson's trichrome ×400.

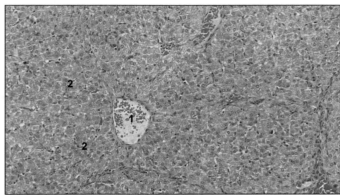

Figure 7.10 Liver (pig). (1) Central vein. (2) Liver sinusoids. H&E ×200.

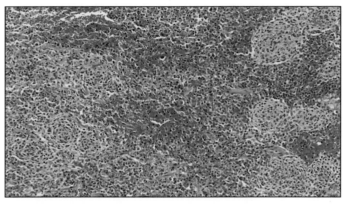

Figure 7.11 Spleen (pig). Sinusoids filled with erythrocytes. H&E ×200.

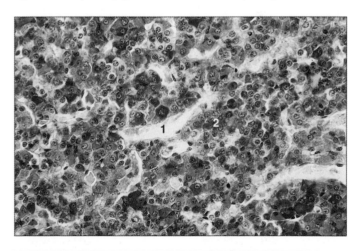

Figure 7.12 Pars distalis of the adenohypophysis (cat). (1) Sinusoids. (2) Cords of hypophyseal cells. Orange G ×400.

Capillaries and Venules

The capillary is the smallest unit of the vascular system. The diameter of the lumen is no larger than an erythrocyte and permits these cells to pass in single file only (**Figure 7.9**). The capillary wall is two-layered: a tunica intima of one or two squamous endothelial cells resting on a basal lamina. The endothelial cells usually form a continuous layer, but fenestrated capillaries occur in the renal glomerulus, the endocrine glands, intestinal villi, and the choroid plexus, where gaps are present between adjoining cells closed by a pore diaphragm. The wall also contains pericytes: undifferentiated cells believed to be capable of becoming fibroblasts, muscle cells, or endothelial cells after injury. Venules collect the blood from the capillaries. Their lumina are wider than those of the arterioles. The tunica intima (there is no tunica media and adventitia) in each venule consists of a continuous layer of endothelial cells and areolar connective tissue. Pericytes are also present.

Sinusoids

Sinusoids are found in the liver, spleen, bone marrow, and adenohypophysis. Their wide lumina are lined by discontinuous endothelial cells and basal lamina, contain large fenestrae without diaphragms, and the wall is interspersed with fixed macrophages of the mononuclear phagocyte scavenging and defence system of the body (**Figures 7.10** and **7.11**). Similar thin-walled venous sinuses are found in endocrine glands (**Figure 7.12**).

Veins

Veins are lined by a continuous layer of endothelial cells and areolar connective tissue. The tunica media, which is always narrow, contains a few circular smooth muscle fibres and some elastic fibres, but no elastic lamina. The tunica adventitia, the thickest layer in veins, consists of

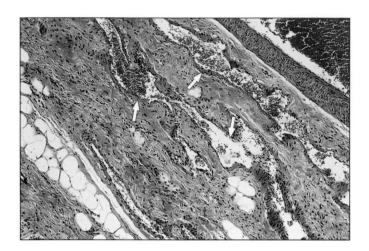

Figure 7.13 Spermatic cord (horse). Venous plexus (arrowed). H&E ×62.5.

longitudinal collagen fibre bundles, elastic fibres, and, in the larger veins, some smooth muscle (**Figures 7.13–7.15**). The lumen contains valves that are projections of the tunica intima; these allow only unidirectional blood flow (**Figures 7.16** and **7.17**).

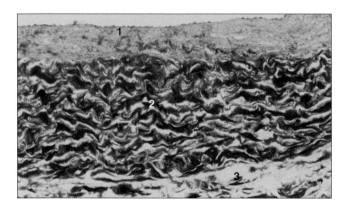

Figure 7.14 Caudal vena cava (dog). (1) Tunica intima. (2) Tunica media. (3) Tunica adventitia. Gomori's trichrome ×125.

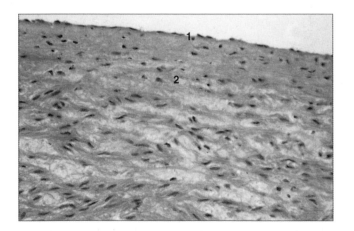

Figure 7.15 Cranial vena cava (horse). (1) Tunica intima. (2) Tunica media. H&E ×125.

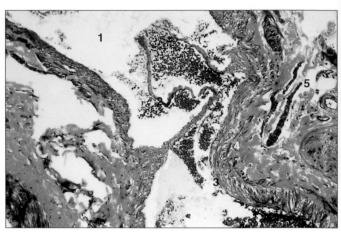

Figure 7.16 Femoral vein with valves (cat). (1) Lumen. (2) Valve. (3) Tunica intima. (4) Tunica media. (5) Tunica adventitia. Masson's trichrome ×100.

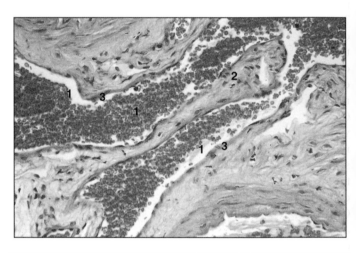

Figure 7.17 Valve in the brachial vein (cat). (1) Lumen filled with erythrocytes. (2) Valve. (3) Tunica intima. H&E ×125.

Arteriovenous Anastomoses

Arteriovenous anastomoses are special areas of the skin, nose, lips, pads, intestine, and male and female reproductive tract, where the arteriole opens directly into a venule without going through the capillary bed (**Figure 7.18**). They are involved in thermoregulation, regulation of blood pressure, erection, and menstruation (in primates).

Heart

The cardiac wall consists of three layers: endocardium (inner), myocardium (middle), and epicardium (outer).

The endocardium contains continuous squamous endothelial cells with an underlying layer of fibroelastic connective tissue and a subendocardial layer with vascular

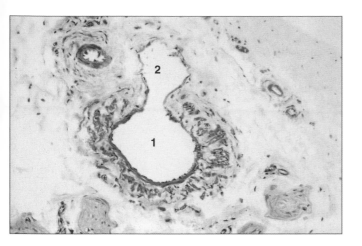

Figure 7.18 Arteriovenous anastomosis in loose connective tissue (cat). (1) Artery. (2) Vein. H&E ×62.5.

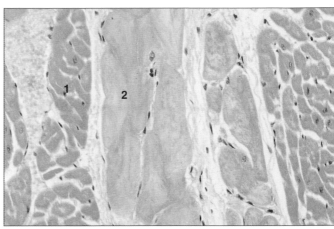

Figure 7.20 Cardiac muscle (horse). (1) Cardiac muscle fibres. (2) Conducting (Purkinje) fibres. H&E ×125.

areolar connective tissue and conducting (Purkinje) fibres. The myocardium is the thicker layer and is composed of cardiac muscle and also contains vascular areolar connective tissue. The epicardium is the serous membrane that covers the external surface of the heart, is thicker than the endocardium, and fat deposits in the rather dense connective tissue and coronary blood vessels are often found (**Figures 7.19** and **7.20**). Fibrous rings support the heart valves (**Figure 7.21**). Between the valves, a triangular connective tissue area (trigonum fibrosum) is found. Those structures, together with the fibrous part of the interventricular septum, provide a means of insertion for the cardiac muscle and may be referred to as the cardiac skeleton (**Figure 7.22**). Depending on the species and age of the animal, the cardiac skeleton may contain dense irregular connective tissue (pig and cat), fibrocartilage (dog), hyaline cartilage (horse), or bone (*ossa* cordis in large ruminants).

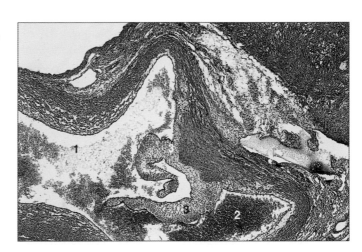

Figure 7.21 Heart (dog). (1) Lumen of the atrium. (2) Lumen of the pulmonary artery. (3) Valve cusps. Masson's trichrome ×62.5.

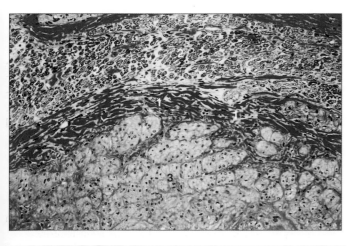

Figure 7.19 Cardiac muscle (ox). (1) Cardiac muscle. (2) Connective tissue of the cardiac skeleton. (3) Atrioventricular node. Masson's trichrome ×125.

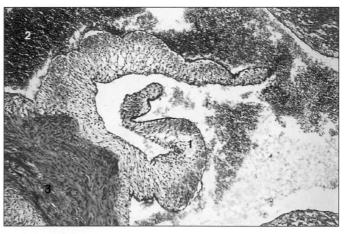

Figure 7.22 Heart (dog). (1) Valve cusps. (2) Blood within atrial lumen. (3) Dense connective tissue part of the fibrous skeleton of the heart. Masson's trichrome ×62.5.

Lymphatic Vessels

Lymphatic vessels drain excess fluid from the tissues and are made up of a thin layer of connective tissue with an endothelial lining with an absent or discontinuous basal lamina. Larger lymphatic vessels, such as the thoracic duct, may have a few smooth muscle fibres in the wall. Lymph vessels contain valves.

Amphibians and reptiles have perilymphatic and endolymphatic systems that are particularly well developed in some species. These lymph-filled structures serve several functions. Perilymphatic pathways encircle the auditory apparatus and may participate in the transmission of sounds. The endolymphatic system consists of receptor organs of the inner ear as well as either bilateral separate or fused thin-walled sacs. These communicate with the skull via narrow ducts and serve as reservoirs for the storage of calcium carbonate microcrystals. In some amphibians, endolymphatic sacs form a ring-like extension of the vertebral canal around the brain.

CLINICAL CORRELATES

A range of cardiovascular diseases is important in veterinary medicine. The heart itself can be affected by congenital, degenerative, inflammatory, and neoplastic conditions with a variety of underlying causes. Disease caused by congenital malformations is naturally recognised most often in young animal.

In adult animals, primary cardiac disease is most often encountered in the form of cardiomyopathy or degenerative change in the muscle of the heart or the valves. Myxomatous degeneration can be observed in the mitral valve from a dog with heart failure (**Figure 7.23**). In **Figures 7.24** and **7.25**, special stains have been used to highlight features of interest. Both are from the same case, a 12-year-old dog with myocardial degeneration. In **Figure 7.24**, a Sirius Red stain, which colours collagen red, demonstrates a large amount of fibrosis replacing muscle bundles, which are stained yellow. In **Figure 7.25**, a Masson's trichrome stain highlights muscle fibres undergoing degeneration.

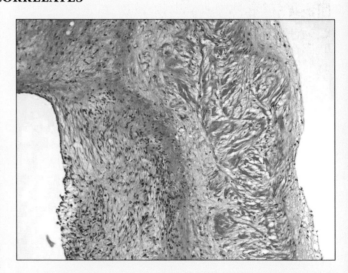

Figure 7.23   Heart (dog) Mitral valve myxomatous degeneration showing a disarray of collagen fibres. H&E ×200.

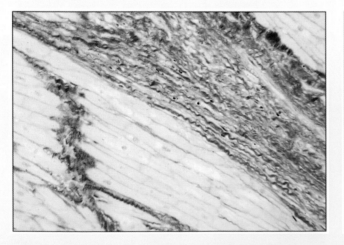

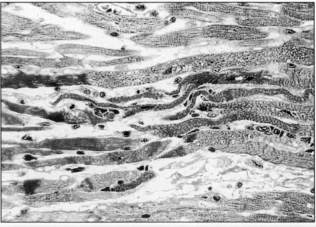

Figures 7.24 and 7.25   Myocardial degeneration and fibrosis in a 12-year-old dog with cardiomyopathy. In **Figure 7.24**, the collagenous tissue which replaces muscle bundle is stained red. Sirius Red ×200. In **Figure 7.25**, degenerative muscle fibres are seen stained strongly orange in the centre. The striations of the muscle fibres are also demonstrated with this stain. Masson's trichrome ×400.

Myocardial degeneration and fibrosis may be found in cases of dilated cardiomyopathy or as a nonspecific response of the cardiac muscle to injury or insult.

One of the most important neoplastic diseases to affect the blood vessels is haemangiosarcoma, a malignant tumour of the endothelial cells that line blood vessels. In dogs, the spleen, right atrium, and skin are common primary sites and metastasis can be very widespread. Other species are also affected. A haemangiosarcoma in the heart from a 5-year-old boxer dog is shown in **Figure 7.26**.

The smooth muscle in the tunica media of blood vessels can be affected by the deposition of calcium salts. This change is frequently induced by oversupplementation of the diet of herbivorous reptiles or amphibians with vitamin D_3 either directly or by the inclusion of commercial dog, cat, or primate food. Similarly, hypervitaminosis D_3 may be induced when supplemented goldfish are fed to fish-eating reptiles and amphibians. Early arteriosclerotic mineralisation of the tunica media in a large pulmonary artery in an iguana is shown in **Figure 7.27**.

Figure 7.26 Haemangiosarcoma in a dog. The tumour cells are spindle-shaped with large, often hyperchromatic, nuclei and are arranged into loosely interlacing bundles that form irregular, blood-filled channels and spaces. H&E ×100.

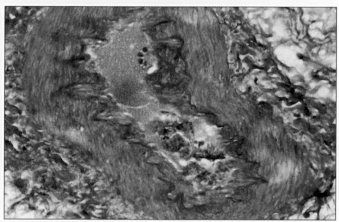

Figure 7.27 Early arteriosclerotic mineralisation in a large pulmonary artery (iguana). The mineral deposition is highlighted in red. Alcian blue/PAS ×125.

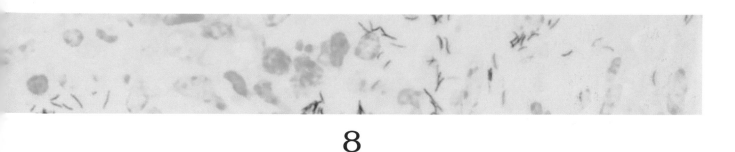

8

RESPIRATORY SYSTEM

The main function of the respiratory system is the conduction of inspired air containing oxygen along the respiratory passages (nasal cavity, nasopharynx, larynx, trachea, and bronchi) to reach the lung alveoli where the gas exchange takes place, and to conduct the expired air containing carbon dioxide out of the body. The air is warmed, moistened, and filtered in these airways before reaching the organs responsible for respiration: the lung parenchyma, the respiratory bronchioles, alveolar sacs, and alveoli. It is between the alveoli and the capillaries that gas exchange takes place.

The paranasal sinuses are air-filled cavities found in skull bones. They are lined with ciliated mucous epithelium continuing with a mucosal membrane of the nasal cavity.

Conduction of Air

The skin around the nostrils has long tactile hairs and numerous sebaceous and sweat glands in many animal species, communicating with the nasal cavities (**Figure 8.1**).

The respiratory mucosa lines all but the finer divisions of the respiratory tract and consists of pseudostratified or simple columnar ciliated epithelial cells and mucus-secreting goblet cells. The lamina propria is continuous with the perichondrium or periosteum where appropriate, is normally highly vascularised, and contains both collagen and elastic fibres. Seromucous glands within the submucosa secrete into the lumen through the epithelium (**Figures 8.2–8.4**).

The mucociliary apparatus consists of three components: the cilia in the apical part of the epithelial cells, a protective mucous layer, and an airway surface liquid layer. These components work in concert to remove inhaled particles from the lung (lower respiratory tract) and the upper respiratory tract. The mucociliary apparatus beats towards the upper respiratory tract, including the pharynx, an area common both to the digestive and the respiratory system.

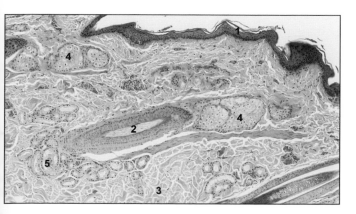

Figure 8.1 Skin. Nostril (horse). (1) Epidermis. (2) Sinus hair. (3) Dermis. (4) Sebaceous glands. (5) Sweat glands. H&E ×100.

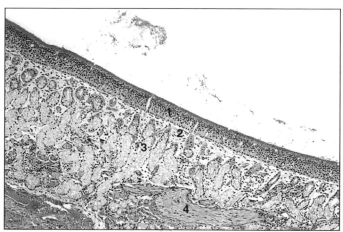

Figure 8.2 Respiratory epithelium (horse). (1) Pseudostratified ciliated columnar epithelium with goblet cells. (2) Lamina propria. (3) Seromucous glands. (4) Nerve. H&E ×62.5.

DOI: 10.1201/9781003333807-8

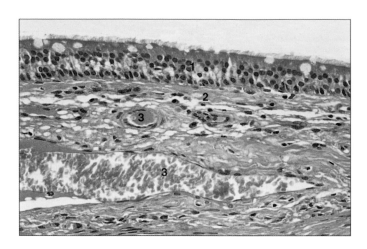

Figure 8.3 Respiratory epithelium (horse). (1) Pseudostratified ciliated columnar epithelium with goblet cells. (2) Lamina propria. (3) Blood vessels. H&E ×250.

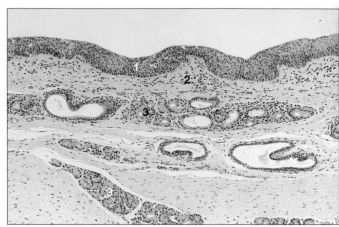

Figure 8.5 Auditory tube (horse). (1) Epithelium. (2) Lamina propria. (3) Seromucous glands. H&E ×100.

Figure 8.4 Respiratory epithelium (horse). The cilia project from the surface of the epithelial cell as fine strands; active mucus-secreting cells lie between the ciliated cells (arrowed). Scanning electron micrograph ×1500.

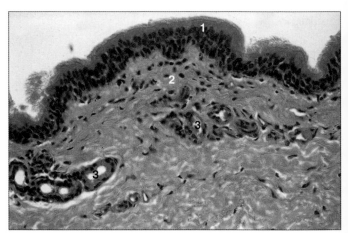

Figure 8.6 Guttural pouch (horse). (1) Respiratory epithelium. (2) Lamina propria. (3) Seromucous glands. H&E ×250.

The auditory tube connects the pharynx to the middle ear and is common to the digestive, respiratory, and auditory systems (**Figure 8.5**). The guttural pouch of equids is a diverticulum of the auditory tube (**Figures 8.6** and **8.7**). The digestive surface of the pharynx is covered by a stratified squamous epithelium that is continuous with the oral cavity and the oesophagus, and the respiratory surface (nasopharynx and laryngopharynx) by respiratory epithelium continuous with the nasal cavities and larynx. The epiglottis is a flap-like structure in the larynx projecting into the pharynx. A plate of elastic cartilage provides internal support. The upper digestive surface mucous membrane is covered by a nonkeratinising stratified squamous epithelium (**Figure 8.8**) and the lower respiratory surface by respiratory epithelium (**Figure 8.9**).

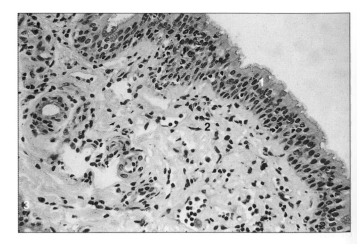

Figure 8.7 Guttural pouch (horse). (1) Respiratory epithelium; the goblet cells are individually stained blue. (2) Lamina propria. (3) Seromucous glands. Alcian blue/periodic acid-Schiff (PAS) ×200.

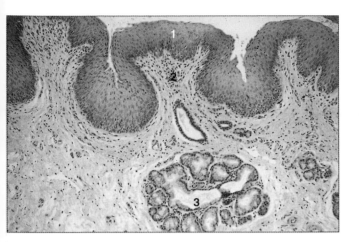

Figure 8.8 Epiglottis (horse). (1) Stratified squamous epithelium. (2) Lamina propria. (3) Seromucous glands. H&E ×200.

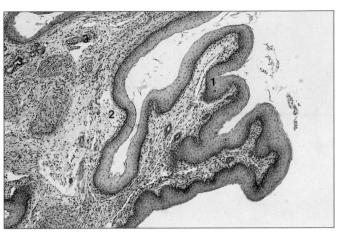

Figure 8.10 Larynx (vocal cord; dog). (1) Stratified squamous epithelium. (2) Lamina propria. (3) Simple tubular glands. H&E ×62.5.

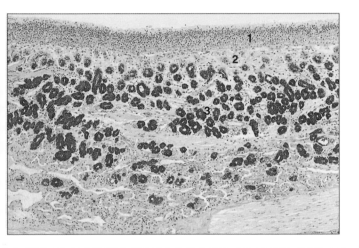

Figure 8.9 Epiglottis (horse). (1) Stratified columnar epithelium with mucus-secreting cells stained blue. (2) Lamina propria. (3) Seromucous glands. Alcian blue/PAS ×125.

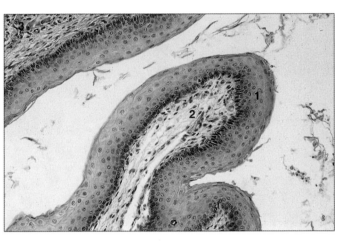

Figure 8.11 Larynx (vocal cord; dog). (1) Stratified squamous epithelium covering the vocal cord. (2) Connective tissue core of the lamina propria. (3) Parasympathetic ganglion. H&E ×250.

The larynx is lined by the respiratory epithelium. The lamina propria is continuous with the perichondrium of the laryngeal cartilages. The vocal cords are covered by stratified squamous epithelium (**Figures 8.10** and **8.11**). The trachea extends from the larynx to the bifurcation of the extrapulmonary bronchi, normally two main bronchi (left and right), but occasionally another large bronchus (the tracheal bronchus) is present in cloven-hoofed mammals. These tubes have the same structure: a lining of respiratory epithelium rising on a lamina propria of loose connective tissue with elastic fibres and mixed seromucous glands opening into the lumen. Rings of hyaline cartilage of different shapes, depending on the species, keep the lumen patent and smooth muscle fibres bridge the gap at the dorsal aspect of the trachea (**Figure 8.12**). In the bronchus, the hyaline cartilage has a plate-like arrangement. The

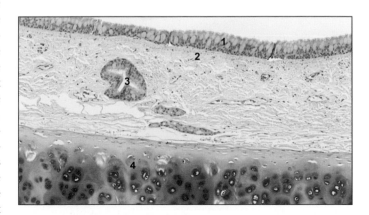

Figure 8.12 Trachea (horse). (1) Respiratory epithelium. (2) Lamina propria. (3) Seromucous glands. (4) Hyaline cartilage. H&E ×250.

smooth muscle forms a spiral and appears as discontinuous blocks in transverse sections (**Figures 8.13** and **8.14**). A fibrous adventitial coat covers the trachea and the extrapulmonary bronchi.

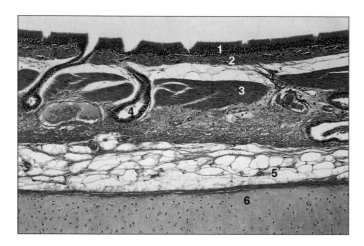

Figure 8.13 Bronchus (sheep). (1) Respiratory epithelium. (2) Lamina propria. (3) Smooth muscle. (4) Simple tubular glands open into the lumen through the epithelium. (5) Connective tissue. (6) Hyaline cartilage. H&E ×25.

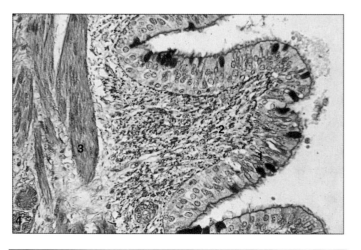

Figure 8.14 Bronchus (ox). (1) Respiratory epithelium; the goblet cells are stained specifically (blue/violet). (2) Lamina propria. (3) Smooth muscle. (4) Simple tubular glands. Gomori/aldehyde fuchsin ×200.

Respiration

The bronchi, both extrapulmonary and intrapulmonary, bring air to the lungs and branch out within the lungs into the bronchioles, which culminate in clusters of minute sacs: the alveoli. In the fetal lung, the duct system is developed, whereas the respiratory part develops slowly (**Figure 8.15**). Expansion begins with the first respiratory movements after birth.

Each lung is covered by elastic connective tissue with an outer layer of the mesothelium, the visceral pleura (**Figure 8.16**). Connective tissue septa divide the

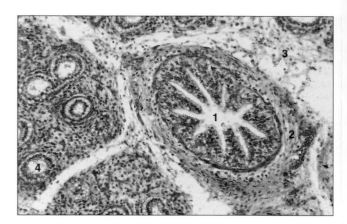

Figure 8.15 Bovine fetal (160-day) lung. (1) Large duct lined by a simple columnar epithelium. (2) Smooth muscle. (3) Vascular mesenchyme. (4) Small ducts. H&E ×200.

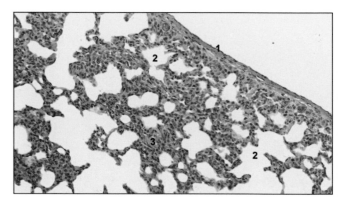

Figure 8.16 Adult lung (pig). (1) Elastic connective tissue with an outer layer of the mesothelium of the visceral pleura. (2) Alveoli. (3) Blood vessels. H&E ×200.

lung into lobes and lobules, and the intrapulmonary bronchi have the same structure as the extrapulmonary bronchi (**Figures 8.17–8.20**). The thickness of these connective tissues septa is highly variable among animal species, to separate the lung areas served by large bronchi (lobation)

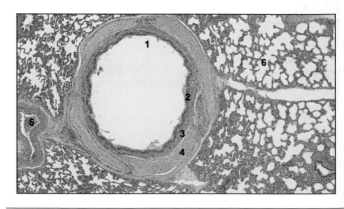

Figure 8.17 Intrapulmonary bronchus. Lung (pig). (1) Lumen lined by the respiratory epithelium. (2) Lamina propria. (3) Smooth muscle. (4) Hyaline cartilage. (5) Blood vessel. (6) Alveoli. H&E ×100.

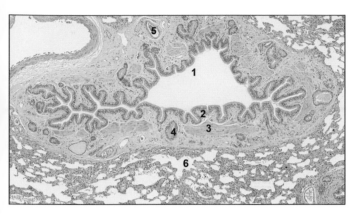

Figure 8.18 Intrapulmonary terminal bronchus. Lung (dog). (1) Lumen lined by respiratory epithelium. (2) Lamina propria. (3) Smooth muscle. (4) Simple tubular glands. (5) Blood vessel. (6) Alveoli. H&E ×60.

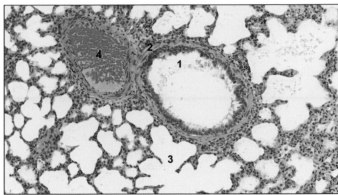

Figure 8.21 Bronchiole. Lung (pig). (1) The bronchiole is lined by cuboidal/low columnar epithelium with ciliated and non-ciliated cells (arrowed). (2) Smooth muscle. (3) Alveoli. (4) Blood vessel. H&E ×200.

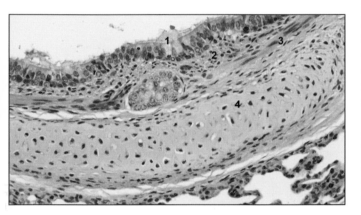

Figure 8.19 Intrapulmonary bronchus. Lung (pig). (1) Respiratory epithelium including goblet cells. (2) Lamina propria. (3) Smooth muscle. (4) Hyaline cartilage. H&E ×400.

or smaller bronchi (lobulation). The bronchi are supported by cartilage, and as the branches become smaller, the cartilaginous support is gradually diminished, and when it disappears completely, the airways are called bronchioles. The epithelium of the bronchioles is columnar or cuboidal and ciliated. In the smaller bronchioles, the epithelium is thinner, the lamina propria is elastic, and the smooth muscle forms a complete ring (**Figures 8.21** and **8.22**).

Club cells are nonciliated epithelial, tall, dome-shaped mostly found in bronchioles, and that protrude into the bronchiolar lumen. They replace the mucus-secreting goblet cells at this level. Both ciliated cells and club cells are present in the terminal and respiratory bronchioles (in the dog and cat, they are lined by the latter exclusively). Club cells divide to form other club or ciliated cells and have an important role in the repair of damaged epithelium after injury. Their secretion also keeps the small airways patent.

Respiratory bronchioles are lined with a low columnar or cuboidal epithelium with ciliated and bronchiolar cells, an elastic lamina propria, and a smooth muscle layer. This opens into the alveolar duct lined by squamous

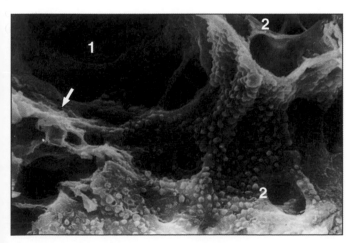

Figure 8.20 Intrapulmonary bronchus (dog). (1) The intrapulmonary bronchus is lined by respiratory epithelium (arrowed). (2) Alveoli. Scanning electron micrograph ×500.

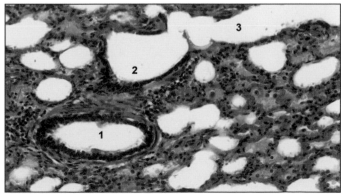

Figure 8.22 Bronchiole. Lung (sheep). (1) The bronchiole is lined by cuboidal epithelium. (2) Respiratory bronchiole. (3) Alveolar duct. (4) Alveoli. H&E ×300.

epithelium, interrupted by atria, and alveoli along its length (**Figure 8.22**). Alveoli are the functional exchange part of the lung. The septa are very thin, with both elastic and collagen fibres, and contain one of the most extensive capillary networks in the body. Cells of the immune system, derived from blood monocytes, are also present and migrate through the alveolar epithelium into the air space, where they phagocytose particulate matter and microorganisms to become dust cells (pulmonary alveolar macrophages). The respiratory membrane where gas exchange takes place consists of capillary endothelial cells, alveolar epithelial cells, and a fused basement membrane (**Figure 8.23**). The squamous alveolar cell, the lining cell responsible for gas exchange, is a type I pneumocyte. The type II pneumocyte is a cuboidal cell that projects into the lumen and secretes surfactant to reduce surface tension. Type II pneumocyte also has a role in the regeneration of damaged alveolar epithelium.

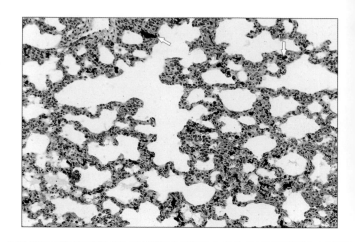

Figure 8.23 Lung (dog). The lung macrophages (dust cells) have phagocytosed the carbon particles (black). The alveolar lining cells are arrowed. Carbon-injected H&E ×256.

CLINICAL CORRELATES

The patterns of disease seen in the respiratory system reflect its structure and function. The respiratory tract is constantly challenged by potentially injurious agents, by both the aerogenous (arrive in inspired air) and haematogenous (arrive in blood supply) routes.

These can include microorganisms such as bacteria, fungi, and viruses or toxic substances or particles in the air. This is especially important in livestock. Adjacent to major airways in many animal species, we can find, typically in bronchi, bronchus-associated lymphoid tissues (BALT), a constitutive mucosal lymphoid tissue with lymphocytes, macrophages, dendritic, and other immune cells. BALT can be activated and proliferate when there is an infection in the lung. A bronchiole surrounded by small cells with dark nuclei and scant cytoplasm is shown in **Figure 8.24**. These cells, which are lymphocytes, also invade the bronchiolar wall. Often termed 'cuffing' pneumonia because of the arrangement of the lymphoid cells around the bronchioles, this is an example of a chronic, non-suppurative pneumonia, commonly seen in calves. Infection with *Mycoplasma* species is quite common.

Respiratory infections are very common in domestic and wild animals. One of the most important infectious diseases that affect humans and animals is tuberculosis. This disease is produced by the infection with a bacillus from the *Mycobacterium tuberculosis* complex. The pathogen induces an inflammatory reaction at the site of infection, with abundant macrophages, neutrophils, lymphocytes, and plasma cells, forming the hallmark lesion of this disease: the granuloma (**Figure 8.25**). The mycobacteria can be identified within the tissue section using a special stain (Ziehl–Nielsen; **Figure 8.26**).

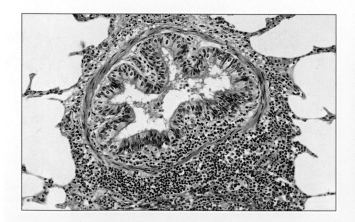

Figure 8.24 Cuffing pneumonia (calf). Lymphocytes cluster around a bronchiole. H&E ×125.

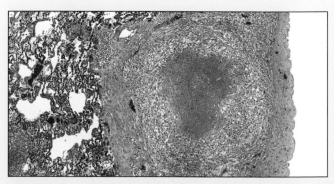

Figure 8.25 Tuberculous granuloma produced by the infection with *Mycobacterium bovis* in the lung from an alpaca (*Lama pacos*). H&E ×100.

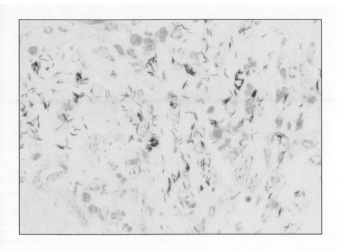

Figure 8.26 Ziehl–Nielsen (ZN) stain for identification of acid-fast bacilli (mycobacteria) within lung granuloma in an infected alpaca (*Lama pacos*). ZN ×400.

The very rich blood supply to the lungs (which have the largest capillary bed in the body) also makes them a common target for haematogenous metastasis from tumours at other sites in the body. Primary lung tumours, both benign and malignant, are recognised in older animals.

Avian Respiratory System

The avian respiratory system contains specialised conducting structures and two lungs containing static structures dedicated to gas exchange connected to air sacs, which expand and contract, moving the air through the static lungs. Unlike mammals, they do not have a diaphragm.

Nasal Cavity

The avian vestibular nasal cavity consists of the nostrils, or the external nares, the operculum, the nasal septum, and turbinates (conchae). The mucosa is lined with a distinctive keratinised stratified squamous epithelium and the conducting passages with pseudostratified ciliated columnar epithelium with simple alveolar mucous glands (**Figures 8.27** and **8.28**).

Trachea

The trachea and tracheal glands are lined with respiratory epithelium and rest on a connective tissue lamina propria. Overlapping rings of ossified hyaline cartilage form the tracheal wall. The trachea is compressed just cranially to the bifurcation into the primary bronchi. Thin vertical bars of cartilage fuse to form the pessulus, a single cartilage rod in the angle of the bifurcation (**Figure 8.29**). This is the tracheobronchial syrinx, the avian sound box. The tympaniform membranes analogous to the mammalian vocal cords are covered with a stratified squamous epithelium (**Figure 8.30**).

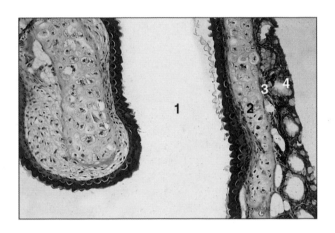

Figure 8.27 Avian vestibule. (1) The vestibule is lined by a distinctive stratified squamous keratinised epithelium. (2) Hyaline cartilage. (3) Perichondrium. (4) Pseudostratified ciliated columnar epithelium with intraepithelial alveolar mucus-secreting glands, respiratory epithelium. H&E ×50.

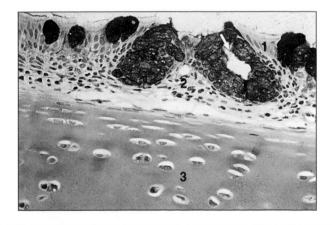

Figure 8.28 Respiratory epithelium (bird). (1) Pseudostratified columnar epithelium with intraepithelial mucus-secreting glands (arrowed). (2) Lamina propria and perichondrium. (3) Hyaline cartilage. Haematoxylin/PAS ×160.

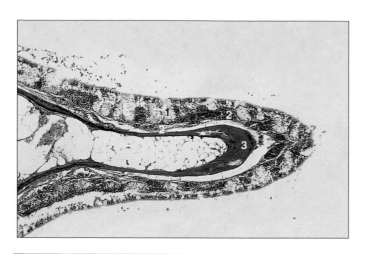

Figure 8.29 Avian syrinx, pessulus. (1) Respiratory epithelium. (2) Vascular lamina propria and perichondrium. (3) Hyaline cartilage. Gomori's trichrome ×5.

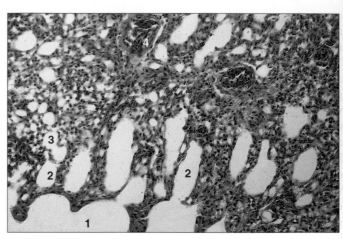

Figure 8.31 Lung (bird). (1) Lumen of the parabronchus (tertiary bronchus). (2) Conical ducts, atria. (3) Air capillaries. (4) Blood vessels filled with nucleated erythrocytes. H&E ×62.5.

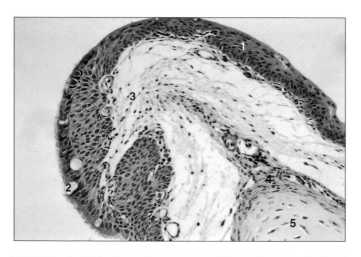

Figure 8.30 Avian bronchotracheal larynx. (1) The tympanic membrane is covered by a stratified squamous epithelium. (2) Respiratory epithelium. (3) Lamina propria. (4) Perichondrium. (5) Hyaline cartilage. H&E ×200.

Bronchi and Parabronchi

The primary bronchi are lined with respiratory epithelium that rests on a connective tissue lamina propria with hemirings of cartilage embedded in fibrous connective tissue and smooth muscle. The primary bronchi pass into the lung, where they give off secondary bronchi. The epithelium of the secondary bronchi contains goblet cells, and cartilage is absent. The secondary bronchi branch into anastomosing parabronchi. Each parabronchus forms the centre of a pulmonary lobule and is lined with simple squamous epithelium. Bundles of smooth muscle form spiral bands that are encased by a thin layer of connective tissue.

The parabronchial wall is perforated with openings leading to spaces lined with squamous or cuboidal epithelium: the atria. The air capillaries arise from the base of the atria via infundibula and radiate towards the periphery of each lobule. The air capillaries are lined with type I epithelial cells forming the respiratory surface and surrounded by a mass of blood capillaries to facilitate the air exchange tissue (**Figure 8.31**).

Air Sacs

Air sacs are thin-walled structures lined by a squamous epithelium (they may be ciliated or columnar) resting on a thin layer of connective tissue. The blood supply is poor, and with the exception of the abdominal air sac (ten branches), they are connected to the secondary bronchi. The humerus and the sternum are some of the bones penetrated by extensions of the air sacs (also called 'pneumatic' bones). Most of the avian species have nine air sacs. The poor vascularisation makes air sacs prone to infection.

Reptilian, Amphibian, and Fish Respiratory Systems

Most fish (except lungfish, the bowfin, some catfish, and a few other teleosts) either lack lungs or possess only primitive elongated sac-like lungs. They must rely upon vascularised gills in order to extract oxygen from and excrete carbon dioxide into their aquatic environment. Gills are composed of parallel rows of gill filaments, the primary lamellae, which are supported by cartilaginous or bony rays forming semilunar folds: the secondary lamellae (**Figures 8.32** and **8.33**). The gill arches contain a fine vascular network of branchial arteries, arterioles, and capillaries, across which respiratory gases are exchanged and osmoregulation (in conjunction with the kidneys) is maintained. In salmonids, eels, and other fish that alternate between freshwater and marine aquatic environments during their life cycles, the electrolyte secreting cells of the gills play major cyclic roles in osmoregulation. Teleost fish also possess a pseudobranch, which is a moderately compressed gill-like structure that is derived during embryological development from the first-gill arch. Its function is believed to be the regulation of blood oxygenation.

The gills of amphibians are similarly structured and function in a similar manner. Some amphibians possess both external gills and internal sac-like lungs, which serve not only as organs of respiratory gas exchange but also have a hydrostatic function. When the sac-like lungs are filled with air, the amphibian becomes more buoyant. When these lungs are empty, buoyancy is lost and the animal sinks to the bottom of the water, thereby requiring little or no effort to remain submerged. Adult plethodontid salamanders lack lungs entirely; their gas respiratory exchange is accomplished solely by diffusion across the well-vascularised moist integument. In some amphibians, lungs are much reduced in size; in others, only a single lung is present. Many amphibians augment their pulmonary and integumentary respiration by buccal movements that help move gases across their oropharyngeal mucosae, where some gas exchange occurs.

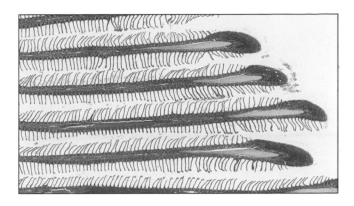

Figure 8.32 Gill from a teleost fish. It is composed of primary lamellae with a cartilaginous 'skeleton' from which secondary lamellae intersect perpendicularly. H&E ×50.

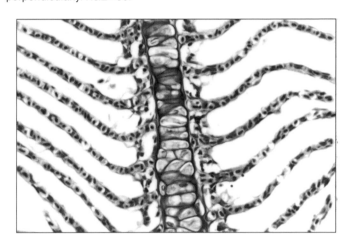

Figure 8.33 Gill from a teleost fish (*Psalidodon anisitsi*). Cartilage can be observed in the centre of the primary lamella from which secondary lamellae branch perpendicularly. Methylene blue ×200.

CLINICAL CORRELATES

The lungs of many diurnal amphibians and reptiles are heavily pigmented with melanin (*see also* Chapter 4). This pigment is believed to confer protection against the effects of solar radiation. A section of lung from a terrestrial frog (*Rana pipiens*) is shown in **Figure 8.34**.

As described on page 74, lungs (and similarly gills) are vulnerable to pathogens present in both the external environment and the blood (**Figure 8.35**). In addition, the delicate capillary bed through which respiratory gases exchange is also predisposed to thromboembolism because of the small cross-sectional area of these vessels.

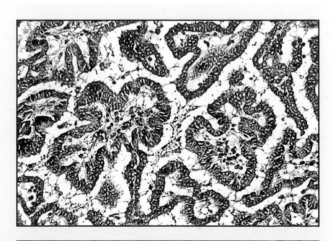

Figure 8.34 Lung of a terrestrial frog (*Rana pipiens*). H&E ×200.

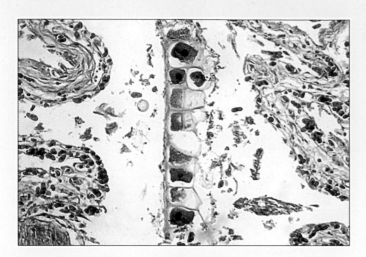

Figure 8.35 Aspiration pneumonia in a lizard. Note the plant fibre within the airspace. H&E ×125.

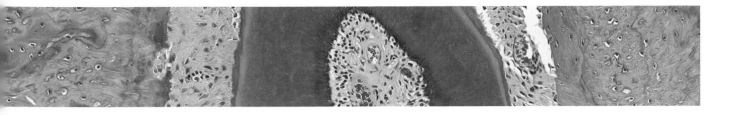

9

DIGESTIVE SYSTEM

The digestive or alimentary tract begins at the entrance to the oral cavity and terminates at the anus. The digestive system consists of a series of tubular organs, including the oral cavity (where we can find the lips, tongue, and teeth), the pharynx, the oesophagus, the stomach, the small intestine (duodenum, jejunum, and ileum), and the large intestine (caecum, colon, rectum, and anus).

There are accessory glands that have important functions in the digestion process, including the salivary glands, pancreas, gall bladder, and liver.

Oral Cavity

The oral (or buccal) cavity is lined by a mucosal membrane, forming the gums in the jawbones. This mucosa covers the hard palate and the soft palate at the back of the oral cavity. Within the oral cavity, we find the tongue and teeth. There is a transition between mucosal membrane and skin forming the lips (**Figure 9.1**). Salivary

glands are adnexal exocrine glands that produce saliva to be secreted into the oral cavity.

Tongue

The tongue lies on the floor of the oral cavity. It is made up of interlacing bundles of skeletal muscle fibres and loose connective tissue. The mucous membrane on the undersurface consists of nonkeratinised stratified squamous epithelium with a lamina propria (**Figure 9.2**). The dorsal surface of the anterior part of the tongue, where the epithelium is keratinised, is rough. The lamina propria is raised in small projections: the lingual papillae, which can have different forms and functions. The filiform, conical, and lenticulate papillae are nonsensory and are heavily keratinised, and

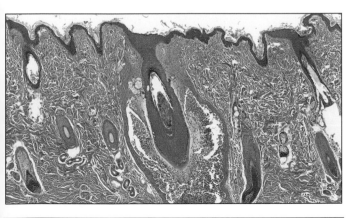

Figure 9.1 Lip (dog). Stratified squamous keratinised epithelium and large sensory pilous follicle. H&E ×75.

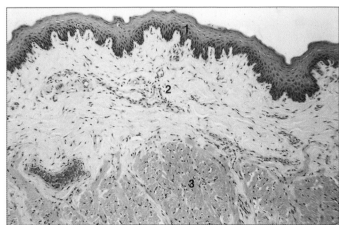

Figure 9.2 Tongue, ventral surface (cat). (1) Stratified squamous non-keratinised epithelium. (2) Vascular lamina propria with small projections. (3) Striated muscle fibres. H&E ×100.

DOI: 10.1201/9781003333807-9

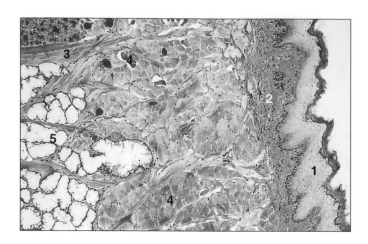

Figure 9.3 Tongue (dog). (1) Stratified squamous keratinised epithelium on the dorsum of the tongue. (2) Lamina propria. (3) Striated muscle fibres cut in longitudinal section and (4) in transverse section. (5) Mixed salivary gland. Masson's trichrome. ×100.

give the tongue a distinctive rough feel (**Figures 9.3–9.5**). The circumvallate, fungiform, and foliate papillae are sensory and are associated with small salivary glands in the lamina propria; the taste buds can be found in the lateral walls. The circumvallate papilla is surrounded by a moat-like trough or vallum and is level with the surface of the tongue (**Figure 9.6**). They are separated from each other and from the exocrine tissue by thin strands of connective tissue that support nonkeratinised and projects above the surface of the tongue (**Figure 9.7**). The foliate papillae are large, nonkeratinised, and leaf-like, crossed by transverse furrows and appear in section as a row of fungiform papillae (**Figure 9.8**).

Taste buds are epithelial structures associated with the terminal fibres of the facial and glossopharyngeal nerve.

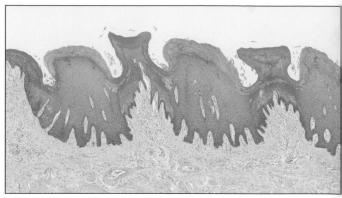

Figure 9.5 Tongue (sheep). Conical papillae. H&E ×60.

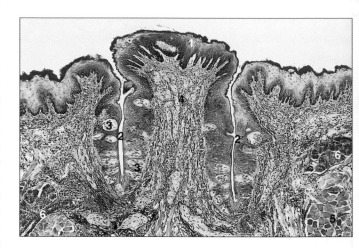

Figure 9.6 Circumvallate papilla. Tongue (cow). (1) Stratified squamous epithelium. (2) Vallum. (3) Taste buds. (4) Lamina propria. (5) Striated muscle. (6) Mixed salivary gland. Gomori's trichrome ×100.

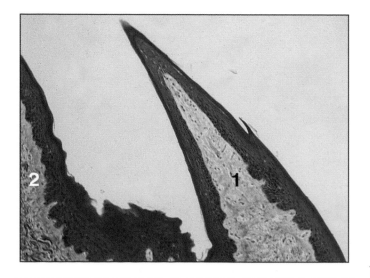

Figure 9.4 Tongue (cow). (1) Filiform papilla. (2) Lenticular papilla. Masson's trichrome ×100.

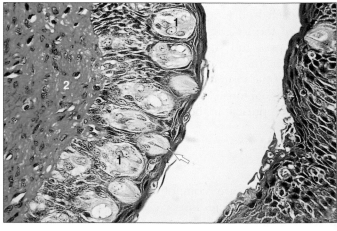

Figure 9.7 Tongue (cow). (1) Taste buds in the stratified squamous epithelium of the lateral wall of the papilla. The taste pore is arrowed. (2) Connective tissue lamina propria. Masson's trichrome ×250.

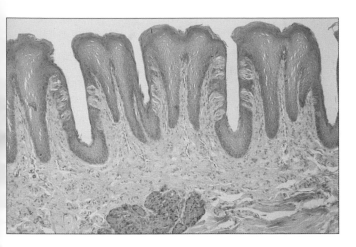

Figure 9.8 Foliate papilla. Tongue (rabbit). This papilla appears as a row of fungiform papillae. The taste buds are arranged along the lateral walls. H&E ×62.5.

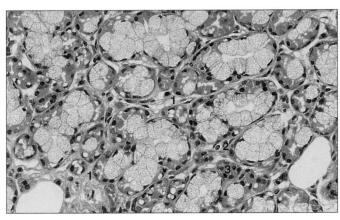

Figure 9.10 Mixed salivary gland (cow). The main mass of tissue is secretory units of seromucous acini. (1) Interlobular connective tissue with (2) blood vessels. (3) Interlobular ducts. H&E ×125.

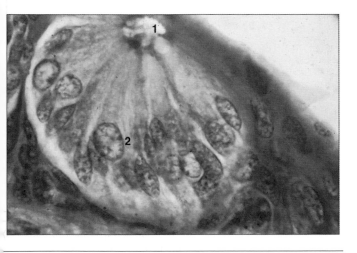

Figure 9.9 Taste bud. Tongue (cow). (1) Taste pore. (2) Taste receptor and sustentacular cells. H&E ×400.

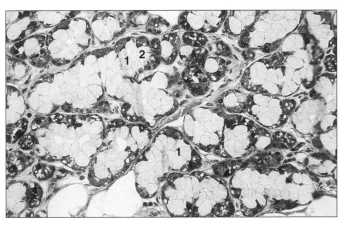

Figure 9.11 Mixed salivary gland (cow). Mixed seromucous acinus with (1) pale staining mucous cells and (2) darkly stained serous cells. Masson's trichrome ×200.

Within each bud is a taste pore, which opens onto the surface of the tongue, and a taste chamber lined with a taste receptor and sustentacular (supporting) cells. Food dissolved in the salivary gland secretion passes into the reservoir of the taste chamber (**Figure 9.9**).

The lingual tonsil is a localised mass of lymphoid tissue that is often present at the base of the tongue.

Salivary Glands

Salivary glands are compound tubuloacinar exocrine glands. They secrete enzymes or, as seromucous or mixed salivary glands, a mixture of enzymes and mucus. The secretory component of each gland is the parenchyma, and the supporting connective tissue is the stroma (**Figures 9.10–9.12**). In mixed salivary glands, the

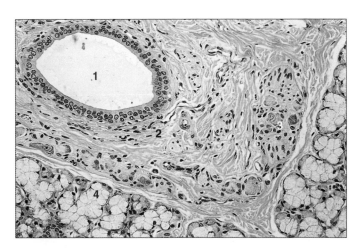

Figure 9.12 Mixed salivary gland (dog). (1) Interlobular duct lined by a stratified columnar epithelium. (2) Connective tissue stroma. (3) Parasympathetic ganglion. (4) Seromucous acini. H&E ×200.

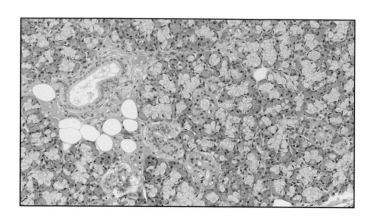

Figure 9.13 Mixed salivary gland (sheep). Pale staining in triangular mucous cells with a basal nucleus and deep staining in serous cells with a round nucleus surrounding the mucous cells. H&E ×200.

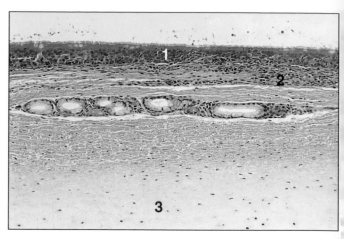

Figure 9.15 Soft palate (ox). (1) Respiratory epithelium. (2) Lymphatic nodule. (3) Bone. H&E ×100.

serous and mucous units may be separate or the serous cells may form a distinctive cap on one side of a mucous unit: a serous demilune (**Figure 9.13**). The serous cell is columnar with a basal nucleus and basal basophilia caused by the presence of abundant rough endoplasmic reticulum. The luminal eosinophilia is caused by the secretory granules accumulating before secretion. The mucous cell is triangular with a basal-flattened nucleus and a pale staining vacuolated cytoplasm. Specialised epithelial cells, the myoepithelial or basket cells, are capable of contracting: these lie between the secretory cells and the basement membrane (*see* Chapter 3).

The diluted salivary secretion leaves the acinus and is concentrated in the first part of the duct system: the striated duct.

Palate

The hard palate is lined with stratified squamous epithelium with the lamina propria continuous with the underlying periosteum (**Figure 9.14**). The oral surface of the soft palate is also lined with stratified squamous epithelium, but the lamina propria has mucus-secreting glands and lymphatic nodules. The nasal surface is covered by the respiratory epithelium (**Figure 9.15**).

Tonsils are lymphoid tissue aggregates present in different areas of the oral cavity and pharynx. Palatine tonsils are normally well developed in all mammals, and they act as a front-line defence against inhaled or ingested pathogens (**Figures 9.16** and **9.17**).

Teeth

In the embryo, teeth develop in the ectoderm as dental papillae within the enamel organs (**Figure 9.18**). The mesoderm invaginates each enamel organ into a bell shape, with an inner enamel epithelium of ameloblasts laying down enamel continuous with the outer enamel epithelium and enclosing

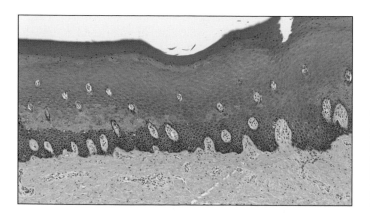

Figure 9.14 Hard palate (sheep). Thick layer of stratified squamous epithelium and lamina propria. H&E ×120.

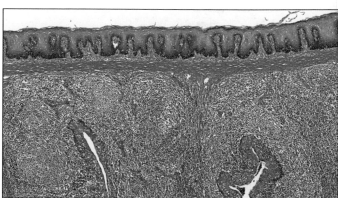

Figure 9.16 Tonsil (pig). Presence of follicular and diffuse lymphoid tissue under the stratified squamous epithelium. H&E ×100.

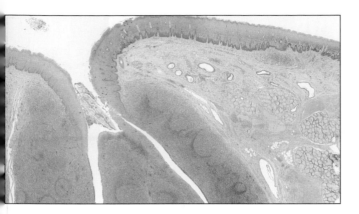

Figure 9.17 Tonsil (sheep). Presence of follicular and diffuse lymphoid tissue and mucus-secreting glands under the keratinised stratified squamous epithelium.

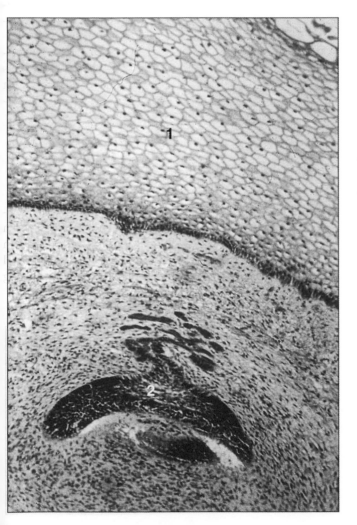

Figure 9.18 Developing tooth (cat embryo). (1) Oral epithelium. (2) Enamel organ surrounded by mesoderm. H&E ×100.

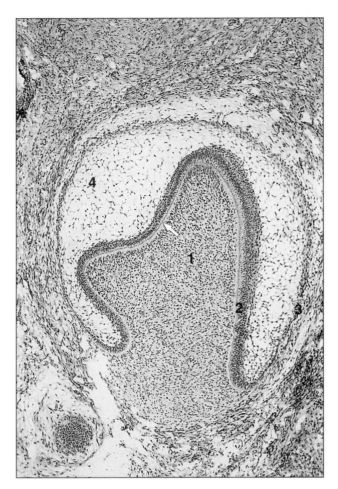

Figure 9.19 Developing tooth (cat embryo). (1) Mesenchymal papilla with a layer of odontoblasts (arrowed). (2) Inner enamel epithelium, continuous with (3) the outer enamel epithelium. (4) Stellate reticulum. H&E ×62.5.

the stellate reticulum (**Figure 9.19**). The mesenchymal cells of the papilla differentiate to become odontoblasts, the dentine-forming cells (**Figure 9.20**) and cementoblasts, and secrete cementum in a similar pattern to that of bone. Enamel and dentine are involved with the creation of the crown; dentine and cementum are involved in the root. The root is formed by an extension of the enamel organ at the junction of the inner and outer enamel epithelium: the root tubule (**Figure 9.21**). The tooth is held in the developing mandible and maxilla by the periodontal membrane of collagen fibres embedded in the cementum. Temporary teeth develop first. The permanent teeth are secondary offshoots on the lingual side of the temporary teeth. The tooth is divided into a crown and a root (**Figures 9.22–9.24**). In the carnivore, teeth cease to grow after eruption and the ameloblast layer is lost: brachydont teeth. In the horse, ruminant, and pig, teeth are much longer and continue to grow for all or part of adult life: hypsodont teeth. In these, the dental sac covers the whole of the tooth before eruption, and the cementum covers the entire tooth, preventing loss of the ameloblasts and allowing continuing deposition

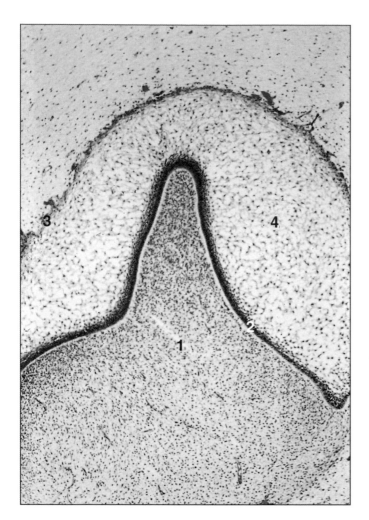

Figure 9.20 Developing tooth (cat embryo). (1) Mesenchymal dental papilla. (2) Ameloblasts in the inner enamel epithelium. (3) Outer enamel epithelium. (4) Stellate reticulum. H&E ×62.5.

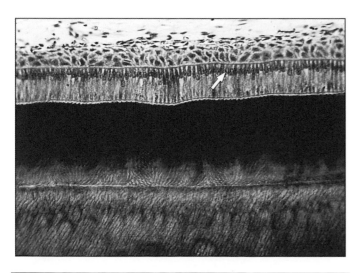

Figure 9.21 Developing tooth (cat). The ameloblasts are tall columnar cells with a basal nucleus (arrowed). H&E ×125.

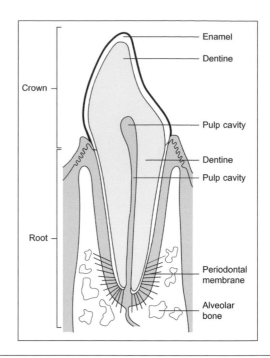

Figure 9.22 Diagram of an incisor.

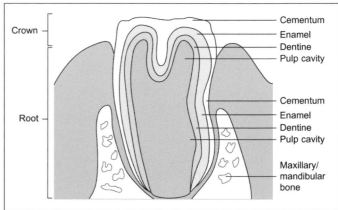

Figure 9.23 Diagram of a molar.

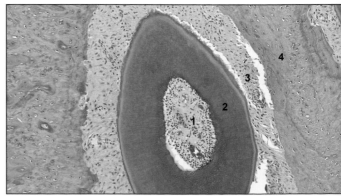

Figure 9.24 Tooth (hamster) within the alveolar bone. Pulp cavity (1), dentin (2), periodontal ligament (3), and maxillary bone (4). H&E ×160.

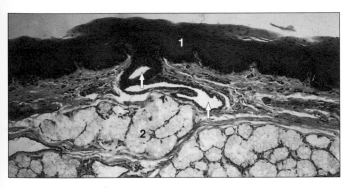

Figure 9.25 Oropharynx. (1) Stratified squamous epithelium. (2) Mucus-secreting glands open onto the surface (arrowed). H&E ×200.

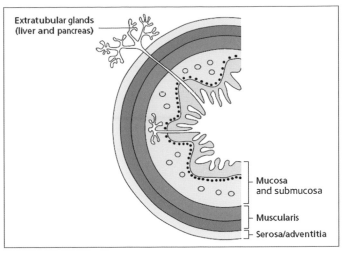

Figure 9.27 Diagram. Alimentary canal.

of enamel and cementum and thus allowing for the wear and tear in these species, which include rodents and lagomorphs.

Oropharynx

This is a short junctional area between the oral cavity and the alimentary canal with some mucus-secreting glands and the presence of tonsils (**Figure 9.25**).

Alimentary Canal

The alimentary canal or digestive tract is a muscular tube extending from the oropharynx to the anus, comprising the oesophagus, stomach, and small and large intestines. Two large glands, the liver and pancreas, are also derived from the embryonic alimentary canal. Each part of the canal has a specific function, and the histology reflects this (**Figure 9.27**). The canal wall is derived from endoderm and mesoderm and consists of four layers:

- Tunica mucosa (mucous membrane), with epithelial lining, supporting vascular lamina propria and lymphatic cells. Mucosal glands, which are derived from the epithelium, are variably present. The outer *muscularis mucosae* (absent from the mouth, pharynx, portions of the oesophagus, and rumen) is smooth muscle.
- Tela submucosa: a connective tissue layer with lymphatic tissue and nerve plexi. Submucosal glands may be present.
- Tunica muscularis: smooth muscle (except in the oesophagus and the anus where the skeletal muscle can also be found).
- Tunica adventitia/serosa: outer layer of connective tissue.

CLINICAL CORRELATES

Any level of the digestive tract can be affected by inflammation. Gingivitis, or inflammation of the gums, is common in dogs and cats, often in association with dental disease.

A gingival biopsy from a dog in which the gingival epithelium is irregularly hyperplastic is shown in **Figure 9.26**. A dense inflammatory infiltrate occupies the superficial submucosa and extends into the mucosal epithelium. Lymphocytes and plasma cells (mature immunoglobulin-secreting cells) predominate in the infiltrate. Their presence indicates a persistent antigenic stimulus. The initiating disease may be local or systemic.

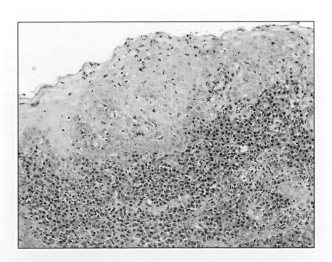

Figure 9.26 Gingivitis (dog). H&E ×125.

Oesophagus

The oesophagus is lined with stratified squamous epithelium, and both mucosal and submucosal mucus- or seromucous-secreting glands may be present. In ruminants, the muscularis externa is skeletal muscle; in the pig and cat, the distal part is a smooth muscle (**Figures 9.28–9.31**).

Stomach

The stomach mucosa may be nonglandular or glandular in domestic animals. In the simple stomach of the dog and cat, the mucosa is basically glandular. In the pig and horse, there is a nonglandular (oesophageal) region and a glandular region (**Figure 9.32**). In the ruminant, the nonglandular forestomach has three compartments: the rumen, the reticulum, and the omasum; the glandular stomach is a separate compartment: the abomasum (**Figures 9.33–9.38**).

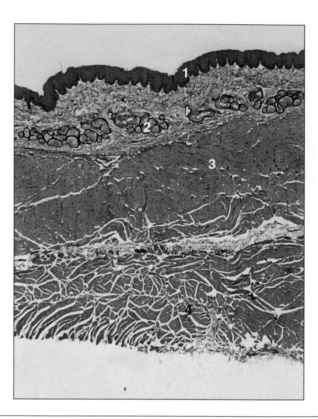

Figure 9.29 Oesophagus (dog). (1) Stratified squamous epithelium. (2) Mucus-secreting glands in the lamina propria. (3) Inner layer of circular skeletal muscle. (4) Outer layer of longitudinal skeletal muscle. H&E ×100.

Figure 9.28 Oesophagus (ox). (1) Stratified squamous keratinised epithelium. (2) Lamina propria. (3) Vein. (4) Inner layer of circular smooth muscle. (5) Outer layer of longitudinal smooth muscle. H&E ×200.

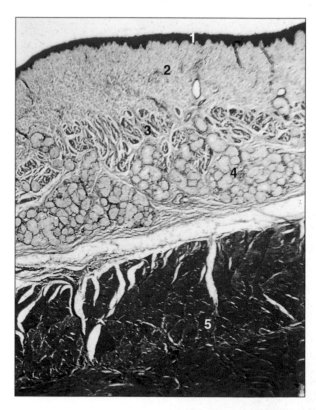

Figure 9.30 Oesophagus (cat). (1) Stratified squamous epithelium. (2) Lamina propria. (3) Muscularis mucosae. (4) Submucosal mucous glands. (5) Muscularis externa. Masson's trichrome ×100.

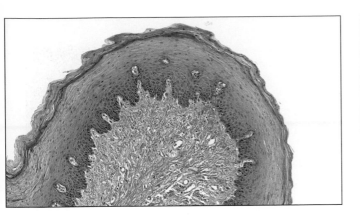

Figure 9.31 Oesophagus (goat). Stratified squamous epithelium and collagen-rich lamina propria with multiple blood vessels. H&E ×150.

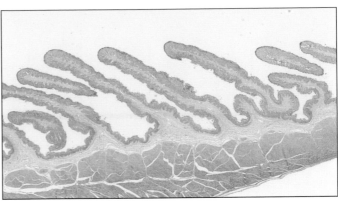

Figure 9.34 Rumen (sheep). The lining epithelium is stratified squamous, the lamina propria is loose connective tissue, and a thick muscular layer is underneath. H&E ×20.

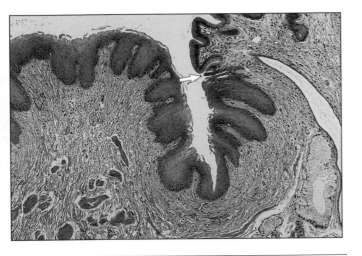

Figure 9.32 Oesophageal/stomach junction (horse). The epithelium changes abruptly from stratified squamous to simple columnar at the junction (arrowed). H&E ×100.

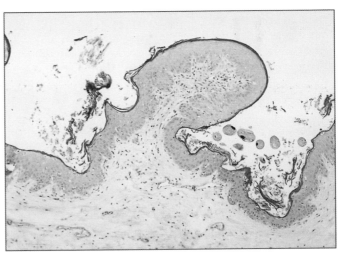

Figure 9.35 Reticular groove (goat). The mucosa is folded; this allows stretching. The lining epithelium is stratified squamous. H&E ×100.

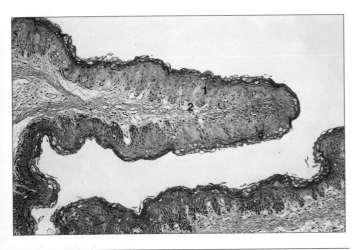

Figure 9.33 Rumen (sheep). (1) Stratified squamous epithelium lines the rumen. (2) Lamina propria. Masson's trichrome ×100.

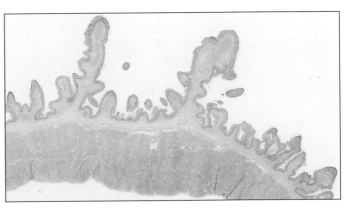

Figure 9.36 Reticulum (sheep). The mucosal folds present conical papillae, and the muscularis mucosae appears at the apical area of the primary folds. H&E ×20.

Figure 9.37 Omasum (sheep). Folds of mucosa forming parallel laminae of variable length with papillae on the surface. H&E ×25.

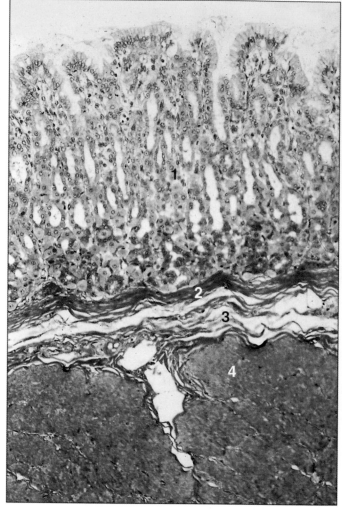

Figure 9.39 Abomasum. Fundic region (goat). (1) Mucosa. (2) Muscularis mucosae. (3) Submucosa. (4) Muscularis externa. Masson's trichrome ×200.

The nonglandular stomach is lined with a stratified squamous epithelium with some keratinisation. In ruminants, clear vacuolated cells in the epithelium give it a distinctive appearance and allow the transfer of water, electrolytes, and short-chain fatty acids. *Muscularis mucosae* layer is present in the omasum and reticulum but absent from the rumen.

The glandular mucosa of the stomach is folded and lined with a simple columnar mucus-secreting epithelium. The gastric pits, the foveoli, are surface depressions continuous with the simple tubular gastric glands (**Figure 9.39**). Three histological regions are recognised: the cardia, the fundus, and the pylorus. Glands are sparse with few cells in the cardia but are abundant and cellular in the fundus (**Figures 9.40–9.45**).

Within the proper gastric (fundic) mucosa are the gastric pits, consisting of three main types of cells responsible for the secretion of gastric juices:

- Mucous neck cells at the neck of the gland secrete mucus.
- Chief cells are the most numerous and secrete the enzyme pepsinogen, that is, converted into pepsin by gastric acid. They are cuboidal or pyramidal with a round basal nucleus surrounded by basophilic cytoplasm (rough endoplasmic reticulum). The apical cytoplasm is eosinophilic (stored secretory granules).
- Parietal cells are large polyhedral cells with a central nucleus and eosinophilic cytoplasm. They secrete hydrochloric acid into canaliculi and elaborate invaginations of the plasma membrane and communicate with the lumen of the gastric gland (**Figure 9.46**).

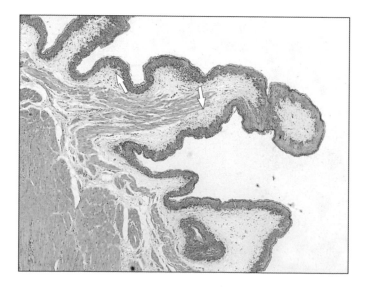

Figure 9.38 Omasum (sheep). The muscularis mucosae is present in the long omasal folds (arrowed). H&E ×50.

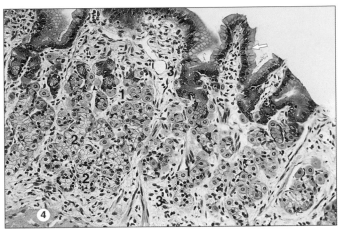

Figure 9.42 Cardiac glands. Stomach (horse). Simple columnar mucus-secreting epithelium lines the stomach (arrowed). (1) Parietal (oxyntic) cell, deep red staining. (2) Zymogen (chief) cell, basophilic staining. (3) Lamina propria. (4) Muscularis mucosae. H&E ×125.

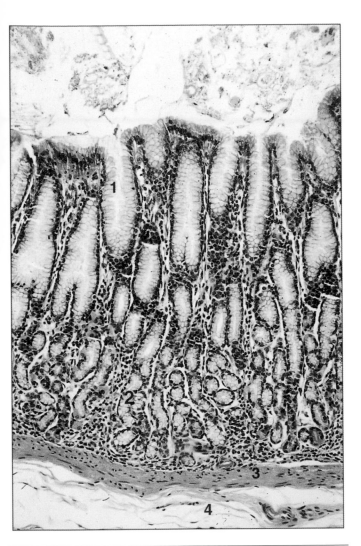

Figure 9.40 Abomasum. Pyloric region (goat). (1) The lining epithelium is simple columnar and mucus secreting. (2) Simple tubular mucus-secreting glands. (3) Muscularis mucosae. (4) Submucosa. H&E ×62.5.

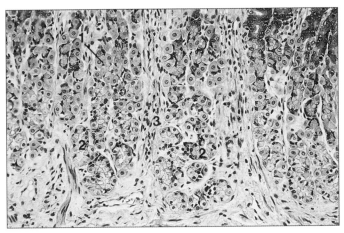

Figure 9.43 Cardiac glands. Stomach (horse). (1) Parietal cell. (2) Chief cell. (3) Lamina propria. H&E ×125.

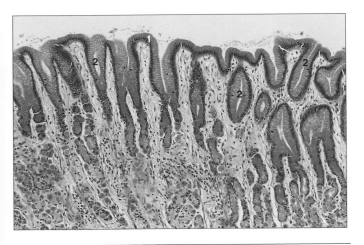

Figure 9.41 Stomach (horse). (1) Simple columnar mucus-secreting epithelium. (2) Gastric pits. H&E ×62.5.

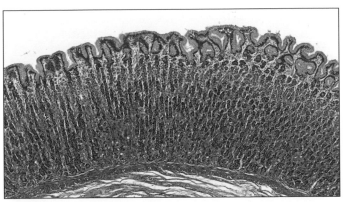

Figure 9.44 Fundic glands. Stomach (pig). Simple mucus-secreting epithelium extends into the gastric pits. Parietal cells and chief cells can be identified. H&E ×100.

Enteroendocrine cells are a diffuse population that are identified with specialised silver stains and are also known as argentaffin and argyrophil cells. Chemical messengers (serotonin, gastrin, somatostatin, and enteroglucagon) are secreted locally to control digestion.

The glands located at the pylorus are branched tubular mucous type (**Figure 9.47**).

The lamina propria is loose cellular connective tissue with lymphatic cells present as a local population and part of the gut-associated lymphoid tissue (GALT). The *muscularis mucosae* is composed of several layers of smooth muscle fibres. The submucosa is aglandular loose connective tissue with parasympathetic nerve plexi (**Figure 9.48**). The muscularis externa consists of three layers of smooth muscle: oblique, circular, and longitudinal. The myenteric parasympathetic nerve plexus (Meissner's) lies between the muscle layers. The outer layer, the serosa, is vascular connective tissue covered with mesothelial cells continuous with the visceral peritoneum.

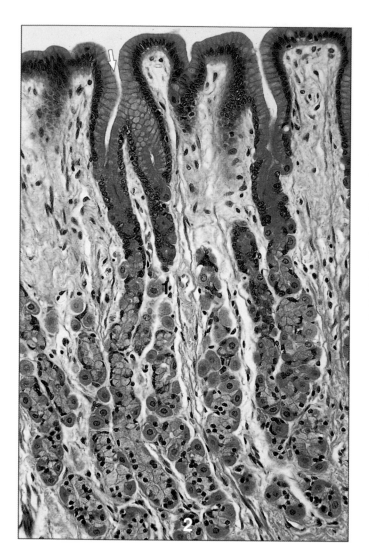

Figure 9.45 Fundic glands. Stomach (dog). Simple columnar mucus-secreting epithelium extends into the gastric pits (arrowed). (1) Parietal cell. (2) Chief cell. H&E ×400.

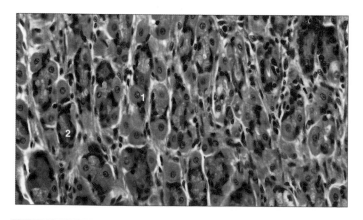

Figure 9.46 Fundic glands. Stomach (cat). (1) Parietal cell. (2) Chief cell. H&E ×250.

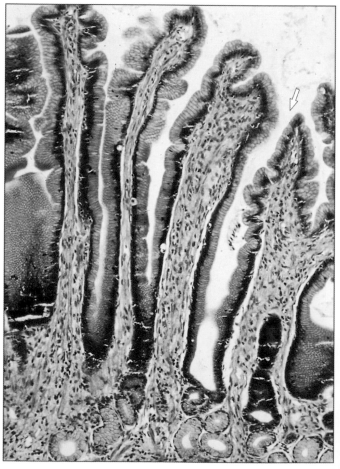

Figure 9.47 Pyloric glands. Stomach (dog). Simple columnar mucus-secreting epithelium extends into the gastric pits (arrowed). Simple columnar epithelium lines the gland tubule seen here cut in transverse section. H&E ×100.

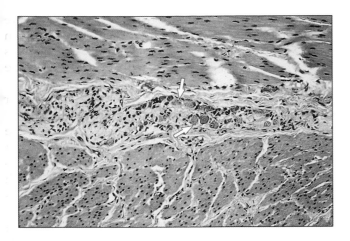

Figure 9.48 Myenteric nerve plexus. Stomach (horse). Parasympathetic neuron cell bodies (arrowed) lie in the connective tissue between the smooth muscle layers. H&E ×200.

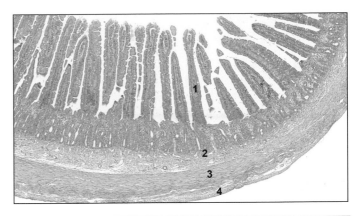

Figure 9.50 Ileum (cat). (1) Mucosa. (2) Submucosa. (3) Muscularis externa. (4) Serosa. H&E ×60.

Small Intestine

The small intestine consists of the duodenum, jejunum, and ileum (**Figures 9.50–9.58**). The main function of the intestinal mucosa is the absorption of nutrients. Finger-like projections of the mucosa (intestinal villi) are long and thin in carnivores and short and thick in ruminants. They increase the surface area for absorption. The core of each villus is formed by the lamina propria, which is vascular, cellular, and reticular, with local aggregations of lymphoid cells. The tall columnar cells that line the intestine have a striated border containing mucus-secreting goblet cells; these increase in number with distance from the stomach towards the rectum. At the bases of the villi, the epithelium dips into the lamina propria to form

CLINICAL CORRELATES

Gastric lesions may be associated with many conditions which have signs that also affect other body systems or other levels of the gastrointestinal tract. Gastric ulceration results from an imbalance between the damaging effects of gastric acid and pepsin and the protective mechanisms of the gastric mucosa, with bacteria like *Helicobacter* spp. involved in the aetiopathogenesis of this process (**Figure 9.49**). Administration of nonsteroidal anti-inflammatory drugs is known to predispose to gastric ulceration by inhibiting prostaglandin metabolism and damaging the gastric epithelium. Systemic disturbances, such as endotoxaemia or uraemia, may produce gastric lesions and complex factors associated with stress can also be implicated. Gastric ulceration can produce abdominal pain, haematemesis (vomit with blood), melaena (blood in faeces), and anaemia.

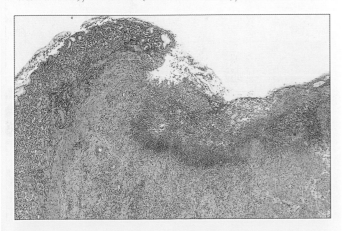

Figure 9.49 Peptic ulceration (dog). Loss of mucosal epithelium is seen, with eosinophilic necrotic debris within the defect. Granulation tissue is developing at the base of the ulcer. H&E ×100.

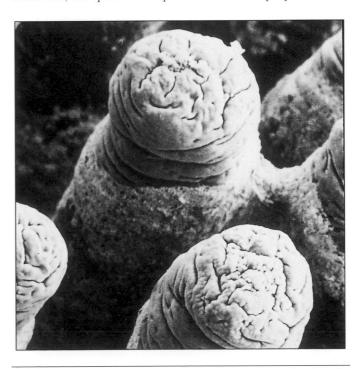

Figure 9.51 Duodenal villi (dog). The finger-like villi project into the lumen of the duodenum. Scanning electron micrograph ×100.

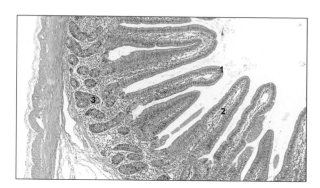

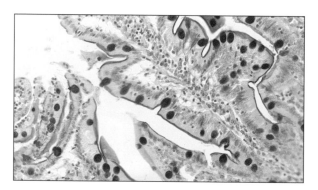

Figure 9.52   Ileum (pig). (1) Villus covered by simple columnar epithelium. (2) Lamina propria forms the core of the villus. (3) Mucosal glands. H&E ×100.

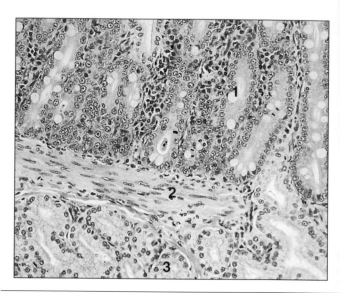

Figure 9.55   Duodenum (dog). (1) Mucosal glands. (2) Muscularis mucosae. (3) Submucosal glands. H&E ×200.

Figure 9.53   Duodenum (horse). Goblet cells in the epithelium are stained deep pink. Haematoxylin/periodic acid-Schiff (PAS) ×200.

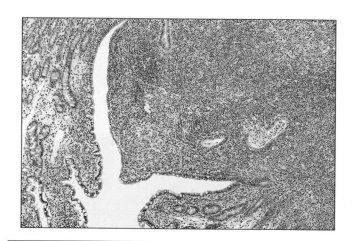

Figure 9.56   Duodenum (dog). The lamina propria is filled with lymphatic tissue, and lymphocytes are seen migrating through the epithelium (a Peyer's patch). H&E ×100.

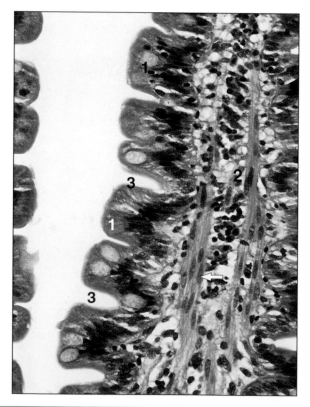

Figure 9.54   Duodenum (horse). (1) Simple columnar epithelium with goblet cells. (2) Lamina propria with smooth muscle fibres (arrowed). (3) Contractile crypts. H&E ×400.

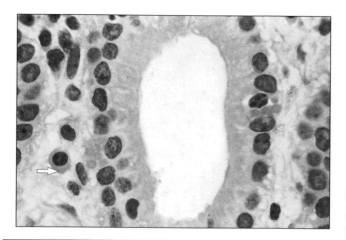

Figure 9.57   Globular leukocyte (horse). The globular leukocyte (function unknown) is in the epithelium of the intestinal gland. A plasma cell is present in the lamina propria (arrowed). H&E ×630.

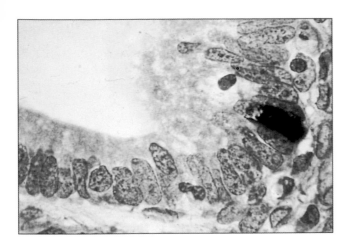

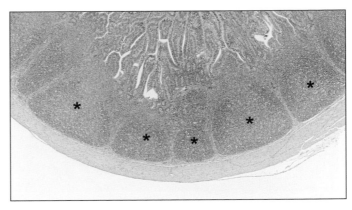

Figure 9.60 Jejunum (pig). Large lymphoid aggregates (Peyer's patches*). H&E ×50.

Figure 9.58 Enteroendocrine argentaffin cell (cat). Argentaffin cell stains black in the intestinal gland epithelium. Methanamine silver/safranin ×500.

mucosal intestinal glands (the crypts of Lieberkühn). The cells lining the crypts are columnar, secreting mucus, enzymes, and local hormones and are the stem cells that are active in the repair and replacement of the epithelium. Paneth cells, which contain secretory granules that contain pepsidase, may also be present in horses and ruminants.

The *muscularis mucosae* consists of two layers of smooth muscle, inner circular and outer longitudinal, and separates the crypts from the underlying sub mucosa. A strip of muscle extends into each villus from the muscularis mucosae; a lacteal (lymphatic that transports chyle) is also present. Indentations on the villi, called contractile crypts, are created by the contraction of the central strip of the muscle.

Brunner's glands are tubular mucous glands found in the mucosa and submucosa of the duodenum of different species, e.g. pigs (**Figure 9.59**).

The *muscularis externa* consists of two layers of smooth muscle dispersed in a gentle spiral, appearing as an inner circular and outer longitudinal layer. As in the stomach, the myenteric parasympathetic nerve plexus (Meissner's) can be found between the layers.

The serosa consists of loose connective tissue, and the mesothelium is continuous with the visceral peritoneum.

Lymphatic nodules can be found throughout the small intestine, isolated and in aggregate, constituting Peyer's patches (**Figure 9.60**), which are more numerous in the ileum.

Large Intestine

There are no villi in the large intestine (caecum, colon, rectum, and rectal canal). Goblet cells are abundant in the surface epithelium and in the mucosal glands. Lymphoid tissue is present in the lamina propria. Large numbers of eosinophils can be found associated with parasitic infestations. Lymphocytes are present in the epithelium when immunoglobulin, which bathes the epithelial cell surface as a defence against luminal antigen, is released. There are no submucosal glands.

A *muscularis mucosae* is present, and the muscularis externa consists of an inner circular and outer longitudinal layer of smooth muscle. The outer layer, or *taenia coli*, is arranged in bands and is characteristic of the colon of the horse and pig. In the horse, elastic fibres replace muscle fibres. The serosa is continuous with the peritoneum (**Figures 9.61–9.63**).

The rectum is lined with simple columnar epithelium. The mucosal glands decrease in number and may disappear entirely as the anus is approached, where there is an abrupt change to a stratified squamous epithelium. The *muscularis externa* is thicker here and becomes striated at the anal

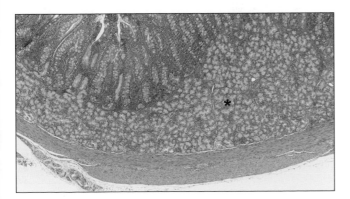

Figure 9.59. Duodenum (pig). Brunner's glands (*) observed in the submucosa. H&E ×100.

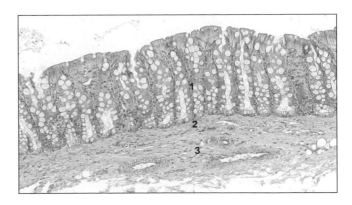

Figure 9.61 Colon (cat). (1) Mucosa. (2) Muscularis mucosae. (3) Submucosa. H&E ×200.

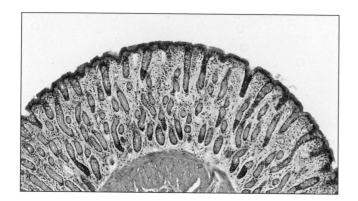

Figure 9.62 Colon (pig). Long crypts in the lamina propria with abundant goblet cells. H&E ×100.

sphincter. Part of the rectum is covered by a serosa and the rest by adventitia. Tubuloacinar anal glands are present at the cutaneous–rectal junction, where they secrete lipids in carnivores and mucus in the pig. In carnivores, circumanal, sebaceous-secreting glands are found in the anal canals. Anal sacs, opened by small tubular alveolar glands and lined with stratified squamous epithelium,

open into the perianal region (**Figure 9.65**). Large sebaceous glands are frequently observed in the external area of the rectum–anal junction (**Figure 9.64**), together with the presence of abundant lymphoid tissue in the internal part (**Figure 9.66**), as part of the well-developed GALT within the large intestine.

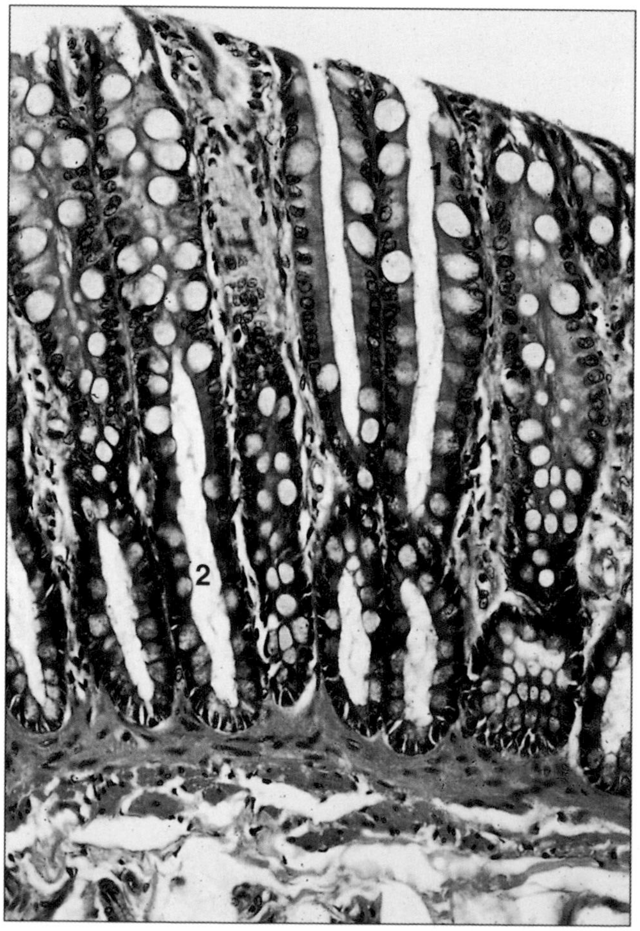

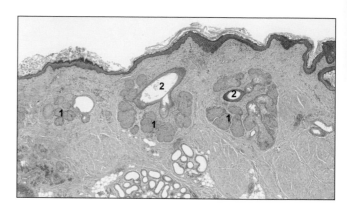

Figure 9.64 Rectum–anal junction (sheep). Abundant large sebaceous glands (1) surrounding hair follicles (2). H&E ×50.

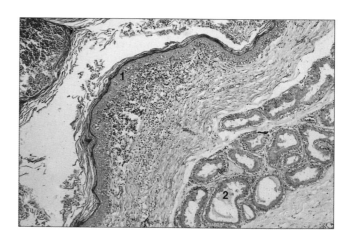

Figure 9.65 Anal sac (dog). (1) Stratified squamous keratinised epithelium lines the sac. (2) Tubuloalveolar glands in the lamina propria. H&E ×100.

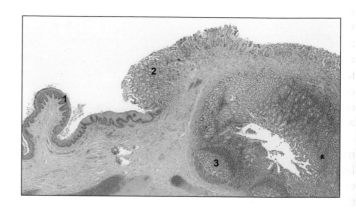

Figure 9.66 Rectum–anal junction (sheep). (1) External skin. (2) Rectum mucosa. (3) Lymphoid aggregates (GALT). H&E ×50.

Figure 9.63 Colon (horse). (1) Simple columnar epithelium with goblet cells. (2) Intestinal mucosal glands. H&E ×400.

CLINICAL CORRELATES

Small and large intestines can suffer from infection from a variety of agents: bacterial, viral, parasitic, or fungal. A highly prevalent disease in ruminants in many areas of the world is Johne's disease, also known as paratuberculosis. The disease is caused by the bacterium *Mycobacterium avium paratuberculosis*. The bacteria induce a granulomatous inflammatory reaction within the intestinal wall layers (**Figures 9.67** and **9.68**), decreasing dramatically the absorption of nutrients and the animals develop emaciation and cachexia.

Different types of neoplasia can also be found in small and large intestines. Lymphoma is not uncommon in dogs and cats (**Figure 9.69**). This tumour infiltrates the intestinal wall and can also produce a malabsorption syndrome, not being able to absorb nutrients and showing clinical signs such as diarrhoea and emaciation.

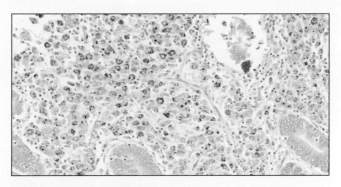

Figure 9.68 Identification of mycobacteria (Acid fast bacilli) in the same tissue from **Figure 9.67**. Ziehl–Nielsen stain ×300.

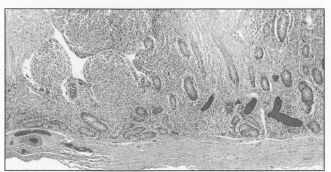

Figure 9.67 Granulomatous enteritis (cow) produced by *Mycobacterium avium paratuberculosis* (Johne's disease). High infiltration of inflammatory cells, mainly macrophages, within the mucosa and submucosa. H&E ×100.

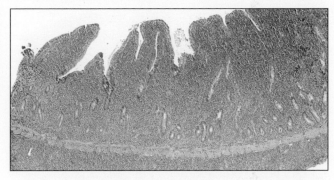

Figure 9.69 High infiltration of neoplastic lymphocytes within all the layers of the jejunum from a cat with intestinal lymphoma. H&E ×70.

Alimentary System of Reptiles and Amphibians

When amphibians metamorphose from larvae to adults, significant changes take place in form and function. Many larval amphibians are facultative (or obligatory) herbivores, the alimentary tracts of which are elongated and often tightly coiled (particularly in frog and toad larvae). During the latter stages of metamorphosis, postlarval amphibians usually cease eating and, therefore, must subsist on their tails and other sources of readily catabolised tissue. As adults, most amphibians are carnivorous.

The lingual apparatus of many amphibians and reptiles is modified for the apprehension of prey: some contain glandular acini that secrete sticky mucus (**Figures 9.70** and **9.71**); others are characterised by numerous papillary projections at the lingual tip to which food particles stick and are then brought into the mouth. When not being used, the tongue of snakes retracts into a lingual sheath that is lined

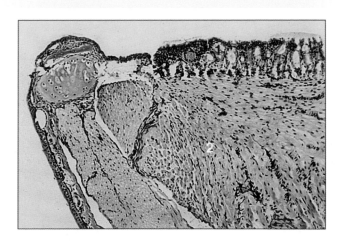

Figure 9.70 Tongue of a poison-arrow frog (*Dendrobates* spp.). The dorsal lingual surface (1) is covered by an unusual and complex epithelium composed of small, dark-staining cuboidal cells and acini of sticky mucin-secreting columnar cells that maintain a coating of adhesive mucus. The muscle fibres (2) are primarily arranged in a longitudinal direction and are attached at the front of the mandible. This facilitates the tongue being rapidly protruded and retracted in order to catch small invertebrates. H&E ×100.

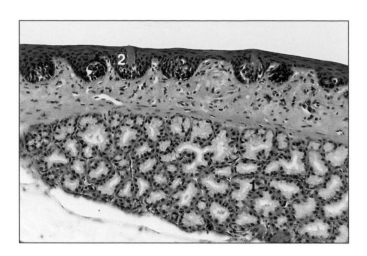

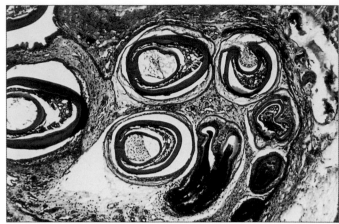

Figure 9.71 The tongue of some lizards overlies a sublingual salivary gland (1), as is illustrated by this longitudinal section of the tongue of a small skink (*Scincella lateralis*). The dorsal surface is covered by a nonkeratinised stratified squamous epithelium (2) in which cup-shaped taste receptors are embedded. Some lizards (for example, many iguanines) possess tongues with a terminal tip composed of papillary projections that are kept moist and sticky with mucus secreted by goblet cells and several salivary glands. H&E ×200.

Figure 9.72 The fangs of venomous snakes are continually being renewed. Illustrated are several teeth primordia of a juvenile rattlesnake (*Crotalus* spp.), forming modified fangs with a central enamel-lined channel through which venom is conducted. H&E ×100.

by mucus-secreting glands. It does not contain glands but is lubricated when it comes into contact with the luminal surface of the lingual sheath. The tongues of other reptiles contain taste receptors that are similar to taste buds found in the tongues of mammals.

The dental histology of amphibians and reptiles is similar to that in mammals, although the teeth of these animals are periodically and continually shed throughout life. Chelonians (turtles, tortoises, and terrapins) lack teeth entirely. Their premaxillae, maxillae, and mandibles are covered with hard and horny keratinous surfaces, called ramithecae, with which these animals cut their food items.

The salivary glands of amphibians and reptiles are similar to those found in mammals. They may be either entirely serous, entirely mucus-secreting, or seromucous.

In venomous snakes and helodermatid lizards (the Gila monster lizard, *Heloderma suspectum,* and the Mexican beaded lizard, *Heloderma horridum*) some salivary glands are greatly modified into structures (*see* Chapter 3) that secrete extremely toxic secretions that help these animals capture their prey and defend themselves. In venomous snakes, the secretions from these glands are conducted to the hollow needle-like fangs through coiled venom ducts. The passage of venom through these ducts is aided by the contraction of the temporal and masseter skeletal muscles that surround the glands and myoepithelial cells that surround the ducts (*see* Chapter 3). The fangs are replaced periodically throughout a snake's life. They are formed with a separate hollow channel

(**Figure 9.72**). Some nominally nonvenomous snakes, especially many colubrids, possess modified salivary (Duvernoy's) glands (**Figure 9.73**), the secretions of which induce a toxic reaction when injected into particularly sensitive prey and humans.

Generally, the alimentary system of the lower vertebrates shows a similar structure to that found in mammals, but major variations exist in species that are highly adapted to a particular diet. Folivorous (leaf-eating) reptilian herbivores utilise hindgut rather than foregut

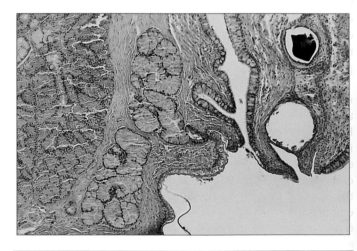

Figure 9.73 Some nonvenomous snakes possess modified (Duvernoy's) maxillary and premaxillary salivary glands connected to short ducts that empty into the oral cavity. Current studies indicate that the secretions from some of these glands manifest venom-like bioactivity on the lower vertebrate prey of these snakes. Also, mild clinical envenomation of sensitive humans bitten by these snakes has been reported. Illustrated are two lobules of the gland from a watersnake (*Natrix cyclopion*). H&E ×62.5.

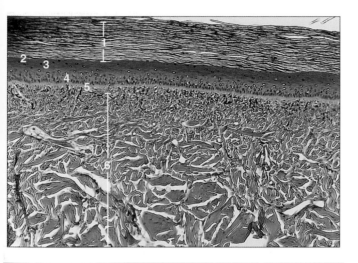

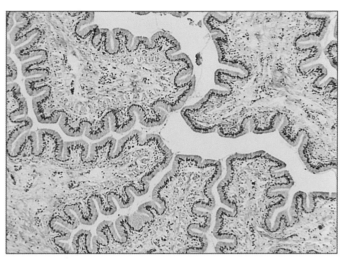

Figure 9.74 The oesophageal lumen of many chelonians, such as this green sea turtle (*Chelonia mydas*), is heavily keratinised and lacks mucus-secreting goblet cells. These characteristics reflect the scabrous diet of these marine animals. (1) Stratum corneum. (2) Stratum lucidum. (3) Stratum granulosum. (4) Stratum spinosum. (5) Stratum basale. (6) Muscularis externa. H&E ×100.

Figure 9.75 The oesophagus of most snakes and many lizards is characterised by its extensive plaiting which permits the oesophagus to stretch to accommodate enormous prey. Illustrated is a cross-section of the oesophagus of a kingsnake (*Lampropeltis triangulum*). Because of the necessity for abundant lubrication during the swallowing of furry, feathered or scaly prey, the oesophageal lumen is lined by a mucous epithelium composed of simple nonkeratinised columnar cells bearing basal nuclei. H&E ×100.

fermentation to accomplish the processing of cellulose and other complex carbohydrates. Modifications that aid in this process are an expanded sacculated colon, which is similar in function to the sacculus rotundus of lagomorphs (rabbits and hares) and some herbivorous rodents, and to the massive caecum and colon of equids. In all of these organs, the surface of the luminal lining is augmented by numerous mucosal villous projections, which greatly increase the area available for microbial digestion and nutrient absorption. Thus, the sacculated colon of reptilian folivores serves the same purpose as the large rumen complex of ruminants, even though it is part of the hindgut rather than the foregut.

The anterior alimentary tracts of various reptiles are modified. The oropharynx and oesophagus of some sea turtles have a heavily keratinised lining (**Figure 9.74**) that helps to protect the lumen from trauma when scabrous food items such as rocky and silica-rich coral are swallowed. The egg-eating snake (*Dasypeltis scabra*) ingests eggs with calcareous shells. As the egg enters the cranial oesophagus, the snake contracts its throat and thereby compresses the egg against multiple horny ridges that extend from the ventral region of the cervical vertebrae. After the eggshell is slit, the snake swallows the fluid and/or embryonic contents and regurgitates the shell fragments *en masse*. Most snakes and many lizards possess an oesophagus with walls formed into multiple longitudinal plaits that permit the swallowing of enormous meals (**Figure 9.75**), many times the diameter of their necks. Other reptiles, such as most crocodilians, have thick-walled muscular stomachs in which their prey are macerated with the aid of ingested stones.

The gastric mucosa of reptiles is similar to that found in mammals, except that only chief and clear cells are present; parietal cells are lacking (**Figures 9.76** and **9.77**). The small intestine lacks Brunner's glands. The serosa covering

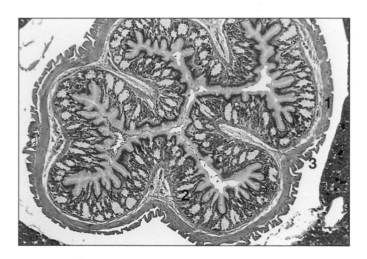

Figure 9.76 Cross-section of the fundic stomach of a small skink (*Scincella lateralis*). A very thin serosa covers the outermost visceral surface. The gastric wall is composed of an outer external longitudinal muscularis externa (1), a circular muscularis externa, the muscularis mucosa, and immediately beneath is the glandular mucosa (2), which is composed of pink staining granular chief cells and clear cells. The lumen is lined by tall mucus-secreting columnar cells. Parietal cells, present in mammalian gastric mucosae, are lacking in amphibians and reptiles. The outermost surface of the stomach is covered by a delicate serosa (3) formed of nonkeratinised squamous cells. H&E ×50.

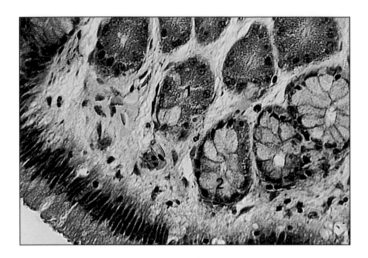

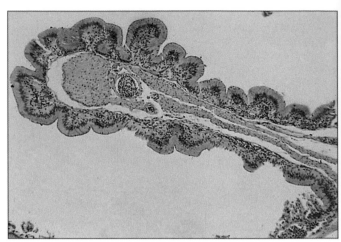

Figure 9.77 Gastric mucosa of a boa constrictor (*Boa constrictor*). The lumen is covered by tall columnar epithelium. The gastric glands consist of only granular, cuboidal, pink staining chief cells (1) with large vesicular nuclei and pale staining clear cells (2) whose nuclei are dark and basal. Some gastric pits are lined by both cell types. H&E ×400.

Figure 9.78 The colon of some lizards, particularly folivorous species, is a highly modified sacculated organ divided into multiple chambers that are functionally analogous to the hindgut of lagomorphs and some (herbivorous/folivorous) rodents, and the forestomachs of ruminants. Digestion is enhanced because the villous surface of the colon is covered by a highly absorptive columnar mucosa across which nutrients processed from cellulose-digesting microorganisms are assimilated. The elongated villi that cover the surface are supported, and stiffened, by thin cores of smooth muscle. Illustrated is the sacculated colon of a green iguana (*Iguana iguana*). H&E ×100.

most or all of the coelomic viscera of many diurnal lizards is heavily pigmented (**Figure 9.78**). Lymphoid patches or aggregates are scattered throughout the length of the alimentary tract.

Many lizards and some snakes possess salt-secreting glands through which hyperosmolar solutions containing sodium, potassium, and chloride ions are secreted. In many lizards, these glands are situated in the nasal cavity. Some sea snakes possess sublingual salt glands. In

some crocodilians, particularly crocodiles that inhabit salt marshes and travel between oceanic islands, salt-secreting glands are located on the dorsal surface of the tongue. All of these aforementioned glands permit the nonrenal secretion of electrolytes without the appreciable loss of water.

CLINICAL CORRELATES

ALIMENTARY SYSTEM OF REPTILES AND AMPHIBIANS

Squamous metaplasia of the nasal and pharyngeal mucosa (**Figures 9.79** and **9.80**) is a frequent clinical condition in reptiles fed diets deficient in vitamin A or β-carotene. Once this alteration occurs, the lubricative mucoid glandular secretion ceases, and the affected animal becomes more susceptible to respiratory and oropharyngeal disorders.

Ulcerative stomatitis is one of the most common conditions found in the cranial alimentary tract of captive snakes. This infectious inflammatory disease is caused by a variety of pathogenic Gram-negative and some Gram-positive bacteria. Depending upon the aetiologic agent, the inflammatory response may be suppurative or nonsuppurative. In suppurative lesions, heterophil granulocytes predominate; in nonsuppurative inflammations, heterophils may be entirely absent.

Glossitis, pharyngitis, oesophagitis, and gastritis also occur in captive amphibians and reptiles.

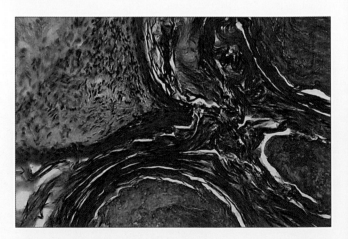

Figure 9.79 Severe pharyngeal hyperkeratosis in a desert tortoise (*Xerobates agassizi*). The pharyngeal glands are replaced by pearl-like masses of desquamated keratin. The luminal epithelial surface is thickened and covered by dense keratin debris. A similar alteration is seen in birds and mammals suffering from vitamin A deficiency. H&E ×100.

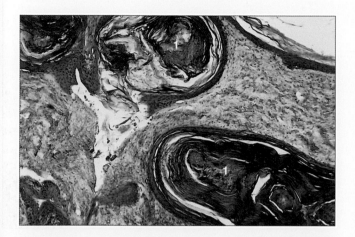

Figure 9.80 Cross-section of the pharynx of a red-eared slider turtle that was fed a diet seriously deficient in β-carotene or preformed vitamin A. The pharyngeal glands display squamous metaplasia and, as a result, have lost their mucus-secreting, goblet-cell-rich glands, which have been replaced by masses of desquamated keratin debris (1). The stratified squamous epithelium lining the pharyngeal lumen is thickened and hyperplastic. H&E ×100.

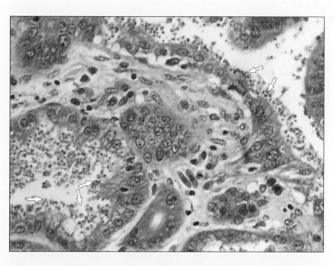

Figure 9.81 Gastric cryptosporidiosis in an Australian tiger snake (*Notechis scutatus*). A myriad number of round organisms (arrowed) are attached to the brush border of the mucosal cells lining the gastric lumen and gastric pits. H&E ×250.

These inflammatory conditions are often caused by items in the diet that injure the delicate mucous membranes that cover the tongue or line these cavities. In snakes, and to a lesser extent in lizards, gastric cryptosporidiosis is a serious clinical problem. The typical lesions induced by *Cryptosporidium serpentis* in snakes include gastric hyperplasia and fibrosis, a decreased diameter of the lumen, and oedema of the mucosa. Petechial haemorrhages are occasionally found together with necrotic foci. Gastric biopsy (or gastric lavage) specimens of infected snakes reveal myriad numbers of protozoan organisms attached to the brush border of the epithelial cells lining the gastric lumen and gastric pits (**Figure 9.81**).

Enteritis (inflammation of the intestine) is usually accompanied by an overproduction of protective mucus by the goblet cells. The inflammation may be suppurative, in which heterophils are easily identified, or nonsuppurative, in which the predominant leukocytes are mononuclear (**Figure 9.82**). The aetiologic agent may or may not be immediately apparent.

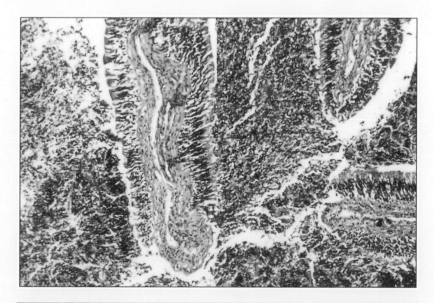

Figure 9.82 Nonsuppurative enteritis in a desert tortoise. Most of the leukocytes are lymphoplasmacytic. H&E ×100.

Intussusception (the telescoping of one segment of the intestine into another, or into the stomach) occurs relatively frequently in some reptiles, particularly in iguanas and Old World chameleons (**Figures 9.83** and **9.84**). The reasons for this high incidence are unknown, but endoparasitism and dietary problems, especially hypocalcaemia, are suspected as predisposing factors.

Benign and malignant neoplasia of the stomach and small intestine are relatively common in captive reptiles, particularly snakes and lizards. This may be a consequence of living considerably longer while in captivity than under natural (wild) conditions.

Preneoplastic leukoplakia and invasive squamous cell carcinoma have been described in chelonians. These proliferative lesions are similar to those observed in mammals.

Adult green iguanas (which are folivorous herbivores) have a simple stomach and a short small intestine that transports the partially processed leafy ingesta into the sacculated and much expanded colon. Villous projections (**Figure 9.78**), covered with pseudostratified, nonciliated columnar epithelium overlying a thin lamina propria and a core of smooth muscle and blood vessels, extend into the colonic lumen and create a larger surface area for the processing of cellulose and absorption of nutrients.

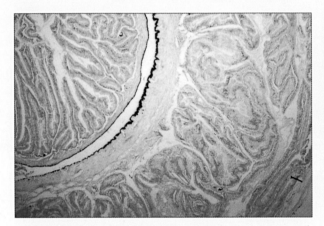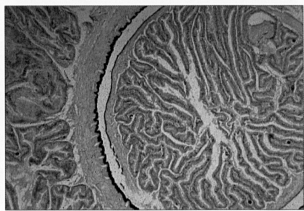

Figures 9.83 and 9.84 Duodenal-jejunal intussusception in a Fisher's chameleon (*Chamaeleo fisheri*; **Figure 9.83**) and an iguana (**Figure 9.84**). A segment of the duodenum has telescoped into the jejunum causing the two serosal layers to lie adjacent to each other. H&E ×100.

Liver

The liver is the largest gland in the body. Blood drains to the liver from the intestines via the hepatic portal vein, and the products of digestion are metabolised, harmful material is detoxified, and bile is secreted. The liver is surrounded by a connective tissue capsule that extends into the gland and divides it into lobes and lobules. The structure of the classic lobule is most clearly visualised in the pig because of its plentiful array of connective tissue dividing the liver into discrete hexagonal lobules with a portal area at the corners of each hexagon (**Figure 9.87**). This is not the case in other domestic animals, except under pathological conditions. The portal areas (triads) occur between three or more lobules, and each contains one or more branches of a hepatic artery, a hepatic portal vein, a lymphatic vessel, and a bile duct. The parenchyma consists of polyhedral epithelial cells of endodermal origin, the hepatocytes, arranged in anastomosing rows separated by sinusoids converging on the central vein. The sinusoids are lined with fenestrated endothelial cells and macrophages (called Kupffer cells in the liver). Blood flows through the sinusoids to the central vein. This, in turn, leaves the liver lobule to travel separately as branches of the hepatic vein.

Bile is secreted by each hepatocyte into the bile canaliculi, channels that are lined with the plasma membranes of the hepatocytes, between adjoining liver cells. It flows from there to a small bile duct in the portal area. Bile ducts are lined with cuboidal epithelium in the portal areas; the larger interlobular ducts are lined with columnar epithelium. Where the bile duct is the central functional axis of the lobule instead of the central vein, the term 'portal lobule' is used. A different functional division of the liver is the liver acinus. It consists of parenchyma served by a terminal branch of the portal vein and the hepatic artery and is drained by two central veins and terminal branches of the bile duct. It has functional and pathological significance (**Figures 9.85–9.90**).

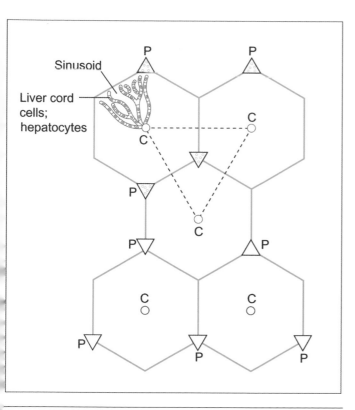

Figure 9.87 Liver (pig). The hexagonal liver lobule is delineated by the interlobular connective tissue. The central vein can be observed in the centre and portal areas at the periphery of the hexagon. H&E ×100.

Figure 9.85 Hepatic lobule. The classic lobule is the hexagon, clearly seen in the figure by the outline (green) of connective tissue. Portal areas (P) occur between the lobules. A portal lobule is defined as the central functional axis of the lobule (the black dotted triangle).

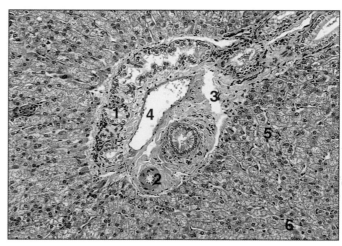

Figure 9.88 Portal canal (triad) (sheep). (1) Branch of the bile duct. (2) Hepatic artery. (3) Lymphatic vessel. (4) Hepatic portal vein. (5) Liver cords. (6) Sinusoids. H&E ×125.

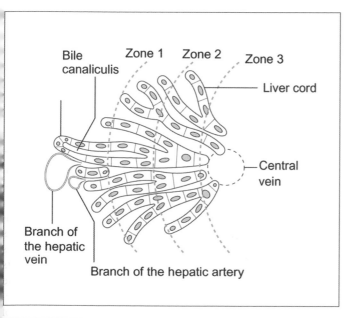

Figure 9.86 Liver acinus (functional unit). The liver parenchyma is served by the terminal branch of the portal vein and the hepatic artery. The acinus is divided into three zones, which indicate the relative position of the cells in relation to the oxygen gradient. Hepatocytes in Zone 1 are closest to the fresh, oxygen-rich hepatic arterial blood, and those in Zone 3 are further away. Equally, cells in Zone 1 are first in line for toxins, etc., carried in the portal blood, with Zone 3 the cells the least affected by these.

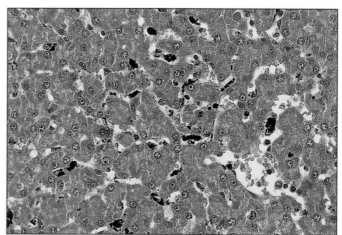

Figure 9.89 Liver (dog). The macrophages have taken up the injected carbon and appear as black areas between the cords of hepatocytes. Carbon-injected with H&E counterstain ×200.

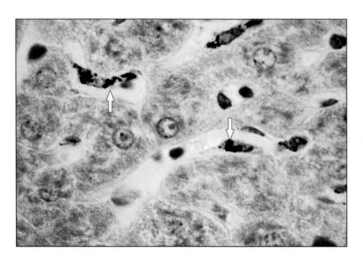

Figure 9.90 Liver (dog). The hepatocytes lie in anastomosing cords separated by sinusoids lined by macrophages (arrowed). Safranin/haematoxylin after carbon injection ×630.

Reptiles, Amphibians, and Fish

The livers of fish, amphibians, and reptiles are superficially similar to those of mammals. However, there are some differences. Many of the lower vertebrates have abundant melanin pigment scattered throughout the hepatocellular parenchyma (*see* Chapter 4). Usually, this pigment is contained in melanophages that are aggregated together in packet-like groups of cells bearing fine dark-brown granules. In some species, the liver is arranged in narrow cords radiating outward from a thin-walled central vein. It has one or more portal triads consisting of an arteriolar branch of the hepatic artery and one or more small bile ducts (as in mammals). In other species, the central veins are scattered randomly throughout the liver and more than one portal triad or triads with multiple arterioles and bile canaliculi or ducts are present.

CLINICAL CORRELATES

With its pivotal role in processing material carried from the intestine via the portal system, the liver is exposed to toxic factors and potentially harmful pathogens reaching from the intestines. Metabolic or nutritional disease (**Figures 9.91** and **9.92**), infectious disease (*see* **Figure 9.98**), and neoplasia, both local and metastatic, can also affect the liver. Inflammation of the liver is termed hepatitis (**Figure 9.93**). Hepatic lipidosis,

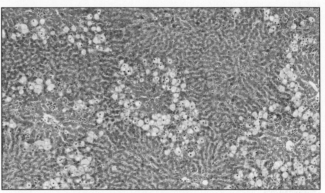

Figure 9.92 Hydropic degeneration in a rat (*Rattus norvegicus*). The cytoplasm of some hepatopcytes appear vacuolated due to an increase of intracellular water. H&E ×100.

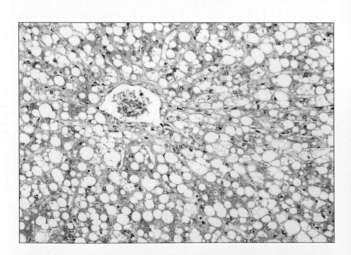

Figure 9.91 Hepatic lipidosis (horse). This micrograph of horse liver shows a central vein surrounded by radiating cords of hepatocytes that contain large, smoothly round vacuoles which occupy most of the cell and displace the nucleus to the periphery. These are fat vacuoles. The horse was hyperlipidaemic with hepatic lipidosis. H&E ×250.

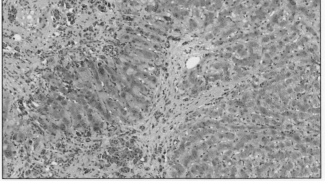

Figure 9.93 Chronic hepatitis in a horse. Hepatocytes show signs of degeneration and the presence of hemosiderin in the cytoplasm and within macrophages. Fibrous tissue is present, separating the hepatic lobules. H&E: Haemosiderin is present ×150.

an excess fat storage in the liver, is seen as a clinical problem in obese animals under physiological stress, e.g. pregnant pony mares, dairy cattle after parturition, and ewes carrying twins in late pregnancy (pregnancy toxaemia). Mobilisation of large amounts of triglycerides causes fatty acids to be presented to the liver in excess of its capacity to handle them. This problem may be quite rapidly fatal, and cases of sudden death are not uncommon.

The liver has a great capacity for regeneration of hepatocytes, but fibrosis is also a characteristic reaction of the liver to chronic injury. Any hepatic injury severe enough to result in hepatic necrosis results in some fibrogenesis, but progressive fibrosis can develop when the insult persists or when the initial damage is severe and provokes an extensive reaction. An equine liver with chronic hepatitis (**Figure 9.93**) shows progressive fibrosis in which the normal architecture of the liver is lost and large strands of fibrous tissue are present, together with large quantities of hemosiderin and hepatocyte degeneration.

AMPHIBIANS AND REPTILES

As with domestic animals, numerous chemical and viral agents induce severe liver disease (**Figure 9.94**). The hepatic parenchyma is sensitive to changes in calcium, and other minerals in the blood and, under conditions of hypervitaminosis D_3, may undergo severe mineralisation and even ossification (**Figure 9.95**).

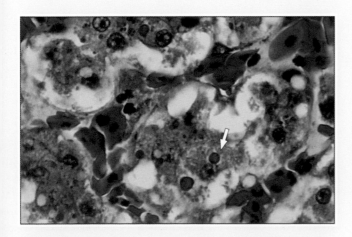

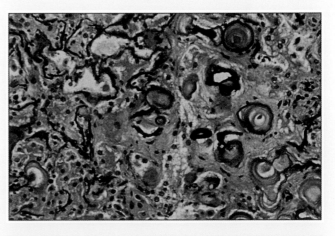

Figure 9.95 Nutrition-related, massive hepatic mineralisation secondary to hypervitaminosis D_3 in an African leopard tortoise (*Geochelone pardalis*). Most of the hepatocellular parenchyma has undergone gross alteration and is replaced by bone. As a consequence, few normal hepatocytes remain. Several osteocytes surrounded by concentric lamellae of compact bone are present. H&E ×125.

Figure 9.94 Viral hepatitis in a Colombian boa constrictor (*Boa C. constrictor*). Most of the hepatocytes contain eosinophilic, intracytoplasmic viral inclusion bodies (arrowed), most of which are surrounded by narrow clear 'haloes'. H&E ×400.

An admixture of hepatocellular and pancreatic tissues, thus forming a hepatopancreas, is present in many fish and in some amphibians and reptiles.

Gall Bladder

The gall bladder (absent in the horse, deer, rats, dolphins, rhinos, and hippos and some avian species like pigeons and some psittacine birds) is a reservoir for bile and is attached to the visceral surface or between the lobes of the liver. The mucous membrane is folded in the flaccid state, and the epithelial lining consists of tall columnar cells with a striated border (**Figure 9.96**). Goblet cells and mucus- and serous-secreting glands may be present in ruminants. The muscularis externa is a circular layer of smooth muscle, and the serosa is continuous with the peritoneum.

The gall bladder of lizards and chelonians is embedded in or surrounded by the liver, as it is in mammals and birds. The gall bladder of snakes is located at a variable distance from the liver and is contiguous with the spleen and

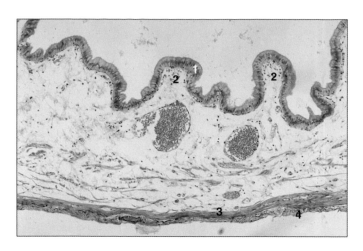

Figure 9.96 Gall bladder (cow). (1) Tall columnar epithelium lining the lumen. (2) Mucosal folds. (3) Muscularis. (4) Serosa. H&E ×50.

pancreas. A long bile duct transports bile from the intrahepatic bile duct(s) to the gall bladder for storage and eventual release into the duodenum. In lower vertebrates, hepatic and pancreatic tissue can be mixed together in one single organ, called the hepatopancreas (**Figure 9.97**).

Pancreas

A fine connective tissue capsule extends into the gland and divides it into lobules. The parenchyma is composed of exocrine and endocrine tissue; both are derived from the endoderm of the foregut. The exocrine portion of the

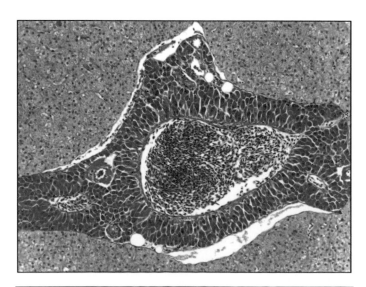

Figure 9.97 The liver and pancreas of some fish, amphibians, and reptiles are fused or admixed with one another and form a hepatopancreas. Illustrated is a section of such a mixed organ in a teleost fish (*Ctenopharyngodon Idella*). H&E ×200.

CLINICAL CORRELATES

The gall bladder can be a site of inflammation (cholangitis), calculi (choleliths), neoplasia, and presence of parasites as trematodes (e.g. *Fasciola hepatica*) (**Figure 9.98**).

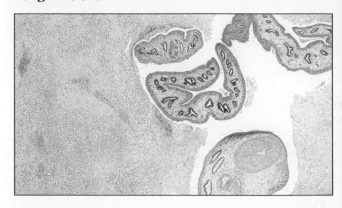

Figure 9.98 Liver (sheep) with parasites within the bile ducts and causing inflammatory cell infiltration (hepatitis) within the surrounding tissue. H&E ×50.

pancreas is a compound tubuloacinar gland that secretes enzymes into the duodenum. The acinar cells are tall columnar with a basal nucleus in basophilic cytoplasm. Where the secretory granules are stored, the luminal cytoplasm is eosinophilic. Projections of duct cells are commonly seen in the acinus; these are the centroacinar cells that are typical of the pancreas. Smaller ducts are lined with cuboidal epithelium and larger ducts with columnar epithelium (**Figures 9.99–9.102**).

The endocrine pancreas is responsible for the control of blood sugar concentrations and isolated groups of pale staining islet cells (pancreatic cells or the islets of Langerhans) are found scattered among the secretory units (**Figure 9.102**). These have two main cell types: A,

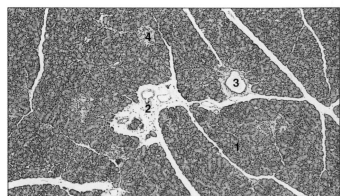

Figure 9.99 Pancreas (dog). (1) Serous acini of the exocrine pancreas. (2) Interlobular connective tissue. (3) Interlobular duct. (4) Pancreatic islet, the endocrine pancreas. H&E ×100.

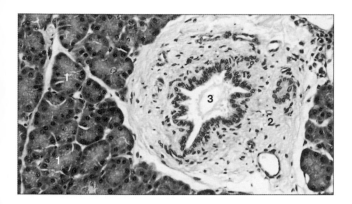

Figure 9.100 Pancreas (dog). (1) Serous acini. (2) Interlobular connective tissue. (3) Interlobular duct. H&E ×200.

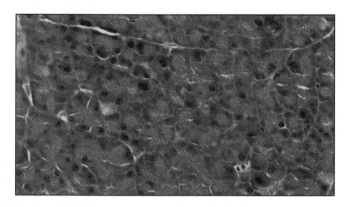

Figure 9.101 Pancreas (mouse). Acinar cell nuclei lie in the basal basophilic cytoplasm of the serous cell. Eosinophilic granules (the secretion) lie in the luminal cytoplasm. H&E ×200.

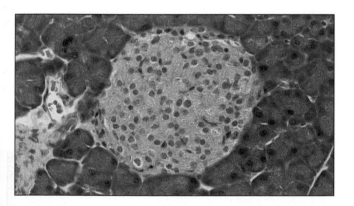

Figure 9.102 Pancreatic islet. Pancreas (mouse). Pale staining endocrine cells forming an islet and surrounded by exocrine acini. H&E ×400.

or alpha, cells secreting glucagon (a polypeptide hormone secreted in response to hypoglycaemia or to stimulation by growth hormone); and B, or beta, cells secreting insulin (a peptide hormone released into the blood in response to a rise in the concentration of blood glucose or amino acids). Rare D, or delta, cells secrete somatostatin and F cells secrete pancreatic polypeptide. These cells belong to the diffuse neuroendocrine system (DNES) cell group (*see* enteroendocrine cells of the stomach on page 90).

CLINICAL CORRELATES

Essentially, all of the various pancreatic disorders that occur in humans also occur in domestic mammals and in the so-called 'lower' vertebrates. Polycystic deformities, diabetes mellitus, pancreatic amyloidosis (**Figure 9.103**), acute and chronic pancreatitis, intraductal calculosis, and both benign and malignant neoplasms are recognised in diverse species.

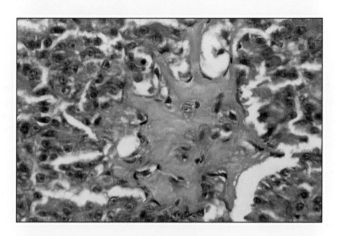

Figure 9.103 Pancreatic islet amyloidosis in a neutered female ocelot (*Felis padalis*). Essentially, this cat's islets are hyalinised and replaced with amorphous, eosinophilic amyloid. H&E ×125.

Reptilian, Amphibian, and Fish Pancreas

Morphologically, the pancreas of teleost fish, amphibians, and reptiles is similar to that found in mammals (**Figure 9.104**), but two major differences are observed in some species. Many fish, and some amphibians and reptiles, possess a pancreas that has an intimate association, and admixture of cells, with the spleen or liver (**Figure 9.105**). This combined organ is termed a 'spleno-pancreas' or 'hepato-pancreas', respectively, and the cells and tissues of each organ receive blood from their respective splenic, pancreatic, or hepatic branches of the splanchnic arteries and veins. Whereas the islet tissue in most reptiles tends to be conventionally arranged and evenly distributed, 'giant' islets of Langerhans are characteristic in the pancreatic tissues of some snakes, particularly members of the family Boidae (pythons and boas); rather than being scattered in a more or less random manner throughout the pancreatic exocrine tissue, these huge islets of endocrine cells tend to be localised in specific areas of pancreatic tissue (**Figure 9.106**).

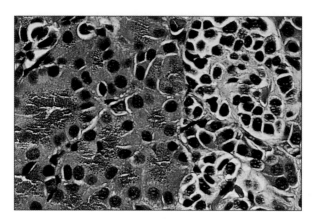

Figure 9.104 Pancreas of a salamander (*Amphiuma tridactyla*). Exocrine pancreatic cells characterised by their fine granular eosinophilic cytoplasm (on the left) extend a finger-like isthmus into the islet of paler staining endocrine cells bearing dense nuclei (on the right). The islet cells are arranged into nest-like lobules that are separated from each other and from the exocrine tissue by thin strands of connective tissue that support small blood vessels. H&E ×630.

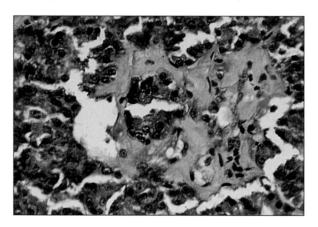

Figure 9.105 The splenopancreas of a milksnake (*Lampropeltis triangulum*). The spleen is on the right, a large aggregate of islet tissue is in the middle, and a portion of the exocrine pancreatic tissue is on the left. The islet displays hyalinisation. H&E ×400.

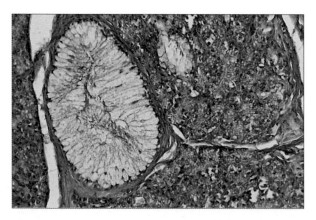

Figure 9.106 The pancreas of some snakes, particularly many pythons, is characterised by possessing endocrine cells formed into giant islets (often found at the edge of the lobule) rather than into many small nests of cells scattered randomly throughout the parenchyma. Illustrated is a section of the pancreas from a regal (ball) python (*Python regius*). H&E ×125.

CLINICAL CORRELATES

REPTILIAN PANCREAS

Certain species of reptiles appear to have a higher than expected incidence of some tumours. In captivity, some lizards, especially savannah monitors (*Varanus exanthematicus*), seem to show a high incidence of adult-onset diabetes mellitus and exocrine deficiency. No evidence suggests that diabetes mellitus or exocrine deficiency are as prevalent in wild savannah monitors. Therefore, these disorders seem to be artefacts of captive husbandry (caused particularly by overfeeding and lack of adequate exercise) resulting in spontaneous acute and chronic pancreatitis with subsequent autodigestion of the pancreatic parenchyma.

The migration of helminth larvae can also induce pancreatitis with secondary fibrosis and loss of secretory function both of exocrine and of endocrine components.

Avian Digestive System

The horny beak replaces functionally the lips and teeth of mammals.

Oral Cavity and Oesophagus

The oral cavity is lined with stratified squamous epithelium. The tongue is also lined with this type of epithelium, with some keratinised areas. The main mass of the tongue consists of striated muscle and a small bar of cartilage or bone: the entoglossal bone. There are no teeth. The glands in the lamina propria of the oral cavity, tongue, and pharynx are simple-branched and mucus secreting.

The oesophagus is lined with stratified squamous nonkeratinised epithelium, with simple mucous glands in the connective tissue lamina propria, with the presence of intraepithelial mucous glands (**Figure 9.107**). The

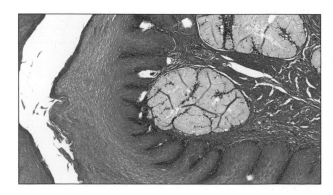

Figure 9.107 Oesophagus (bird). Stratified squamous nonkeratinising epithelium with simple tubular mucosal glands and intraepithelial mucous glands. H&E ×100.

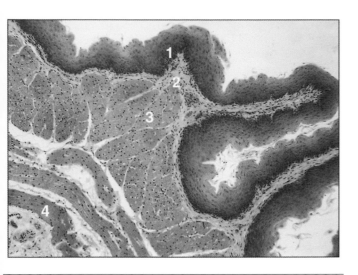

Figure 9.108   Crop (bird). (1) Stratified squamous nonkeratinised epithelium. (2) Lamina propria. (3) Muscularis mucosae. (4) Muscularis externa. H&E ×100.

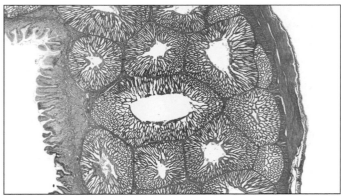

Figure 9.109   Proventriculus (chicken). The gastric epithelium is simple columnar and mucus secreting. A thin lamina propria separates it from the submucosal glands. Each submucosal gland lobule contains a central cavity with secretory tubules radiating to the interlobular connective tissue. An external muscular layer is present underlying the organ. H&E ×30.

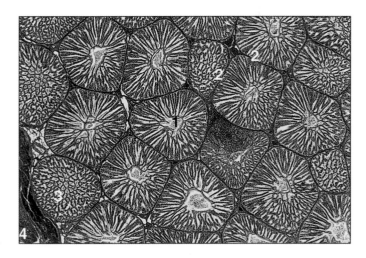

Figure 9.110   Proventriculus (bird). (1) Simple columnar mucus-secreting epithelium. (2) Lamina propria. (3) Submucosal glands. (4) Muscularis externa. Masson's trichrome ×25.

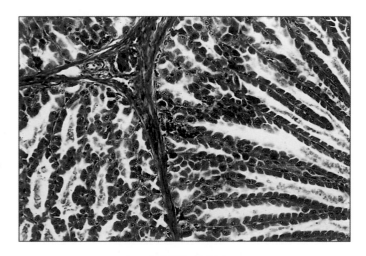

Figure 9.111   Proventriculus (bird). The submucosal gland lobules are separated by thin strands of connective tissue. Masson's trichrome ×250.

muscularis externa consists of a thick inner layer of circular and a thin outer layer of longitudinal smooth muscle. Lymphoid tissue accumulates in the caudal oesophagus as the oesophageal tonsil. The crop is an aglandular caudal diverticulum situated two-thirds of the way down the oesophagus (**Figure 9.108**). In the pigeon, two lateral glomerular sacs secrete crop milk.

Stomach

The stomach consists of the glandular proventriculus and a muscular ventriculus (gizzard). The gastric epithelium of the proventriculus is simple columnar and mucus secreting. A thin lamina propria separates it from the lobules of the submucosal glands. These glands form an almost continuous mass of tissue, with adjacent lobules separated by fine strands of connective tissue. Each gland lobule contains a central cavity with straight secretory tubules radiating to the interlobular connective tissue. An excretory duct drains onto the gastric mucosal surface. The glands contain only one type of cell, which secretes acids and pepsinogen, thus combining the functions of both the chief and parietal cells of the mammal. The *muscularis externa* is arranged as inner circular and outer longitudinal layers of smooth muscle (**Figures 9.109–9.111**).

The ventriculus is the aglandular stomach or gizzard. The luminal surface is lined with a secretory product of the mucosal glands, which solidifies at the surface to form a hard cuticle of koilin. The epithelium is low columnar and continues within the simple straight tubular mucosal glands in the lamina propria. A submucosa is present, and the *muscularis externa* is a thick layer of smooth muscle (**Figures 9.112** and **9.113**). There is no *muscularis mucosae*.

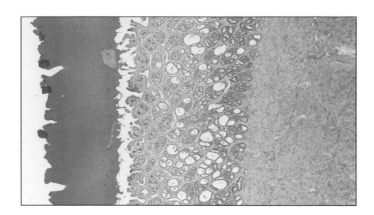

Figure 9.112 Gizzard (chicken). A thick cornified layer of koilin adjacent to the lumen, with underlying epithelium and glands and a thick muscular layer. H&E ×100.

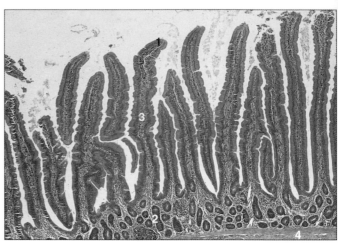

Figure 9.114 Duodenum (bird). (1) Simple columnar epithelium. (2) Intestinal mucosal glands. (3) Connective tissue core of the villus. (4) *Muscularis mucosae.* H&E ×62.5.

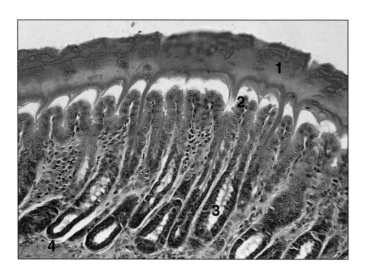

Figure 9.113 Ventriculus, gizzard (bird). (1) Cornified lining. (2) Epithelium lining the gizzard. (3) Mucosal glands. (4) Lamina propria. H&E ×250.

Intestine

The small intestine is similar to that of mammals but is more uniform throughout its length. Diffuse lymphatic tissue is present in the lamina propria and the submucosa, and the third layer of circular smooth muscle may be present in the muscularis externa (**Figures 9.114** and **9.115**).

The caeca are two blind sacs at the junction of the small and large intestines and are of considerable size in domestic birds. The epithelium is a simple columnar with mucous cells. Lymphatic tissue is particularly abundant, forming the caecal tonsil in the narrow proximal part of the caecum (**Figures 9.116** and **9.117**).

The large intestine has the same histological appearance as the caeca. The cloaca is lined with tall columnar

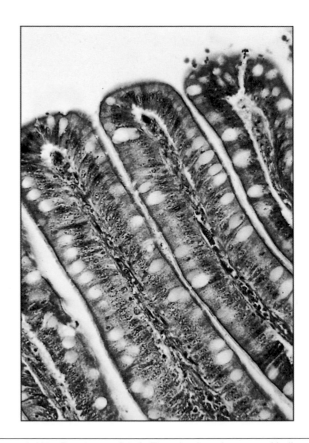

Figure 9.115 Duodenum (bird). The simple columnar epithelium lining the duodenum has a striated border to allow absorption and goblet cells. Masson's trichrome. ×400.

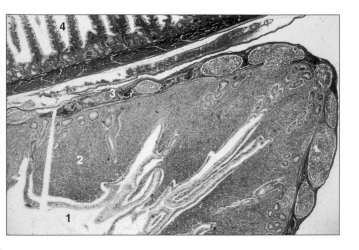

Figure 9.116 Caecum (bird). (1) Lumen of the caecum. (2) Mucosa consists of a columnar epithelium and a lamina propria with extensive deposits of lymphatic tissue. (3) Muscularis. (4) Lumen of the duodenum. Masson's trichrome ×50.

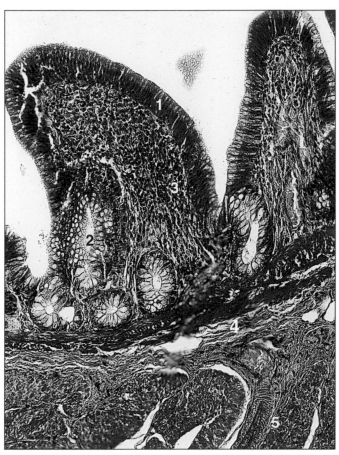

Figure 9.118 Cloaca (bird). (1) Simple columnar lining epithelium. (2) Simple tubular glands. (3) Lamina propria with lymphatic tissue. (4) Muscularis mucosae. (5) Muscularis externa. Alcian blue/PAS ×400.

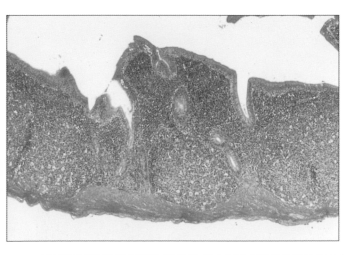

Figure 9.117 Caecum (bird). Dense masses of lymphatic tissue fill the lamina propria. H&E ×100.

epithelium with a variable number of mucus-secreting cells. A vascular lamina propria separates it from the muscularis mucosae and externa (**Figure 9.118**). (*See* Chapter 16 for the cloacal bursa.)

The avian liver is very similar to the mammalian; the connective tissue capsule extends into the gland and divides it into lobes and lobules. The hepatocytes are arranged in rows, often two cells thick, separated by sinusoids.

The avian exocrine pancreas is similar to the mammallian but has less interlobular connective tissue (**Figure 9.119**). The endocrine pancreatic islets are of three types: light (beta) islets, dark (alpha) islets, and mixed.

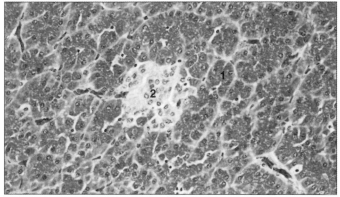

Figure 9.119 Pancreas (bird). (1) Serous units of the exocrine gland. (2) Pancreatic islet, the endocrine gland. H&E ×200.

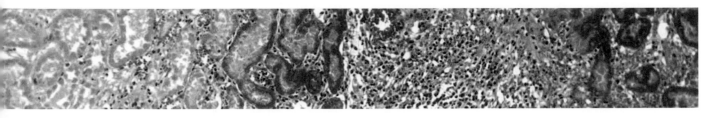

10

URINARY SYSTEM

The urinary system of mammals is composed of two kidneys, two ureters, a urinary bladder, and a urethra. The main function is to excrete the nitrogenated wastes from the body while regulating the volume and composition of the body fluids.

Kidneys

The kidneys are highly vascularised organs that filter the blood and excrete waste materials, excess water, and electrolytes via the ureters to the bladder as urine (*see* **Appendix Figures A1–A7**). The kidneys have an endocrine function: they secrete erythropoietin, which stimulates erythrocyte production in the bone marrow, and renin, which helps to regulate blood pressure. Each kidney is enclosed in a tough connective tissue capsule extending into the parenchyma and has two regions – the cortex and the medulla (**Figures 10.1–10.4**). Smooth

muscle may be present in the capsule. The hilus is a deep fissure on the medial border of the kidney and contains the renal artery, the renal vein, lymph vessels, and the ureter.

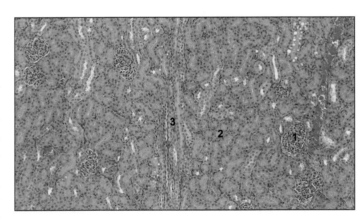

Figure 10.2 Kidney cortex (pig). (1) Renal corpuscle. (2) Uriniferous tubules. (3) Medullary ray H&E ×150.

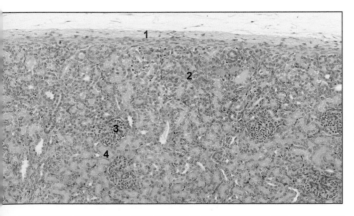

Figure 10.1 Kidney (horse). (1) Capsule. (2) Outer area of the cortex. (3) Renal corpuscle. (4) Uriniferous tubules. H&E ×100.

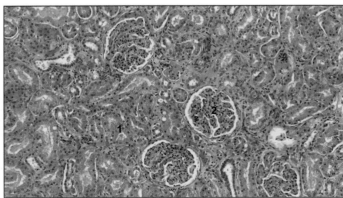

Figure 10.3 Kidney cortex (dog). (1) Uriniferous tubules. (2) Renal corpuscle. H&E ×150.

DOI: 10.1201/9781003333807-10

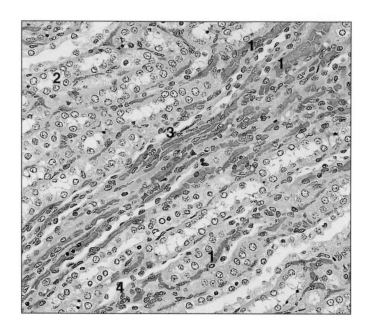

Figure 10.4 Kidney medulla (dog). (1) Blood vessels. (2) Collecting tubules. (3) Ascending thin limb. (4) Descending thin limb. H&E ×400.

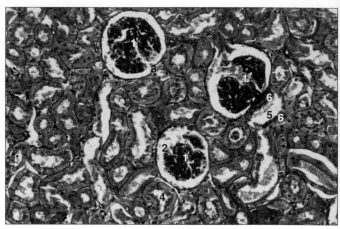

Figure 10.5 Kidney (horse). (1) Renal corpuscle with (2) capsular space (3) Urinary pole. (4) Proximal convoluted tubule. (5) Distal convoluted tubule with (6) the macula densa. H/periodic acid-Schiff (PAS) ×250.

Renal (Uriniferous) Tubules

Uriniferous tubules are the structural and functional units of the kidney. Each tubule has two components, the nephron and the collecting duct system.

Nephron

The nephron is composed of the renal corpuscle, the proximal tubule (with the convoluted and straight portions), the thin tubule, and the distal tubule (with the straight and convoluted portions).

The blind end of the proximal tubule is indented with a network of capillaries and supporting cells to form a filtering system: the renal corpuscle. Each renal corpuscle consists of a glomerulus (a tuft of capillaries) and a glomerular (Bowman's) capsule. The outer layer of the glomerular capsule is the capsular (parietal) wall, composed of simple squamous epithelial cells and separated from the glomerular (visceral) layer by the capsular (urinary or also called Bowman's) space (**Figure 10.5**). The capillaries are lined with a fenestrated endothelium resting on a basal lamina. The visceral layer of the capsule is composed of modified epithelial cells, the podocytes, that closely invest the capillary endothelium of the glomerulus and develop primary processes wrapped around each capillary. These processes develop secondary foot processes called pedicels. The foot processes of adjacent podocytes interdigitate, resulting in the formation of small gaps called filtration slits (**Figure 10.6**). The podocyte basal lamina is fused with the endothelial basal lamina, and blood passing through the capillary is filtered through this common

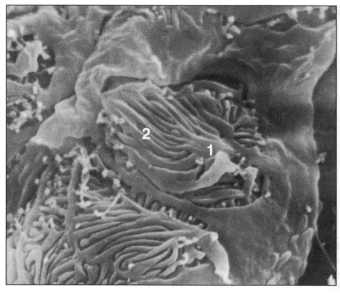

Figure 10.6 Kidney (dog). (1) Part of the podocyte. (2) Secondary foot processes, the pedicels – the filtration slits are the spaces between the solid pedicles. Scanning electron micrograph ×1500.

basal lamina into the capsular space (**Figures 10.6** and **10.7**). Intraglomerular mesangial cells are present between the endothelium and the basal lamina. The capillaries of the glomerulus are served by an afferent and an efferent arteriole, entering and leaving the renal corpuscle at the vascular pole (**Figure 10.8**). At the opposite pole is the capsular space, where the filtrate passes into the proximal tubule at the urinary pole of the renal corpuscle (**Figures 10.8** and **10.9**).

The proximal convoluted tubule (PCT) is long and lined with low columnar cells with a basal nucleus. The cytoplasm is deeply stained with eosin, has a granular appearance, and the apical surface is a continuous brush border. The basal plasma membrane is folded, with mitochondria in the cytoplasm giving a striated effect, and functions to increase the

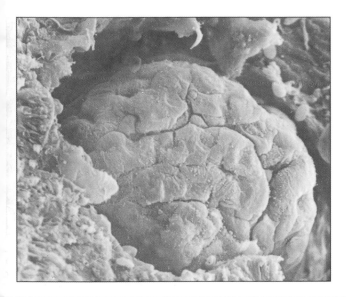

Figure 10.7 Kidney (dog). The renal corpuscle projects from the surrounding tissue. Scanning electron micrograph ×500.

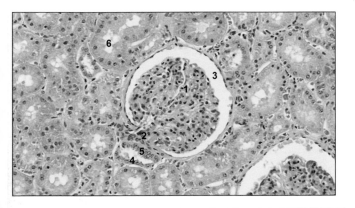

Figure 10.8 Kidney (horse). The width of the glomerular capsule is shown by the line. (1) Renal corpuscle with epithelial cells and mesangium. (2) Vascular pole. (3) Capsular space. (4) Distal convoluted tubule. (5) The macula densa. (6) Proximal convoluted tubule. H&E ×300.

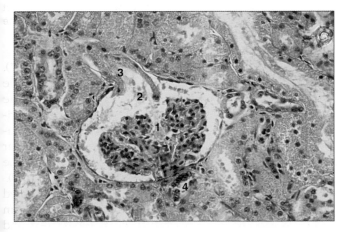

Figure 10.9 Kidney (horse). (1) Renal corpuscle with epithelial cells and mesangium. (2) Urinary pole opening into (3) the proximal convoluted tubule. (4) Distal convoluted tubule and macula densa. H&E ×250.

surface for transport. The PCT is continued with the proximal straight tubule. It is similar in appearance and extends towards the medulla where the epithelium changes abruptly to simple squamous. The tubule descends into the medulla as the descending limb and bends sharply to return to the cortex as the ascending limb. In the cortex, the epithelium becomes cuboidal or columnar and forms the distal straight tubule and coils near the glomerulus to become the distal convoluted tubule (DCT). The thin tubule, together with the proximal and distal straight tubules, forms the loop of Henle. The DCT is shorter than the PCT, the epithelium is cuboidal, the cytoplasm is paler, and there is no brush border.

The DCT approaches the glomerulus at the vascular pole, where it thickens, and the cell nuclei of the tubule wall become crowded together to form the macula densa, part of the juxtaglomerular apparatus. The organelles of these cells are polarised towards the basal cell surface, and this structure acts as a sensor monitoring the tubular concentration of sodium and chloride. Juxtaglomerular cells are modified smooth muscle cells in the walls of afferent arterioles close to the glomerulus. The cells are epithelioid, contain granules, and produce renin (which plays a role in the regulation of blood pressure) and angiotensin (a vasoconstrictor and stimulus of aldosterone secretion) (**Figures 10.1–10.3** and **10.9**). Extraglomerular mesangial cells, which function remains unknown, are located between the macula densa and the two glomerular arterioles (afferent and efferent).

Collecting Duct System

The collecting duct system (lined with poorly staining cuboidal epithelium) begins with the arched collecting ducts (connecting tubules), a continuation of the DCT within the cortex (**Figures 10.4** and **10.10**); it continues with the straight collecting ducts in the medulla, formed by the union of several connecting tubules, joining several of them to form the papillary ducts (of Bellini). Here, the epithelium becomes

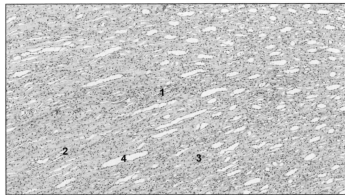

Figure 10.10 Kidney medulla (dog). (1) Capillaries lined by endothelium. (2) Collecting tubules lined by cuboidal cells. (3) Ascending limb lined by cuboidal epithelium. (4) Descending limb lined by squamous epithelium. H&E ×125.

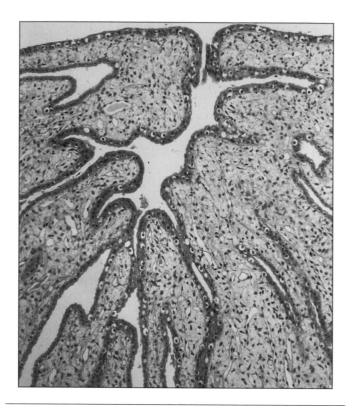

Figure 10.11 Kidney. Renal pelvis (sheep). The renal pelvis is lined by urothelium resting on a vascular lamina propria. H&E ×62.5.

columnar and then becomes urothelium towards the opening into the renal pelvis (**Figure 10.11**).

Renal Pelvis

This is the funnel-like dilatation at the cranial end of the ureter. It is lined with urothelium resting on a loose connective tissue lamina propria-submucosa. In the horse, there are numerous mucus-secreting glands (**Figures 10.12** and **10.13**).

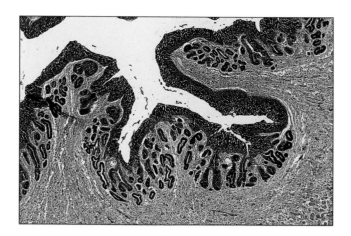

Figure 10.12 Kidney. Renal pelvis (horse). The renal pelvis is lined by urothelium; simple mucus-secreting glands are present in the lamina propria. H/PAS ×62.5.

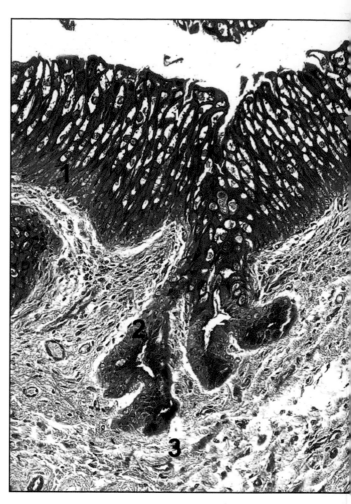

Figure 10.13 Ureter (horse). (1) Urothelium. (2) Simple mucus-secreting tubular glands. (3) Vascular lamina propria. H/PAS ×250.

The *muscularis* is three ill-defined layers of smooth muscle (longitudinal, circular, and longitudinal). The tunica adventitia is loose connective tissue.

Ureter

The ureter leaves the kidney at the pelvis and runs to the bladder. The mucosa is formed into plait-like longitudinal folds, and elastic fibres allow stretching. The urothelium consists of at least five to six cell layers, and in the horse, simple tubuloalveolar mucous glands are present in the lamina propria-submucosa. The *muscularis externa* is ill-defined with connective tissue between the bundles of smooth muscle (**Figures 10.14** and **10.15**). The outer coat may be loose connective tissue adventitia or a serosa, depending on the part of the ureter that is examined.

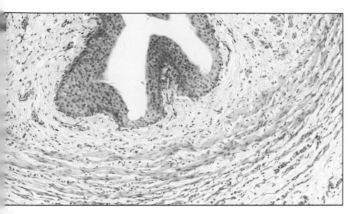

Figure 10.14 Ureter (sheep). The urothelium lines the lumen of the ureter, with underlying lamina propria and muscular layer. H&E ×100.

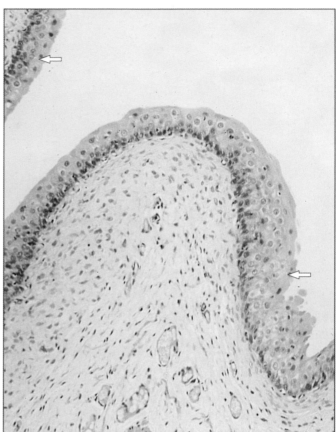

Figure 10.16 Urinary bladder (dog). The bladder is relaxed, and the surface cells are rounded adjacent to the lumen (arrowed). H&E ×200.

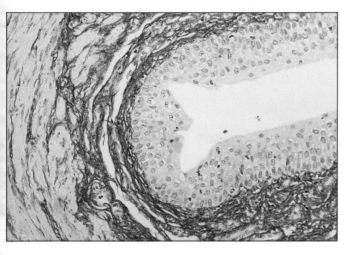

Figure 10.15 Ureter (dog). The elastic fibres are stained reddish-orange. Van Gieson ×200.

Urinary Bladder

The urinary bladder is lined by urothelium, a specialised epithelium with the number of cell layers depending upon whether the bladder is stretched or unstretched. There are elastic fibres in the lamina propria. Small bundles of smooth muscle form a discontinuous lamina muscularis except in the cat. The muscularis (detrusor muscle) is the same as in the ureter, and the outer layer may similarly be adventitia or serosa. Parasympathetic ganglia and nerve receptors are present (**Figures 10.16** and **10.17**).

Urethra

The male urethra serves a genital function and is discussed in Chapter 12. The female urethra is short, running from the bladder to the external urethral orifice, and has a purely urinary function. The mucosa is folded

longitudinally, and the epithelium varies from urothelium at the bladder to stratified squamous at the urethral orifice. Endothelium-lined caverns form an erectile plexus in the propria-submucosa varying their distribution in each species (*see* **Figures 12.15–12.18** and **13.26**).

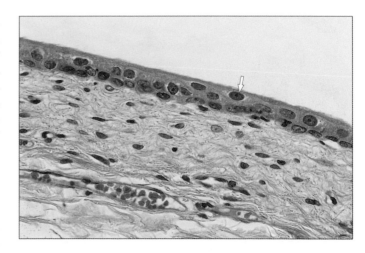

Figure 10.17 Urinary bladder (dog). The bladder is stretched, and the surface cells are flattened (arrowed). H&E ×400.

Avian Urinary System

Each kidney consists of three pyramidal divisions: cranial, middle, and caudal. These are not comparable to the lobes of the mammalian kidney. Each division receives a branch of the renal artery, a branch of the great renal vein, and in the renal portal vein, this is a branch of the internal iliac. The avian renal division is composed of a number of indistinct lobes made up of lobules, the structural unit of the kidney. Lobules that drain into a single branch of the ureter constitute a lobe. Each lobule is pear-shaped; the wider part is cortical tissue, and the tapering part is medullary tissue. In histological sections, the appearance is of the larger cortical areas surrounding cone-shaped islands of medullary tissue, called medullary tracts. The interlobular veins are wedged between the lobules.

There is no renal pelvis or urinary bladder.

There are two types of uriniferous tubules: the mammalian (metanephric) tubule extends into the medullary tissue and the shorter 'reptilian' (mesonephric), or cortical, tubule lacks a loop (of Henle). Both types begin with a renal corpuscle. The reptilian renal corpuscle is smaller than the mammalian, but more numerous, and has a prominent central mass of mesangial cells. The PCT is about half of the tubule and is connected by a short intermediate segment to the DCT. In the medullary tubule, the intermediate segment is the loop descending into the medullary tissue. A juxtaglomerular complex is present as in the mammal. The DCTs are joined by collecting tubules, lined with mucus-secreting cuboidal to the low columnar epithelium, to the perilobular collecting ducts. These fuse with other ducts to form larger ducts and lead to a secondary branch of the ureter. Five or six secondary branches fuse to form a primary ureteral branch (**Figures 10.18** and **10.19**).

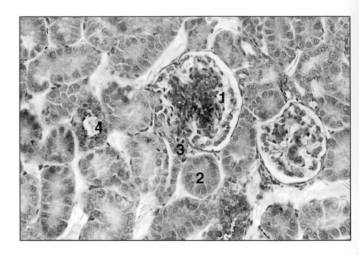

Figure 10.19 Kidney cortex (bird). (1) The renal corpuscle has a central mass of epithelial cells and mesangial cells. (2) Proximal convoluted tubule. (3) Distal convoluted tubule. (4) Small collecting tubule with low columnar mucus-secreting epithelium. H&E ×250.

The ureters are lined with a mucus-secreting, pseudostratified epithelium supported by a cellular lamina propria with variable amounts of diffuse lymphoid tissue. The thick muscularis consists of an inner longitudinal and an outer circular layer of smooth muscle (**Figure 10.20**). The ureter drains into the middle compartment of the cloaca: the urodeum.

Reptilian Urinary System

The renal tissues of reptiles are superficially very similar to those of birds and mammals, but there are some notable differences. In adult males of some species of squamate reptiles (snakes and lizards), there is an obvious and seasonal alteration in the size, shape, staining characteristics, and cytoplasmic granularity of the cells comprising the DCTs. This change

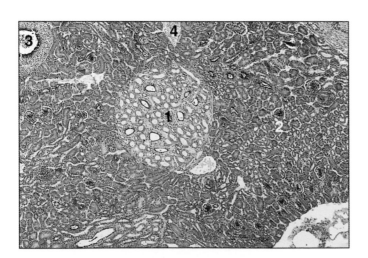

Figure 10.18 Kidney (bird). (1) The central, pale staining medullary area is surrounded by (2) the much denser staining cortical area. (3) Lobar duct. (4) Renal vein. H&E ×25.

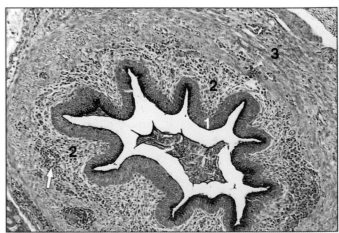

Figure 10.20 Ureter (bird). (1) Pseudostratified mucus-secreting epithelium lines the lumen. (2) Cellular lamina propria with groups of lymphocytes (arrowed). (3) Muscularis externa. H/PAS ×125.

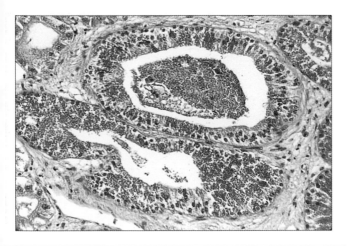

Figure 10.21 The kidneys of some species of sexually mature male snakes and lizards possess a characteristic 'sexual segment granulation', which is discerned by a marked hypertrophy and eosinophilic granularity of the distal convoluted tubules. This makes the kidneys of sexually mature males readily distinguishable from the kidneys of sexually mature females of the same species. Illustrated is a section of kidney from a mature male timber rattlesnake (*Crotalus horridus*). H&E ×125.

is termed the 'sexual segment' and is seen in sexually mature male crotalids (pit vipers), teiid, and some iguanid lizards, to name but a few (**Figure 10.21**). The change is so striking that the sex of the animal from which the renal tissue was obtained is immediately apparent to the histologist. The epithelial cells of the DCTs of mature males become markedly hypertrophied and become packed with small, round, and highly eosinophilic cytoplasmic granules that are extruded into the urinary wastes. They are believed to contribute a pheromone or similar chemical cue to the urates that are passed during periods of sexual courtship and mating activity. The sexual segment does not occur in females.

The histological characteristics of the amphibian and reptilian ureter and urinary bladder are similar to those of mammals. The lumen is lined by transitional epithelium (urothelium) that is several layers thick when the bladder is empty but becomes flatter and thinner when the bladder is distended. The bladder wall contains wispy strands of smooth muscle. Not all reptiles possess urinary bladders. Instead, the urinary wastes are passed from the twin ureters into the urodeum portion of the cloaca and thence out through the cloacal vent. In some reptiles, the urine is retained for a variable length of time, during which water is actively removed and 'recycled' before the now concentrated, urate-rich wastes are discharged as a pasty white, relatively water-insoluble microcrystalline substance.

Amphibian Urinary System

There is a substantial change in the renal histology and function before, during, and after metamorphosis in most amphibians. The pronephric kidney of early larval

amphibians changes to become fully metanephric in the adult stage of most amphibians. However, in the caecilian amphibians, which are elongated, legless creatures, the kidneys are opisthonephric.

Some amphibians without tails (frogs and toads) excrete most of their nitrogenous wastes as toxic ammonia. Others excrete urea, yet others produce urate salts of uric acid similar to those produced by terrestrial reptiles. The mode of nitrogen excretion is dictated mainly by the living habits; aquatic forms tend to be more ammonotelic, whereas terrestrial amphibians tend towards ureotelism and uricotelism. In addition, the season of the year and hydration affect the means by which these animals excrete their nitrogenous wastes.

The kidneys are one of the major sites of haematopoiesis in larval amphibians. During and after metamorphosis, the kidneys lose most of this ability and, as a result, assume a more familiar histological pattern consisting of nephrons that superficially resemble those of reptilian and avian species (**Figure 10.21**).

There is no loop interposed between the PCT and DCT and, therefore, the degree of concentration of the glomerular filtrate is variably limited.

Fish Urinary System

Most fish excrete toxic ammonia-rich urinary wastes that are converted via nitrification by the bacteria *Nitrosomonas* spp. to nitrite and then by *Nitrobacter* spp. to nitrate, a less toxic ionic product that can be metabolised by aquatic plants. Thus, the potential toxicity of water containing urinary wastes is prevented by the action of these two essential microbial organisms and aquatic flora, which, in turn, yield oxygen via photosynthetic pathways.

The kidneys of fish are divided into a cranial pronephric (head) kidney and caudal mesonephric (tail) kidney. The cranial portion is the major site of erythropoiesis. Erythropoietic tissue occupies the interstitial spaces between adjacent glomeruli and renal tubules. The histology of fish kidneys varies widely between species and between marine and freshwater fish: the kidneys of some marine teleosts lack glomeruli. Structurally, the renal tissue of freshwater fish is readily recognisable as kidneys at low magnification, but the intervening erythropoietic component may seem to be a cellular inflammatory infiltrate to histologists unfamiliar with fish kidneys.

The caudal portion functions in a conventional renal manner as a site of proteinaceous waste filtration and removal. The glomerulus is easily recognisable by the tuft of capillaries and its parietal and visceral capsule. The renal tubules are composed of cuboidal to low columnar epithelial cells, similar to those seen in other vertebrates. In teleosts, the kidney also plays a role in the osmoregulation of sodium and chloride. The gills also participate in osmoregulation.

CLINICAL CORRELATES

In all animals, the main role of the kidney is the homeostatic control of extracellular fluid composition. This involves maintenance of normal concentrations of salt and water in the body, control of acid–base balance, and excretion of waste products.

To function normally, the kidney requires adequate perfusion with blood, sufficient functional renal tissue, and unimpeded urinary outflow. Failure of kidney function can therefore be related to inadequate perfusion (prerenal), to inadequate processing in the kidney (renal), or to blockage of urinary outflow (postrenal).

In animals, renal disease is often subclinical. Clinical disease may be divided into acute and chronic renal failure. Acute renal failure involves a sudden onset of oliguria or anuria and azotaemia and is often the result of acute glomerular, interstitial damage or acute tubular necrosis. This form of renal disease is often reversible. Once the kidney is fully developed, new nephrons (functional units) are not produced and chronic renal failure with progressive destruction of functional tissue, regardless of initiating cause, leads to a syndrome of salt and water imbalance, acid–base disturbance, and accumulation of wastes. Chronic renal failure results in irreversible changes that produce shrunken, fibrosed 'end-stage' kidneys.

A wide variety of developmental, circulatory, metabolic, inflammatory, and neoplastic conditions can affect the kidneys. Renal infarcts can be found in the kidneys of the swine in cases of artery occlusion due to thrombosis and aseptic emboli or cases of renal vasculitis (**Figure 10.22**). Familial nephropathies are recognised in several dog breeds, and renal cysts are quite common in pigs and cattle. Glomerulonephritis, often of immune origin, is a common cause of chronic renal failure in both dogs and cats (**Figures 10.23** and **10.24**).

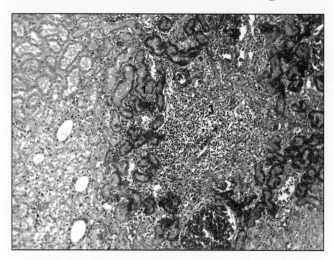

Figure 10.22   Renal infarct caused by an artery occlusion in a pig. Coagulative necrosis (1), hyperaemic area (2), and inflammatory infiltrate (3). H&E ×100.

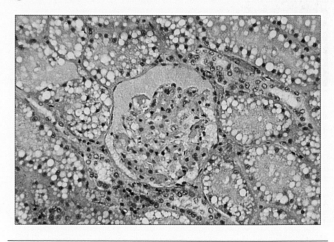

Figure 10.23   Feline infectious fibrinoperitonitis is caused by a feline coronavirus and can produce an immune-mediated glomerulonephritis with severe proteinaemia. The glomerular space and tubular lumina contain eosinophilic proteinaceous material. The numerous intracytoplasmic lipid vacuoles seen in the tubular epithelial cells are normal in the feline kidney. H&E ×250.

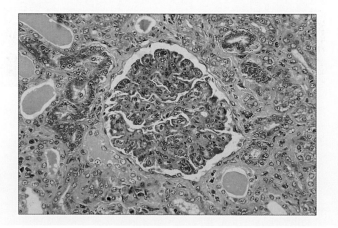

Figure 10.24   Chronic glomerulonephritis (dog). This high-power micrograph shows thickening of the glomerular capillary loops, fine interstitial fibrosis, and accumulation of proteinaceous fluid in the tubules. This disease is usually of immune origin and is a common precursor to the 'end-stage' kidney of renal failure. Masson's Trichrome & Orange G ×250.

Interstitial nephritis characterised by foci of nonsuppurative inflammation can be caused in pigs by systemic viral infections such as porcine reproductive and respiratory syndrome (PRRS) virus (**Figure 10.25**). In chronic cases of interstitial nephritis, areas of interstitial fibrosis can occur (**Figure 10.26**). Porcine dermatitis and nephropathy syndrome (PDNS) is an immune-mediated disease associated with porcine circovirus type 2 (PCV2) and type

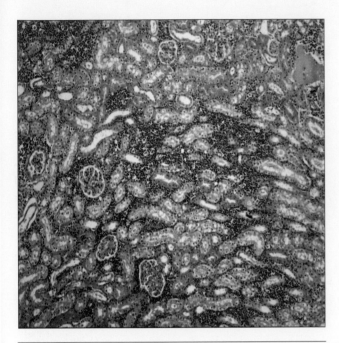

Figure 10.25   Acute interstitial nephritis (pig). This micrograph is of a kidney section from a pig with acute interstitial nephritis. Large numbers of inflammatory cells, mostly lymphocytes, are present between the tubules and around glomeruli in a case of infection by PRRS virus. H&E ×100.

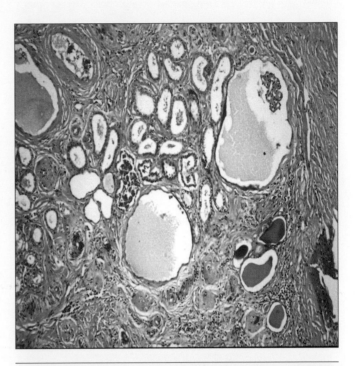

Figure 10.26   Chronic interstitial nephritis (pig). Presence of interstitial fibrosis, destruction of glomeruli, and accumulation of proteinaceous fluid in the tubules. H&E ×100.

3 (PCV3) infection, although the role of those viruses in the aetiology has not been demonstrated so far, that affects the kidney and skin, causing a fibrinonecrotising glomerulitis with fibrin and neutrophils filling and occluding Bowman's spaces and nonpurulent interstitial nephritis in the kidney, and a necrotising vasculitis in the skin (**Figure 10.27**).

As glomerular damage leads to significant protein loss, this can result in the development of nephrotic syndrome, characterised by hypoalbuminaemia, generalised oedema and hypercholesterolaemia. Primary renal neoplasms are uncommon in domestic animals, with renal carcinoma the most commonly recognised tumour in dogs, sheep, and cattle, while in pigs, nephroblastomas (true embryonal tumours which arise in primitive nephrogenic tissue) are more frequently seen, especially in younger animals.

Renal tumours are common in salmonid fish and some amphibians, particularly leopard frogs (Rana pipiens) in which a specific virally induced adenocarcinoma (of Lucke) is found.

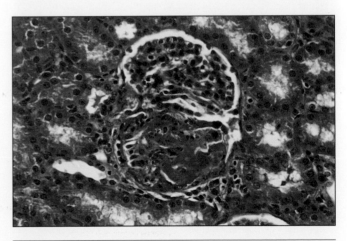

Figure 10.27   Glomerulonephritis in a case of PDNS in a pig. Abundant fibrin and cell debris can be observed within the glomerulus. H&E ×400.

Purulent interstitial glomerulonephritis in a kingsnake (**Figure 10.28**), renal gout in a boa constrictor (**Figure 10.29**), cholesterol nephrosis in a Galapagos tortoise (**Figure 10.30**), and renal trematodiasis in an Argentine horned frog (**Figure 10.31**) are shown.

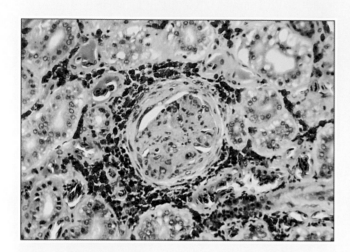

Figure 10.28 Purulent interstitial glomerulonephritis in a kingsnake (*Lampropeltis getulus*). The glomerular capsule and glomerular tuft are thickened. The renal interstitial connective tissue is infiltrated by heterophil granulocytic and mixed small mononuclear leukocytes. H&E ×180.

Figure 10.29 Renal gout. Illustrated is a section from the kidney of a boa constrictor that had been treated with several aminoglycoside antibiotic drugs without receiving adequate parenteral fluid therapy. A large 'star burs'-shaped tophus has replaced the glomerulus, and the periglomerular interstitial connective tissue is infiltrated by mixed small mononuclear leukocytes. H&E ×180.

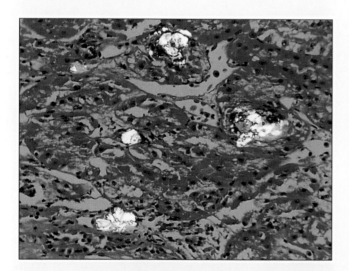

Figure 10.30 Cholesterol nephrosis. Illustrated is a section of a kidney from a Galapagos tortoise (*Geochelone elephantopus*), which displays multifocal deposits of cholesterol crystals that are obstructing some glomeruli and renal tubules. This condition is seen in captive herbivorous reptiles that have been fed abnormal diets such as commercial dog and cat food. H&E, photographed with cross-polarised illumination ×200.

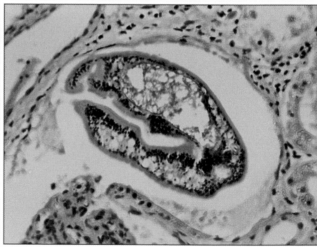

Figure 10.31 Renal trematodiasis. The renal pelvis of this Argentine horned frog (*Ceratophrys ornata*) is dilated and contains a large fluke. H&E ×200.

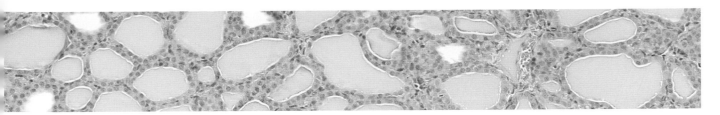

11

ENDOCRINE SYSTEM

The endocrine system is composed of a series of ductless glands that secrete their products, the hormones, into the intercellular compartments and diffuse cells in different body structures.

Endocrine tissue is derived from epithelioid parenchymal cells and may form discrete glands, such as hypophysis cerebri (pituitary gland), thyroid, parathyroid, adrenal, and epiphysis cerebri (pineal gland). Groups of endocrine cells are also active in the interstitial cells of the testis, the granulosa and luteal cells of the ovary, the transient endocrine cells of the placenta, the pancreatic islets (of Langerhans), which are responsible for insulin production and release (*see* **Figure 9.102**), the juxtaglomerular apparatus of the kidney, endocrine cells in the heart, adipocytes, and diffuse neuroendocrine system (DNES) cells, formerly known as amine precursor uptake and decarboxylation (APUD) cells. Hormones can be secreted directly

into a blood or a lymphatic vessel, or tissue fluid to influence the activity of target organs.

Hypophysis Cerebri (Pituitary Gland)

The hypophysis cerebri consists of a glandular lobe, the adenohypophysis, and a fibrous lobe, the neurohypophysis (**Figures 11.2** and **11.3**).

Adenohypophysis

The adenohypophysis (also called pars anterior of the hypophysis cerebri) is derived from an invagination of the ectoderm of the dorsal portion of the oral cavity, the

CLINICAL CORRELATES

Diabetes mellitus, one of the most common endocrinopathies of the dog, may be associated with degenerative change and fibrosis in the pancreas (**Figure 11.1**). However, the diagnosis of this condition rests on clinical testing rather than biopsy, as the pancreas of nondiabetic elderly dogs may appear similar and, conversely, some diabetic dogs will have a histologically normal pancreas.

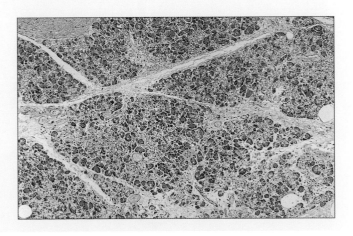

Figure 11.1 Diabetes mellitus (dog). The section of pancreas shown here was taken from an 11-year-old, neutered female English Springer Spaniel with a 2-year history of diabetes mellitus. There are pale eosinophilic areas of replacement fibrosis and the remaining epithelial cells of the exocrine pancreas, with their dark basal nuclei and strongly eosinophilic cytoplasm, can be seen. No islet tissue is discernible. H&E ×62.5.

DOI: 10.1201/9781003333807-11

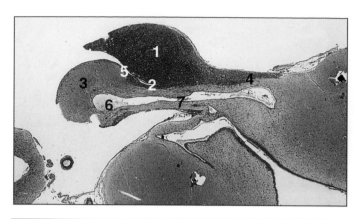

Figure 11.2 Hypophysis cerebri (cat). (1) Pars distalis. (2) Pars intermedia. (3) Pars nervosa. (4) Pars tuberalis. (5) Residual lumen (Rathke's pouch). (6) Infundibular recess (third ventricle). (7) Infundibular stalk. H&E ×5.

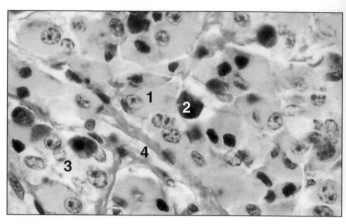

Figure 11.4 Pars distalis (cat). (1) Eosinophilic alpha cells. (2) Basophilic beta cells. (3) Chromophobes. (4) Sinusoidal blood vessels lie along the cords and clusters of secretory cells. Orange G/acid fuchsin ×250.

hypophyseal (Rathke's) pouch. It has three subdivisions: the pars distalis, the pars intermedia, and the pars tuberalis.

Pars Distalis The pars distalis is the major constituent of the glandular pituitary. The dense connective tissue capsule is continuous with a fine network of reticular fibres supporting the cords and clusters of parenchymal cells and the capillary/sinusoidal blood vessels. There are two broad

categories of cells based on staining affinity: the chromophobes and the chromophils (**Figures 11.4–11.8**). The chromophobe cytoplasm has a few granules that are nonreactive to dyes. These may be reserve cells or exhausted degranulated cells. The chromophils have a strong affinity for dyes and are divided into acidophils (alpha cells) and

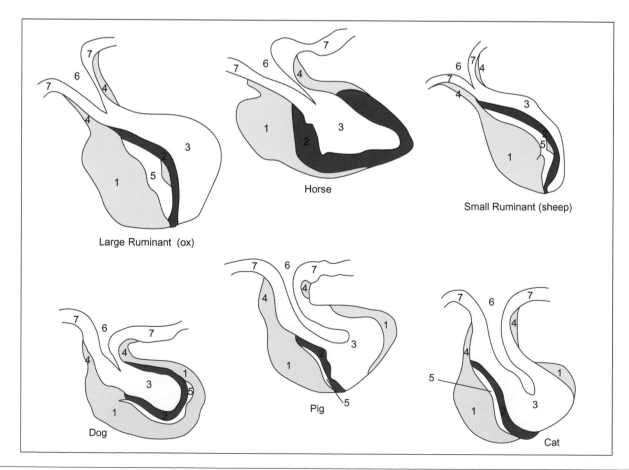

Large Ruminant (ox)

Horse

Small Ruminant (sheep)

Dog

Pig

Cat

Figure 11.3 Diagram of the hypophysis cerebri (pituitary) in some domestic species. (1) Pars distalis. (2) Pars intermedia. (3) Pars nervosa. (4) Pars tuberalis. (5) Hypophyseal cleft (residual lumen of Rathke's pouch). (6) Infundibular recess (third ventricle). (7) Infundibular stalk.

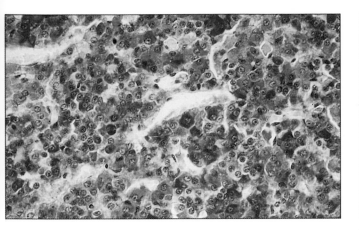

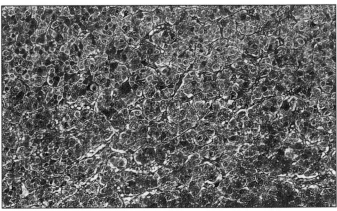

Figure 11.5 Pars distalis (horse). The orange staining A cells are easily distinguished from the deep pink staining B cells. Orange G/PAS ×125.

Figure 11.8 Pars distalis (horse). H&E ×100.

basophils (beta cells) in H&E stained sections. The acidophils secrete the growth hormone (GH) and prolactin (PRL). In some species, somatomammotroph or bihormonal cells are described, producing both GH and PRL. Basophils are larger than acidophils, fewer in number, and the cytoplasmic granules stain strongly with periodic acid-Schiff (PAS). They secrete thyroid-stimulating hormone (TSH), follicle-stimulating hormone (FSH), luteinising hormone (LH), and adrenocorticotropic hormone (ACTH).

Pars Intermedia The pars intermedia is well developed in domestic animals and lies between the pars distalis and the pars nervosa (neurohypophysis). In the horse, these regions lie closely together. In other domestic animals, the pars intermedia and pars distalis are partially separated by the hypophyseal cavity, lined by a cubical to pseudostratified epithelium. The parenchymal cells are basophilic with a few acidophilic cells and are arranged in columns or in follicles (**Figure 11.9**). The hormone secreted by these cells

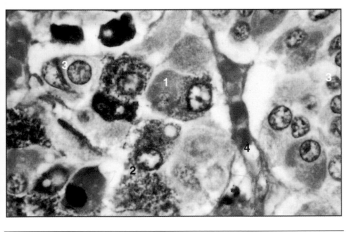

Figure 11.6 Pars distalis (cow). (1) Eosinophilic alpha cells. (2) Basophilic beta cells. (3) Chromophobe. (4) Sinusoid. H&E ×500.

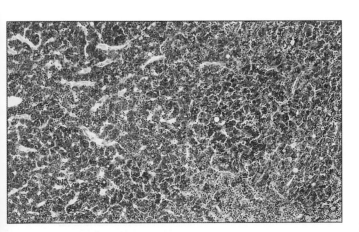

Figure 11.7 Pars distalis (dog). A cells stain with orange G and the B cells with acid fuchsin. Orange G/acid fuchsin ×100.

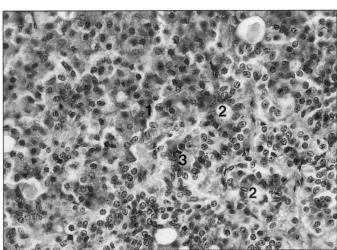

Figure 11.9 Pars intermedia (cat). The parenchymal cells are arranged in (1) columns or in (2) follicles. (3) Blood vessels. PAS/orange G/haematoxylin ×250.

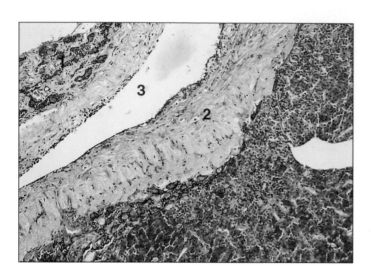

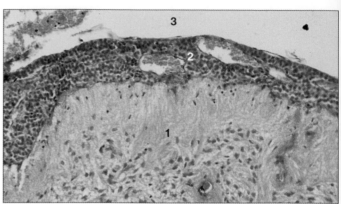

Figure 11.12 Pars nervosa (cat). (1) Pars nervosa. (2) Pars intermedia. (3) Infundibular recess. H&E ×100.

Figure 11.10 Pars intermedia/infundibular stalk (horse). (1) The parenchymal cells are separated by blood vessels (orange-staining erythrocytes). (2) The infundibular stalk is fibrous in appearance with small blood vessels of the portal system. (3) The third ventricle is lined by ependyma. PAS/orange G/haematoxylin ×50.

Neurohypophysis

The neurohypophysis is derived from a ventral invagination of the diencephalon (the caudal part of the forebrain) and is divisible into a pars nervosa, median eminence, and infundibular stalk. The neurohypophysis is composed of numerous unmyelinated nerve fibres with cell bodies that are located in the supraoptic and paraventricular nuclei of the hypothalamus. The axons converge at the median eminence to form the hypothalamic–hypophyseal tract and pass through the infundibular stalk to terminate on the endothelial lining of the capillaries of the pars nervosa (**Figures 11.12** and **11.13**). The neurosecretions of these cells pass down the axons and accumulate at the terminal regions of the nerve fibres as neurosecretory (Herring) bodies (**Figure 11.14**). The axons are supported by pituicytes (neuroglial cells). Oxytocin (OT) and antidiuretic hormone (ADH) are the major hormones produced by the pars nervosa.

stimulates melanocytes and controls the degree of skin pigmentation.

Pars Tuberalis The pars tuberalis surrounds the infundibular stalk, and together they form the hypophyseal stalk (**Figure 11.10**). It is highly vascularised. The major blood vessels of the hypothalamic–hypophyseal portal system lie in the stalk and allow the transfer of releasing factors secreted in the brain to the target cells in the pars distalis. The tuberalis cells are small and basophilic (**Figure 11.11**).

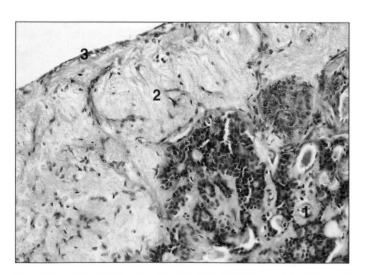

Figure 11.11 Pars nervosa/infundibular stalk (horse). (1) Pars tuberalis. (2) Infundibular stalk with portal vessels. (3) Ependyma lining the third ventricle. PAS/orange G/haematoxylin ×125.

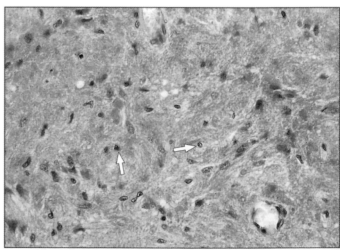

Figure 11.13 Pars nervosa (cat). The pars nervosa has a distinctive fibrous appearance; the nuclei of the pituicytes are arrowed. H&E ×250.

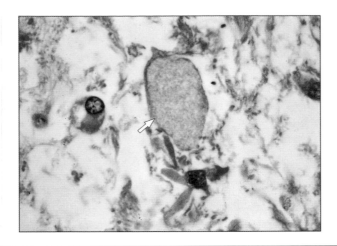

Figure 11.14 Pars nervosa (cat). The neurosecretion accumulates in the terminal part of the axons of the hypothalamic–hypophyseal tract to form neurosecretory bodies (arrowed). H&E ×400.

CLINICAL CORRELATES

Neoplasms involving one or more lobes of the pituitary are more common. Because of the location of the pituitary gland and the limited space available for an expanding lesion to occupy within the skull, the clinical signs that attend pituitary tumours may be referable to pressure on adjacent structures. If the optic chiasm is affected, deterioration of vision extending to blindness will ensue. More specifically, the signs manifested by an animal with a hypophyseal neoplasm refer to hyper- or hyposecretion of one or more pituitary tropic hormones.

Hormonally active tumours of the adenohypophysis result in a clinical picture of corticosteroid excess known as Cushing's disease. Neoplasms involving neurohypophysis usually result in either the hypo- or hypersecretion of ADH or OT. Therefore, the first clinical signs may be substantial changes in renal function, water loss or retention, and electrolyte regulation.

Thyroid Gland

The thyroid gland is derived from an endodermal outgrowth from the floor of the embryonic pharynx. The thin capsule of dense irregular connective tissue is continuous with the fine reticular fibres of the vascular stroma. It is partially divided into lobules by thin trabeculae. Each lobule consists of numerous follicles lined with simple cuboidal epithelium; the principal follicular cells secrete the thyroid hormones (**Figures 11.15** and **11.16**). The secretion is stored in the follicles as homogenous eosinophilic colloid. The hormones cross back through the follicular cells as required and enter the capillaries as tri-iodothyronine (T3) and thyroxine (tetra-iodothyronine; T4). These regulate the metabolic activity of all of the body cells and tissues. Large

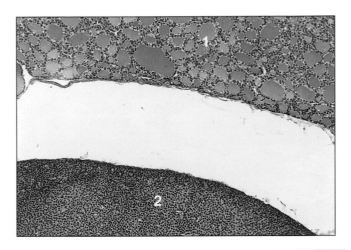

Figure 11.15 Thyroid/parathyroid gland (cat). (1) The thyroid follicles are filled with colloid. (2) Densely basophilic cords of parathyroid chief cells lie in the thyroid capsule. H&E ×100.

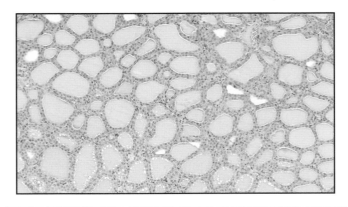

Figure 11.16 Thyroid gland (goat). The thyroid follicles show considerable variation in size and are full of eosinophilic colloid. H&E ×100.

pale cells with an eosinophilic cytoplasm, the parafollicular cells, lie between the follicular cells (**Figure 11.17**). These cells produce calcitonin, an antagonist of parathyroid hormone controlling blood–calcium concentrations. The secretory activity of the thyroid is controlled by the TSH or

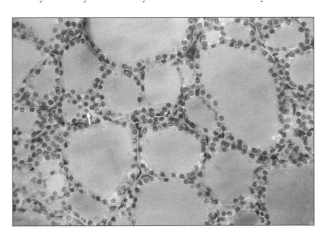

Figure 11.17 Thyroid gland (sheep). The thyroid follicle is lined by a simple cuboidal epithelium. The clear parafollicular cells are arrowed. H&E ×160.

CLINICAL CORRELATES

Feline hyperthyroidism (**Figure 11.18**) is one of the most common endocrine diseases encountered in veterinary practice. Typically, an elderly cat presents with weight loss and polyphagia. Vomiting and diarrhoea are also often reported. Hyperthyroidism results from hypersecretion of thyroid hormone by a hyperplastic or adenomatous thyroid gland which is often palpably enlarged (goitre).

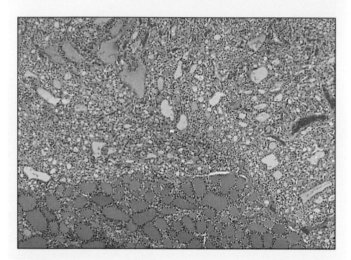

Figure 11.18 Thyroid adenoma (cat). Normal thyroid tissue, at the base of the image, is compressed by an adenomatous growth that is more cellular and has less colloid production. The cells are quite uniform in appearance and form recognisable acinar patterns. The mitotic rate is low. H&E ×50.

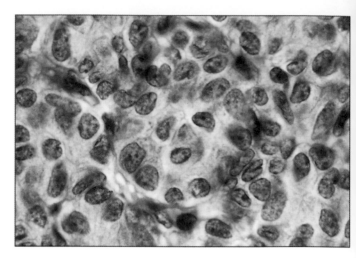

Figure 11.19 Parathyroid gland (dog). Darkly staining basophilic principal (chief) cells are ranged along the capillary bed. H&E ×500.

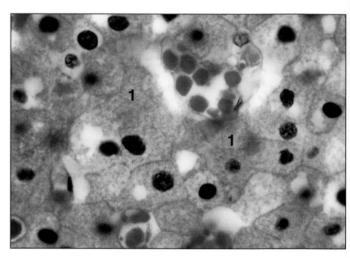

Figure 11.20 Parathyroid gland (cow). Clumps of oxyphil cells (1) are present in the parenchyma associated with small blood vessels. H&E ×500.

thyrotrophin secreted by the pars distalis of the pituitary gland, what is stimulated by the thyrotropin-releasing hormone (TRH) produced in the hypothalamus. Somatostatin, also produced in the hypothalamus, has the opposite effect on the pituitary gland production of TSH.

Parathyroid Gland

The parathyroids are derived from the endoderm of the third and fourth pharyngeal pouches and are either internal (embedded in the capsule of the thyroid) or external (lying a variable distance away). A fine reticular framework supports the cells and the blood vessels. Cords of densely packed basophilic epithelial cells range along the rich capillary bed (**Figures 11.15** and **11.19**). They are of two types: principal (chief) cells and oxyphil cells. Principal cells are the major source of parathyroid

hormone regulating calcium homeostasis. Oxyphil cells are large cells with an acidophilic cytoplasm and a pyknotic nucleus (**Figure 11.20**). They are found in the horse and large ruminants but are rare in other domestic animals. Their function is unknown.

Adrenal Gland

The adrenal glands are paired and lie in the abdominal cavity close to the craniomedial border of the kidneys. The connective tissue capsule extends into the gland as thin trabeculae of loose vascular reticular connective tissue. The adrenal is divided into an outer cortex derived from mesenchyme and an inner medulla derived from neural crest cells (**Figures 11.21** and **11.22**).

Figure 11.21 Adrenal gland (cat). The adrenal gland is divided into an outer cortex (top) and an inner medulla (bottom). H&E ×50.

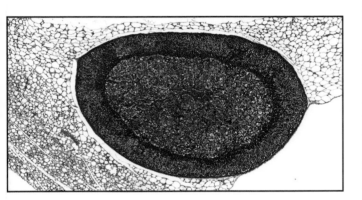

Figure 11.22 Adrenal gland (mouse). The adrenal gland is divided into an outer cortex and an inner medulla. The organ is covered by a connective capsule and perirenal adipose tissue. H&E ×130.

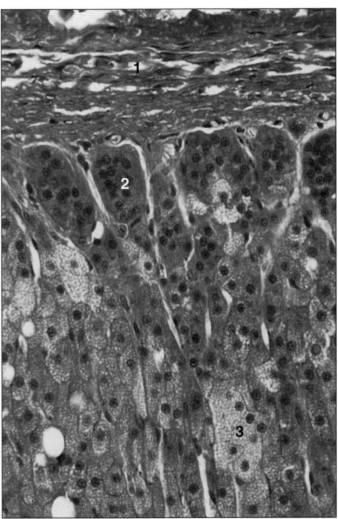

Figure 11.23 Adrenal cortex (cat). (1) Connective tissue capsule. (2) Zona glomerulosa. (3) Zona fasciculata cells are vacuolated spongiocytes. H&E ×400.

The adrenal cortex produces three main groups of hormones: glucocorticoids, which regulate carbohydrate metabolism; mineralocorticoids, which maintain electrolyte concentrations in the extracellular fluid; and androgens, which possess the same masculinising effect as testosterone. It is divided into four zones: the zona glomerulosa, the zona intermedia, the zona fasciculata, and the zona reticularis.

The zona glomerulosa (arcuata, multiformis) is the outer layer immediately beneath the capsule. It consists of curved cords or arcades of columnar cells in horses, carnivores, and pigs (**Figures 11.23** and **11.24**), and as clusters of polyhedral cells in ruminants. The cellular cytoplasm is acidophilic with a small dark nucleus and secretes mineralocorticoids.

The zona intermedia (more common in the horse and carnivores than in other domestic animals) lies between the zona glomerulosa and the zona fasciculata. It is composed of small undifferentiated stem cells that can generate parenchyma for both the inner and the outer cortex.

The zona fasciculata, the most extensive zone, is formed of cuboidal or polyhedral cells arranged in radial cords separated by a sinusoidal network of blood vessels. The cytoplasm of these cells (also called spongiocytes) may appear foamy after routine processing and staining because of the loss of the steroid glucocorticoid hormones (**Figures 11.25** and **11.26**).

The zona reticularis is the innermost zone next to the medulla (**Figures 11.27** and **11.28**). The cells are small, darkly staining anastomosing cords surrounded by sinusoids. They secrete sex hormones.

The adrenal medulla produces adrenalin (epinephrine) and noradrenaline (norepinephrine). Adrenaline is

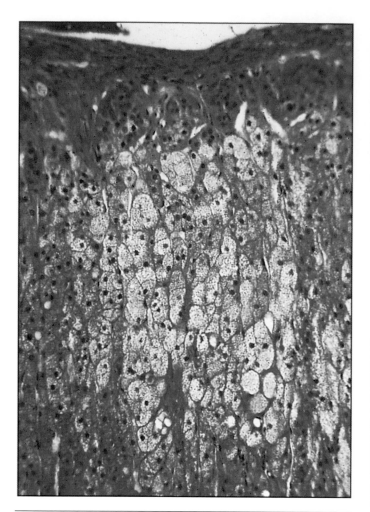

Figure 11.24 Adrenal cortex (pig). The vacuolated appearance of the zona fasciculata cells (spongiocytes) is evident. H&E ×250.

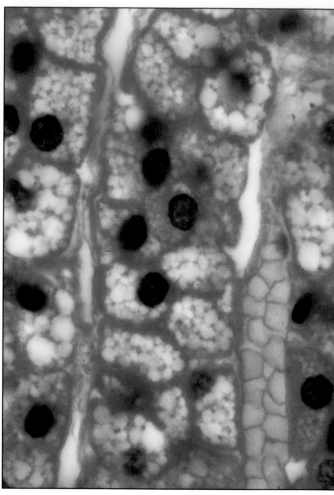

Figure 11.25 Adrenal gland (pig). Spongiocytes line the sinusoids. H&E ×630.

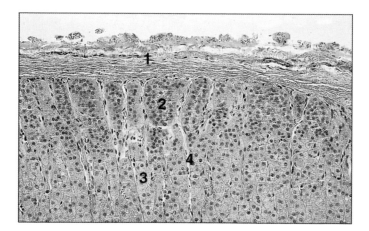

Figure 11.26 Adrenal cortex (cat). (1) Connective tissue capsule. (2) Zona glomerulosa cells are arranged in curved cords. (3) Zona fasciculata cells are polyhedral arranged in long radial cords. (4) Sinusoids. H&E ×100.

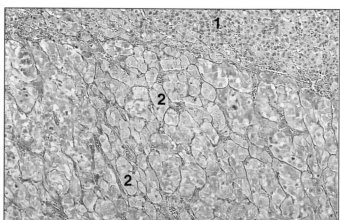

Figure 11.27 Adrenal cortex/medulla (cat). (1) Zona reticularis cells are arranged in anastomosing cords. (2) Adrenal medulla sympathetic neuron cell bodies. The cytoplasm is filled with golden brown granules, the chromaffin reaction. Bouin's fixation and H&E stain ×100.

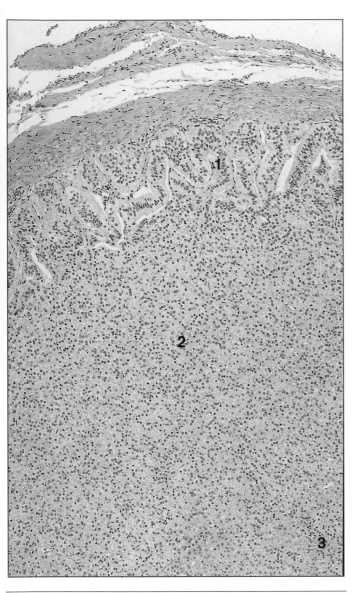

Figure 11.28 Adrenal gland (horse). (1) Zona glomerulosa. (2) Zona fasciculata. (3) Zona reticularis. H&E ×100.

CLINICAL CORRELATES

Clinical manifestations of the adrenal disease depend on which cellular populations within the gland are affected and to what extent. The function of the adrenal cortex can be insufficient, leading to hypoadrenocorticism (Addison's disease). In dogs, the most common lesion of this type is idiopathic bilateral adrenocortical atrophy, in which all layers of the cortex are reduced in thickness, and there is reduced production of all classes of corticosteroids. Inflammation of the adrenal cortex caused by a variety of microbial infections can also lead to insufficiency. Clinically, a patient often exhibits progressive loss of condition, weakness, and gastrointestinal signs, but the presentation in a shock-like state of circulatory collapse is also possible.

Hypersecretion of adrenal corticosteroids produces hyperadrenocorticism (e.g. pituitary-dependent Cushing's syndrome), a clinical syndrome of steroid excess characterised by skin and hair changes, polydipsia, polyphagia, and weight gain with muscle weakness. Causes include a functional adrenal tumour or may be referable to a pituitary tumour that overstimulates the adrenal glands. The typical skin biopsy changes in canine hyperadrenocorticism are illustrated in **Figure 11.29**. A primary adrenal tumour from a Syrian hamster is shown in

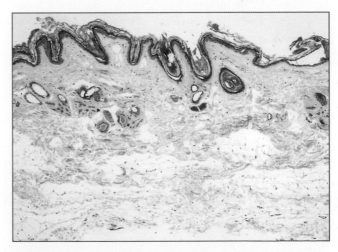

Figure 11.29 Hyperadrenocorticism (dog). This skin section is from an 8-year-old Whippet with hyperadrenocorticism. The epidermis is thin, the hair follicles inactive, and their associated sebaceous glands atrophic. There is dermal atrophy with reduction in the number and thickness of the collagen bundles in the deep dermis. No inflammatory reaction is present. H&E ×50.

a powerful vasopressor, increasing cardiac output when the animal is distressed. It also regulates the sympathetic branch of the autonomic nervous system and stimulates the release of glucose from the liver. The medulla is composed mostly of columnar or polyhedral cells, modified postganglionic sympathetic neurons that take up and stain strongly with chromium salts and have numerous brown granules in the cytoplasm. The chromaffin reaction demonstrates the presence of adrenalin and noradrenalin. In domestic mammals, an outer and inner zone of the medulla can often be distinguished.

Other glands, such as the small carotid and aortic bodies, also demonstrate the chromaffin staining reaction.

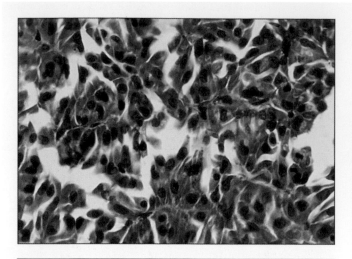

Figure 11.30 Adrenal cortical adenoma in a Syrian hamster (*Mesocricetus auratus*). The corticosteroid-secreting cells are moderately pleomorphic, and a few mitotic figures are seen. H&E ×400.

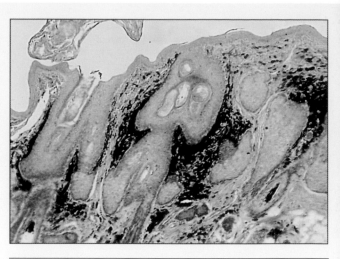

Figure 11.31 A section of skin from the hamster illustrated in **Figure 11.30**. The epidermis is hyperplastic, hyperkeratotic, acanthotic, and heavily pigmented. There is a generalised atrophy of hairs and their precursors. H&E ×50.

Figure 11.30 and is accompanied by a skin section (**Figure 11.31**) from the same animal. In addition to the expected findings of adnexal atrophy, the skin shows marked secondary hyperplastic changes, possibly associated with self-trauma.

Disorders of the adrenal medulla are less commonly described but phaeochromocytomas (**Figure 11.32**), tumours of the chromaffin cells, are the most prevalent. These are usually recognised in dogs or cattle. In dogs, around half of these show malignant behaviour with metastasis to regional lymph nodes and beyond. The majority of phaeochromocytomas are not functionally active, but occasional tumour can secrete catecholamines (adrenalin/noradrenalin).

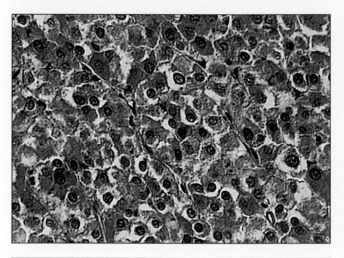

Figure 11.32 Phaeochromocytoma in a dog. The tumour cells contain abundant red–brown granules and their nuclei bear prominent nucleoli. Bouin's fixation; H&E ×400.

Epiphysis Cerebri (Pineal Gland)

The pineal gland (epiphysis cerebri) is a dorsal evagination of the diencephalon, attached by a stalk to the dorsal wall of the third ventricle of the cerebrum. It is covered by a capsule and trabeculae of the pia mater and is divided into lobules by connective tissue. The parenchyma is composed primarily of pinealocytes, small epithelioid cells with round nuclei and acidophilic cytoplasm, supported by astrocyte-like central gliocytes. Corpora arenacea are local calcified deposits present in the gland. These are more numerous

in older animals (**Figures 11.33** and **11.34**). This gland secretes melatonin, which modulates circadian and seasonal cycles.

Avian Endocrine System

The avian pituitary gland is similar to that of other mammals except for the absence of the pars intermedia. In the thyroid gland, follicles are identical to those of mammals, but the parafollicular cells are located in a separate gland,

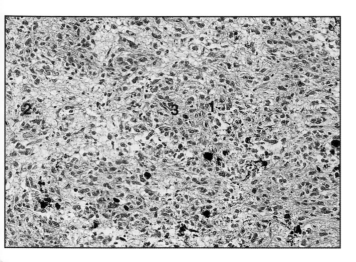

Figure 11.33 Epiphysis cerebri (pineal gland; ox). The pinealocytes are small epithelioid cells arranged in (1) cords, (2) clusters, or (3) follicles. Corpora arenacea (brain sand) is a feature of this gland. H&E ×125.

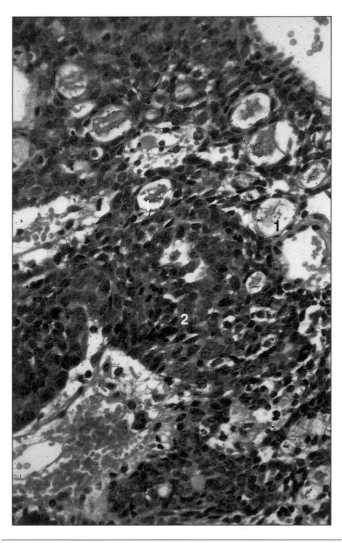

the ultimobranchial body (**Figure 11.35**), which is located close to the origin of the carotid artery. There is no capsule. Vesicles or cords of round basophilic cells lie in the connective tissue surrounding the carotid artery. This is the source of calcitonin in the bird and together with parathyroid hormone; it regulates calcium metabolism.

In the parathyroid gland, the parenchyma is composed of irregular cords of principal/chief cells, separated by connective tissue and numerous sinusoids.

The adrenal gland has no clear division into cortex and medulla. Instead, the parenchyma is composed of intermingled cortical tissue and medullary (chromaffin) tissue. The cortical cells are arranged as anastomosing cords and are steroid-secreting. The chromaffin cells are polygonal (**Figure 11.36**).

Figure 11.35 Ultimobranchial body (bird). The basophilic cells are arranged in vesicles (1) or cords (2) in the connective tissue surrounding the carotid artery. H&E ×250.

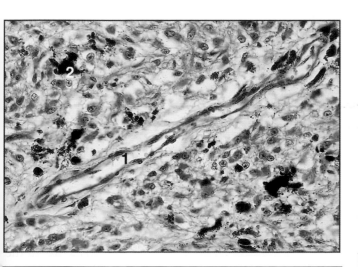

Figure 11.34 Epiphysis cerebri (pineal gland; ox). (1) The pia mater extends into the gland. (2) Corpoa arenacea. H&E ×250.

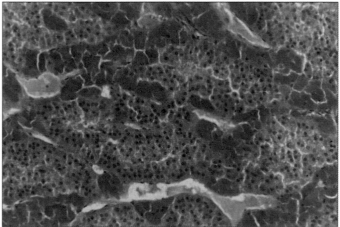

Figure 11.36 Adrenal (inter renal) gland (bird). The chromaffin cells are filled with golden brown granules and lie in islands between the anastomosing cords of steroid-secreting cells. Bouin's fixation. H&E ×125.

Reptilian, Amphibian, and Fish Endocrine System

Pituitary Gland

The pituitary gland of amphibians and reptiles is similar to that of mammals and birds, but there are substantial differences between disparate families of reptiles.

The adenohypophysis is composed of the intermediate lobe (pars intermedia) and distal lobe (pars distalis); the neurohypophysis is composed of the median eminence and neural lobe (pars nervosa). Neurosecretory perikarya of the supraoptic and paraventricular nuclei form the hypothalamoneurohypophyseal tract.

The intermediate lobe is closely juxtaposed to the neural lobe and is joined to the distal lobe by a narrow band of glandular tissue. The distal lobe of most amphibians and reptiles tends to be flattened and lies ventral to the pars nervosa and is caudoventral to the median eminence. It is composed of cellular cords of cuboidal or prismatic cells that are surrounded by a thin fibrovascular connective tissue. This forms fine septa separating adjacent cords of chromophilic cells that are situated peripherally from chromophobic ('principal') cells that are situated more centrally. Capillaries and thin-walled venous sinuses are a variable feature, depending upon the taxon of an animal being studied.

The pars tuberalis (if it is present) consists of small groups of cells forming follicle- or duct-like structures on the ventral surface of the infundibular recess, in the region of the median eminence.

The pars nervosa forms a pit-like depression in the infundibular recess in some reptiles. In others, it lies more lateral or dorsolateral to the distal lobe. It may or may not be divided into lobules by thin connective tissue septa. The cells comprising the pars nervosa tend to be finely granular and stain paler than the pars distalis or pars intermedia (**Figure 11.37**).

The staining characteristics of amphibian and reptilian pituitary glands are generally comparable with those observed in higher vertebrates. Depending upon species and the laboratory staining protocols employed, acidophilic, basophilic, amphophilic (the secretory granules staining both erythrophilic and cyanophilic), chromophilic, or erythrophilic glandular cells can be discerned. The pituitary cells of some reptiles are strongly PAS-positive, whereas they are not in reptiles of different families. In addition, seasonal, age, and sex differences may further complicate the situation.

Thyroid Gland

The thyroid glands of most amphibians and reptiles are located just cranial to the cardiac outflow tract, often lying between or immediately adjacent to the twin aortic arches, cranial vena cava, and pulmonary vasculature (**Figure 11.38**). In amphibians, the thyroid is present as paired lobes. In reptiles, it is usually a single organ, although bilobed glands with a thin isthmus joining the two lobes have been observed. The colloid-filled follicles are highly variable in size and are separated by a fine fibrovascular connective tissue (**Figure 11.39**). The thyroid often displays a marked seasonal change in the shape and height of the follicular epithelial cells and amount of colloid. They are usually flattened and decidedly squamous or may be tall columnar and pallisaded into a parallel 'picket-fence'-like pattern and contain scanty colloid during times of hypoiodine-induced hypothyroidism. Interfollicular 'C' cells are located in the interstitial connective tissue that

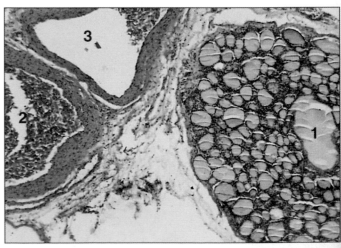

Figure 11.38 Whole mount section of the thyroid (1) of a Central American rattlesnake (*Crotalus durissus bicolor*). The thyroid lies immediately adjacent to the heart base and is readily identified by its characteristic pink, colloid-filled follicles. One aortic arch (2) and a jugular vein (3) are immediately to the left of the thyroid. H&E ×100.

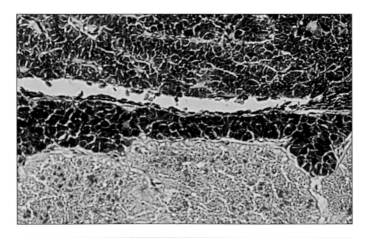

Figure 11.37 Whole mount sagittal section of the pituitary gland of a Children's python (*Liasis childreni*). The pars distalis is at the top; the pars intermedia is represented as a variably narrow band of eosinophilic cells; the pars nervosa is in the lower half. H&E ×200.

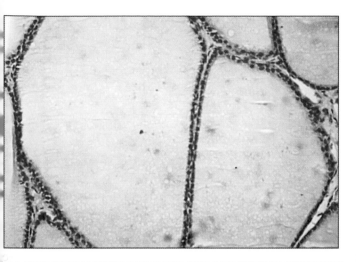

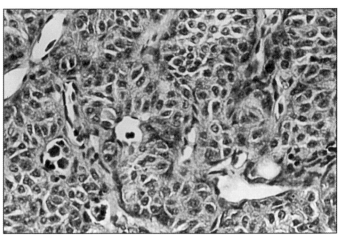

Figure 11.39 Thyroid from a red-eared slider turtle (*Trachemys scripta elegans*). The histological characteristics of the reptilian gland change seasonally and may range from follicles lined by cuboidal to a much flattened squamous epithelium. H&E ×125.

Figure 11.41 Parathyroid gland of a desert tortoise (*Xerobates agassizi*). The secretory cells are formed into small lobules bound by thin fibrovascular connective tissue septa. H&E ×250.

separates adjacent follicles (**Figure 11.40**). Usually, the intrafollicular colloid is amorphous and agranular, but it can also be distinctly granular.

Parathyroid Gland

Fish lack a parathyroid gland. Rather, the ultimobranchial body is highly developed and serves as the major regulator of calcium and phosphorus metabolism. The parathyroid glands of most amphibians and reptiles are present as one or (usually) two pairs of lobes that are located in the caudal cervical region, often adjacent to the thymus or jugular veins. The cells may be cuboidal to prismatic or polyhedral and contain small intracytoplasmic granules

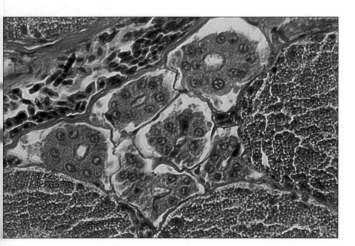

Figure 11.40 Interfollicular ('C') cells in the thyroid of a hog-nosed snake (*Heterodon platyrhinos*). These endocrine cells are arranged into small diameter tubuloacinar follicles with tiny central lumens. The thyroidal follicular colloid in this section is unusually granular. H&E ×400.

that usually stain pale pink with H&E. They are formed into small lobules by thin fibrovascular connective tissue septa that penetrate the gland from the surrounding capsule (**Figure 11.41**). Under some conditions, clear or foamy appearing spaces may be scattered throughout the glandular tissue.

Adrenal Glands

The adrenal tissues of fish, amphibians, and many reptiles are significantly different. The adrenal glands (sometimes referred to as interrenal or intrarenal tissue) of most fish are embedded in the cranial kidneys, where they surround the larger blood vessels and may be intermixed with haematopoietic tissue. There are two kinds of adrenal cells: medullary or chromaffin cells and sympathetic nerve-like paraganglion cells. Amphibians possess discrete paired adrenals composed of three major cell types: chromaffin 'medullary' cells that are of neurectodermal origin; 'cortical' cells; and Stilling cells that are of mesodermal origin. Although it is known that the cortical cells secrete adrenal corticosteroids, the function of Stilling cells is not known. Some investigators consider the Stilling cell to be a distinct glandular organ.

In reptiles, the paired adrenals usually lie immediately adjacent to the kidneys or, in some instances where the kidneys are located within the pelvic canal, they lie medial to the gonads. Two major cell types are found: dark staining chromaffin or medullary cells that secrete catecholamines; and pale staining cortical steroidogenic cells. In mammals, these cells are confined to the cortical and medullary zones. In reptiles, the location of the cells tends to be less circumscribed; the chromaffin cells are in islands or nests scattered throughout the steroidogenic cortical cells. The

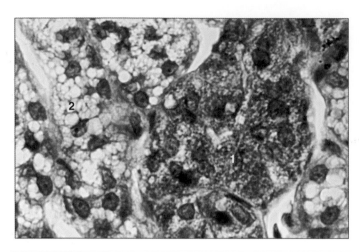

Figure 11.42 Adrenal gland of a desert tortoise (*Xerobates agassizi*). The kidney is in the lower right; the adrenal is on the left. Rather than being confined to a discrete cortex and medulla, the deeply staining chromaffin cells (1) and the pale spongiocytes (2) are grouped into nest-like aggregates that are admixed throughout the gland. H&E ×400.

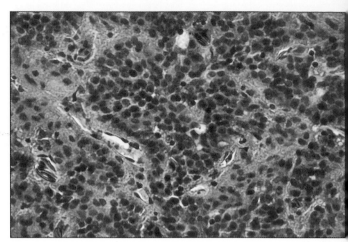

Figure 11.43 Pineal gland of a Burmese python (*Python molurus bivittatus*). The cuboidal epithelial cells have indistinct cell membranes and large round nuclei. They are arranged into lobule-like groups that are separated from each other by delicate fibrovascular septa. H&E ×200.

cytoplasm of cortical cells often contains small clear lipid-like vacuoles (**Figure 11.42**).

Pineal Gland

In teleost fish, the pineal gland is hollow and consists of columnar epithelial cells of three types: sensory, sustentacular, and ganglion-like. It resembles a sensory organ; some primitive cyclostomes actually possess a pineal covered by a relatively clear lens-like structure that admits light.

In amphibians, the pineal gland is more solid than in fish and arises as a median outgrowth on the dorsal surface of the brain. Like the gland in fish, it is photosensitive. In response to darkness, it secretes melatonin, which causes peripheral melanophages to aggregate, thus lightening the integument's colour.

In reptiles, the pineal gland is solid and composed of glandular secretory cuboidal to low columnar epithelial cells that are arranged into lobules separated by delicate fibrovascular septa that are extensions from the surrounding capsule (**Figure 11.43**). Preliminary investigations have revealed at least some inter-relationships between the pineal and other endocrine glands, especially the thyroid. The pineal gland of lizards lies slightly rostral to the parietal foramen. Whether sufficient light can enter the skull through the parietal eye and the parietal foramen (in those species that possess them) is conjectural but appears to be possible.

Ultimobranchial Body

In the lower vertebrates, the ultimobranchial bodies play an important role that is shared in the higher vertebrates by the parathyroid and thyroid. In teleost fish, the ultimobranchial body is well developed and lies in the transverse

septum between the ventral surface of the oesophagus and liver, immediately caudal to the heart. The glandular tissue consists of tall columnar or pseudostratified columnar epithelium formed into spherical vesicle-like structures. They are paired in some amphibians, single in some salamanders and absent in some frogs and in caecilians.

In reptiles, the ultimobranchial bodies are paired and function in calcium and phosphorus regulation. They also have a relationship to thyroid function in some reptilian species. They are believed to secrete calcitonin in response to blood calcium concentrations. Reptilian ultimobranchial bodies are composed of cuboidal to low columnar glandular epithelial cells with finely granular pink staining cytoplasm often containing clear lipid-like vacuoles (**Figure 11.44**).

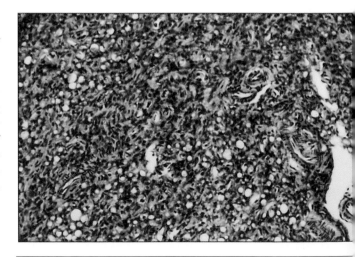

Figure 11.44 Ultimobranchial body of a boa constrictor (*Boa c. constrictor*). Many of the secretory epithelial cells that comprise this glandular organ contain large, clear lipid vacuoles. Typically, these glands are formed from solid sheets of cells with little separation by connective tissue septa. Small blood vessels penetrate the tissue. H&E ×50.

CLINICAL CORRELATES

THYROID GLAND

As is the case with the higher vertebrates, hypothyroidism, goitre, thyroid inflammation, and neoplasia occur occasionally in the lower vertebrates. Dietary hypothyroidism is relatively common in semiaquatic turtles, especially those kept as pets by persons living in so-called 'goitre belts', where humans suffer from chronic hypothyroidism unless they receive supplemental iodine. Herbivorous chelonians, especially those originating from oceanic volcanic islands, appear to be particularly prone to dietary hypothyroidism.

PARATHYROID GLAND

Captive reptiles often display the clinical signs of secondary hypoparathyroidism that are induced by being fed either food containing insufficient available calcium or, more commonly, food containing improperly low calcium: high phosphorus ratio. The parathyroid glands from these animals become hyperplastic and hypertrophied. With chronicity, changes consistent with adenomatous hyperplasia are seen: the glandular cells become pleomorphic, distorted, and may lose their normal granularity. Sometimes, actual primary hyperparathyroidism occurs. The cells then assume frankly pleomorphic shapes, their nuclear chromatin becomes more dense, occasional bipolar cells and mitotic figures may be observed, and the lobules become distorted or break through their septal confines and coalesce (**Figure 11.45**).

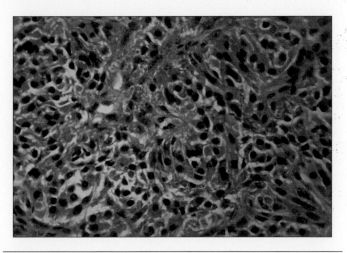

Figure 11.45 Parathyroid adenoma from a desert tortoise that was displaying clinical manifestations of primary hyperparathyroidism, including severe osteopenia. The tumour cells are pleomorphic, hyperchromatic, and arranged irregularly. H&E ×250.

PINEAL GLAND

One benign tumour (pinealoma) has been reported in a green iguana (*Iguana iguana*). It was found at necropsy and was grossly and microscopically pigmented. It was composed of branching papillary fronds of streamer-like, very elongated columnar epithelial cells borne on thin stalks of fibrovascular connective tissue that projected into the lumen of cystic spaces within the mass.

Destruction of the pineal gland might be expected to disrupt the normal diurnal–nocturnal behaviour pattern, and perhaps the colour, of an animal subjected to this experimental protocol.

ULTIMOBRANCHIAL BODY

The paired ultimobranchial bodies of reptiles undergo hypertrophy as a consequence of chronic hypocalcaemia. They have been reported to involute spontaneously with ageing in some normal lizards. Functional hyperplasia (or adenoma) of the ultimobranchial bodies can cause excessive mobilisation of calcium from skeletal stores. Hypersecretion of calcitonin, with subsequent hypercalcaemia, may result in osteopenia and, occasionally, renal tubular calcium urolithiasis.

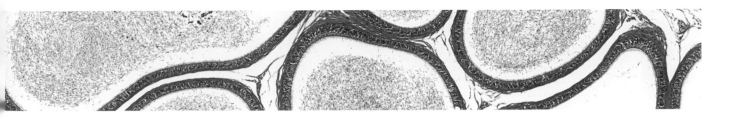

12
MALE REPRODUCTIVE SYSTEM

The male reproductive system consists of the paired testes suspended in the scrotal sacs, an intratesticular and extratesticular continuous duct system for the storage and transport of the male gamete, the spermatozoon, and an ejaculatory organ, the penis. The accessory sex glands secrete into the urethra and provide a suitable fluid medium to transport the spermatozoa during ejaculation in the female reproductive tract.

Testis

The testis is a compound tubular gland with an exocrine and an endocrine function. The exocrine function is the production of spermatozoa by the lining epithelium of the seminiferous tubules. The endocrine function is the production of the male sex hormone, testosterone, by the specific interstitial (Leydig) cells in the intertubular connective tissue. The testis is covered by mesothelium continuous with the visceral layer of the tunica vaginalis. A thick, irregular dense connective tissue capsule, the tunica albuginea, encloses the testis. A variable amount of smooth muscle may be present. The tunica vasculosa is the inner layer of the tunica albuginea and is composed of highly vascularised loose connective tissue.

The capsule is reflected into the median plane of the testis to form a partition, the mediastinum, and gives off loose vascular connective tissue, the septula testis, to divide the testis into lobules to support the seminiferous tubules (**Figures 12.1** and **12.2**). The coiled seminiferous tubules are lined with a multilayered seminiferous epithelium of spermatogenic cells and sustentacular (Sertoli) cells. They rest on a basement membrane and are surrounded by a lamellated connective tissue with myoid elements. The specific interstitial (Leydig) cells are found in the loose vascular connective tissue separating the tubule and arranged in cords or clusters.

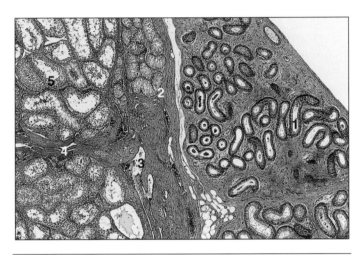

Figure 12.1 Testis (cat). (1) Efferent ducts invested by connective tissue. (2) Tunica albuginea. (3) Rete testis. (4) Mediastinum testis. (5) Seminiferous tubules. H&E ×50.

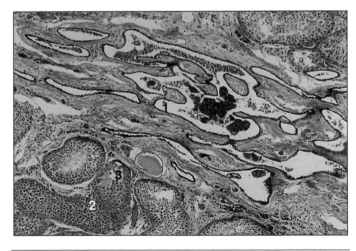

Figure 12.2 Testis (cat). (1) Rete testis lying in the mediastinum. (2) Seminiferous tubules leading into (3) tubuli recti lined by sustentacular cells. H&E ×100.

DOI: 10.1201/9781003333807-12

Production of Spermatozoa

In the prepubertal male, there are two cell types: the sustentacular (Sertoli) cell and the spermatogonium, the immature male germ cell. The sustentacular cells are tall columnar, extending from the basement membrane to the lumen of the tubule, with a pale vesicular basal oval nucleus and a prominent nucleolus. As the name suggests, the sustentacular cells support the later stages in the development of spermatozoa. Spermatogonia lie next to the basement membrane and are small round cells with a dark staining nucleus. At puberty, they move away from the membrane and undergo mitotic divisions. When these cease, the cells go through a period of growth, the deoxyribonucleic acid is replicated, and the cells become tetraploid primary spermatocytes. The transformation process of spermatogonia into primary spermatocytes is known as spermatocytogenesis. The nucleus of primary spermatocytes is granular, and the coiled chromosomes give a dense staining reaction. The primary spermatocyte divides meiotically to form two secondary spermatocytes, which each divide immediately to form two haploid spermatids. These are small cells with a spherical nucleus and lie close to the lumen of the tubule. The spermatids move into recesses in the sustentacular cells and metamorphose into spermatozoa, shedding the excess cytoplasm into the lumen of the tubule. The process of differentiation of spermatids into spermatozoa is known as spermiogenesis (**Figures 12.3–12.5**).

Structure of Spermatozoa

The mature spermatozoon consists of a head and a tail (**Figure 12.6**). The head contains the condensed haploid nucleus. The anterior two-thirds are covered by the acrosomal cap, a derivative of the Golgi apparatus containing hydrolytic enzymes and proteases (acrosin), required for

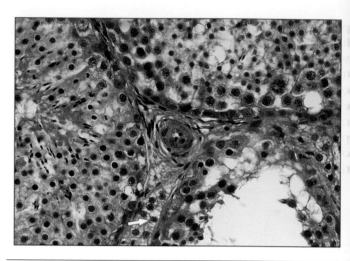

Figure 12.4 Testis (bull). Vascular interstitial tissue lies between the seminiferous tubules; specific interstitial cells are vacuolated (arrowed). H&E ×200.

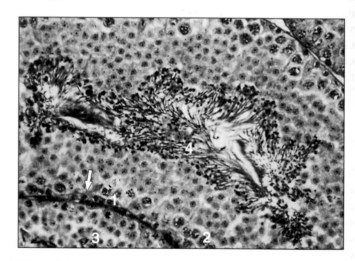

Figure 12.5 Testis (dog). All the stages of spermatogenesis are seen in the lining epithelium. (1) Spermatogonia. (2) Primary spermatocytes. (3) Spermatids. (4) Developing spermatozoa. The sustentacular cells are arrowed. H&E ×200.

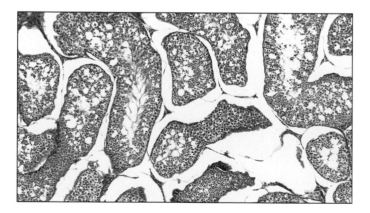

Figure 12.3 Testis (pig). The seminiferous tubules are lined by a multilayered epithelium with spermatogonia, primary spermatocytes, spermatids, and developing spermatozoa. Interstitial tissue separates the tubules. H&E ×100.

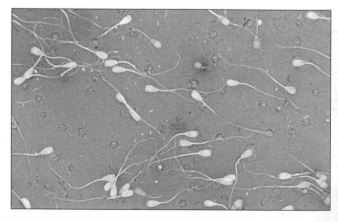

Figure 12.6 Spermatozoa (bull). The spermatozoa consist of a head and a tail (flagellum). Unstained spermatozoa were live, and stained spermatozoa were dead at the time of staining. Nigrosin–eosin ×625.

penetration of the ovum at fertilisation. The tail (a flagellum) has the characteristic structure of flagellae and cilia, with two central microtubules and nine peripheral doublets (the axial filament complex), surrounded by nine thick outer fibres. The proximal part is surrounded by an end-to-end helix of mitochondria to provide energy during movement. Sperm are nonmotile and immature at this stage. The peritubular myoid elements push the mixture of sperm and debris down the seminiferous tubule towards the mediastinum. Spermatozoon size and head shape vary by species.

Tubuli Recti and Rete Testis

The seminiferous tubules, at their terminal segments, are joined by a transitional zone, lined by sustentacular cells, to straight tubules (tubuli recti), lined by simple squamous to columnar epithelium containing numerous macrophages and lymphocytes. Tubuli recti are continuous with a network of anastomosing channels that form the rete testis (**Figure 12.7**). The rete testis has a simple squamous or columnar epithelium. It is surrounded by the loose connective tissue of the mediastinum and is drained by between 7 and 20 ductuli efferentes (*see* **Figure 12.1**).

Epididymis and Ductus Deferens

The epididymis is formed by the ductuli efferentes and the ductus epididymis and is the place where sperm mature and become motile (**Figures 12.8–12.10**). Ductuli efferentes are lined with simple columnar or pseudostratified epithelium, and some of the cells are ciliated. The lumen of each duct is wide, and the lamina propria is sparse connective tissue (with some smooth muscle in the stallion).

The ductus epididymis, divided into head, body, and tail, is a single, long coiled duct lined with pseudostratified

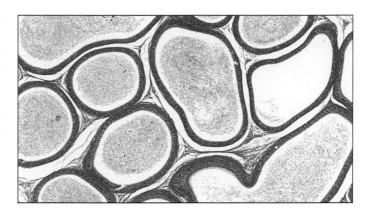

Figure 12.8   Epididymis (pig). The single coiled ductus epididymis is lined by a pseudostratified columnar epithelium resting on a vascular lamina propria continuous with the supporting connective tissue of the head of the epididymis. Spermatozoa are present in the lumen of the tubule. H&E ×15.

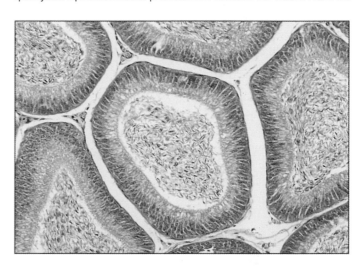

Figure 12.9   Epidiymis (bull). Pseudostratified columnar epithelium lines the tubule; the lumen is filled with spermatozoa. H&E ×100.

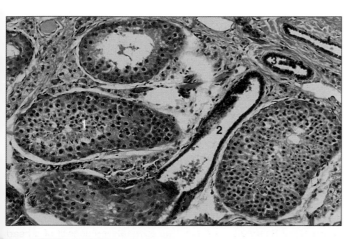

Figure 12.7   Testis (cat). (1) Seminiferous tubules. (2) Straight tubules. (3) Rete testis. H&E ×100.

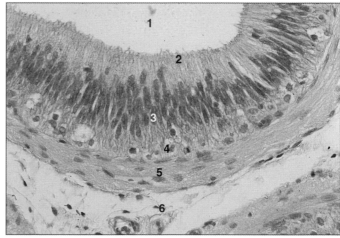

Figure 12.10   Epidiymis (bull). (1) Lumen of the tubule. (2) Luminal cytoplasm extends into the lumen as stereocilia. (3) Nuclear layer of the epithelial cells. (4) Vascular lamina propria. (5) Muscularis. (6) Vascular connective tissue. H&E ×250.

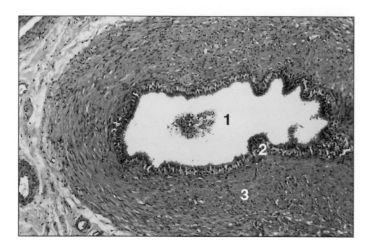

Figure 12.11 Ductus deferens (dog). (1) Lumen of the ductus deferens. (2) Mucosal folds. (3) Thick layer of the smooth muscle of the muscularis externa. H&E ×100.

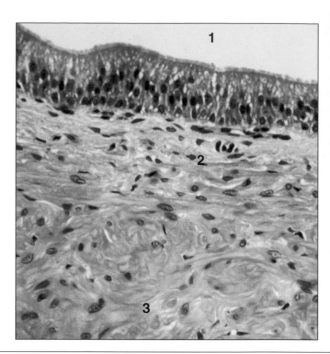

Figure 12.12 Ductus deferens (dog). (1) The lumen is lined by a pseudostratified ciliated epithelium. (2) Mucosal connective tissue. (3) Smooth muscle of the muscularis externa. H&E ×400.

epithelium. Long cytoplasmic processes, the stereocilia, project into the lumen. The epithelium rests on a loose connective tissue lamina propria, and there is a variable amount of circular smooth muscle. The latter increases in volume as the tail of the epididymis approaches the ductus deferens.

The ductus deferens links the ductus epididymis with the urethra. It has a very thick muscular wall and a small lumen, and it is lined with pseudostratified columnar epithelium resting on a delicate lamina propria containing collagen and elastic fibres. The mucosa is arranged in longitudinal folds, and the muscularis is composed of smooth muscle fibres, presenting a variety of arrangements. In domestic animals, it is often difficult to determine specific layers of muscle. Circularly arranged fibres predominate, and a fibrous adventitial coat is present (**Figures 12.11** and **12.12**). The ductus deferens terminates at the colliculus seminalis of the proximal urethra. Near to this junction it dilates to form an ampulla in the stallion, ruminants, and dog, but not in the boar and tomcat. Simple branched tubuloalveolar glands are present in the lamina propria-submucosa of the terminal portion in all domestic mammals. Spermatozoa can be present in the lumen and the glandular openings. The mucosa is arranged in complex folds and shows a glandular appearance to become one of the accessory sex glands (**Figure 12.13**).

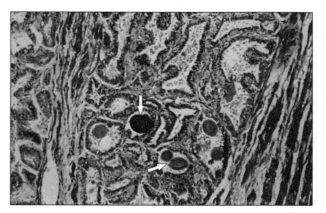

Figure 12.13 Ampulla (bull). The lining epithelium is a simple columnar secretory; the mucosal folds create a glandular appearance. In the lumen of the glands, calcified structures called *corpora arenacea* can be found (arrows). H&E ×100.

Urethra

The urethra is the terminal part of the male duct system and has both a urinary and a reproductive function (**Figure 12.14**). It is a long mucous tube extending from the bladder to the glans penis; it is divided into a prostatic (from the urinary bladder to the prostate), membranous

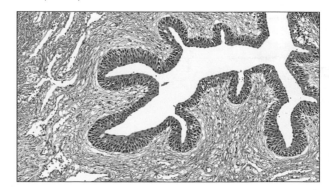

Figure 12.14 Urethra (pig). The lumen of the urethra is lined by urothelium with the underlying vascular connective tissue of the lamina propria. H&E ×100.

(from the prostate to the bulb of the penis), and spongiose (the cranial continuation until the external opening) parts. The epithelial lining varies from urothelium proximal to the bladder, to stratified columnar or cuboidal along the greater part of the urethra, and to stratified squamous at the tip of the penis. Simple tubular mucosal (urethral) glands are present in stallions and tomcats. The tunica muscularis is composed of smooth muscle in the vicinity of the urinary bladder and skeletal muscle in the remainder of the urethra. The adventitia is loose connective tissue carrying blood vessels and nerves; nerve endings are a constant feature. Cavernous blood spaces are present in the lamina propria of the prostatic and membranous urethra and increase markedly in the penile (spongy) region to become the corpus spongiosum urethrae (**Figure 12.15**). This is part of the erectile tissue of the penis, with little or no muscle. In the stallion and ruminants, the urethra extends beyond the tip of the penis as the processus urethrae and is covered by a cutaneous membrane lined with transitional or stratified squamous epithelium.

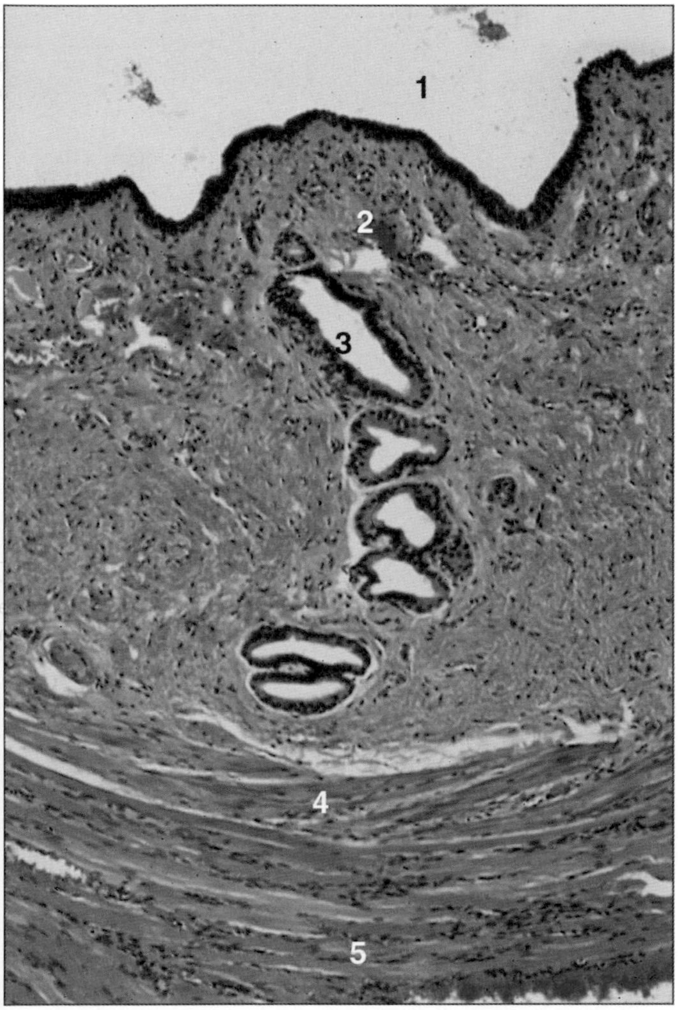

Penis

The male copulatory organ consists of an outer fibroelastic connective tissue capsule, the tunica albuginea, extending into the body of the penis to form trabeculae supporting a network of endothelial lined spaces: the corpus cavernosum (**Figures 12.16** and **12.17**). With the corpus spongiosum urethrae, these spaces fill with blood from the coiled helicine arteries during erection. Intimal cushions prevent the back-flow of blood in these arteries and help to maintain the erection. In the stallion, smooth muscle predominates (vascular-type penis) (**Figure 12.18**); in the boar and ruminants, fibroelastic connective tissue forms the bulk of the organ (fibrous-type penis); carnivores have an intermediate-type penis. In carnivores, the terminal part of the corpus cavernosum ossifies to become the os penis

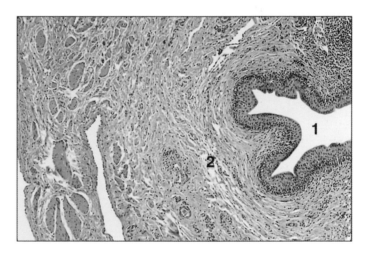

Figure 12.16 Penile urethra (dog). (1) Lumen of the urethra lined by urothelium. (2) Vascular lamina propria: the corpus spongiosum. H&E ×62.5.

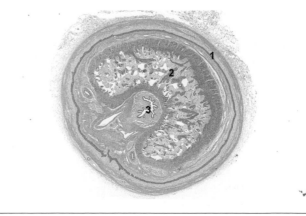

Figure 12.17 Penis (pig). (1) Superficial fascia. (2) Thick connective tissue layer with semierectile tissue. (3) Lumen of the urethra. H&E ×15.

Figure 12.15 Cranial urethra (cat). (1) Lumen of the urethra lined by urothelium. (2) Vascular lamina propria. (3) Simple tubular gland of the disseminate prostate. (4) Inner layer of smooth muscle. (5) Outer layer of skeletal muscle. H&E ×200.

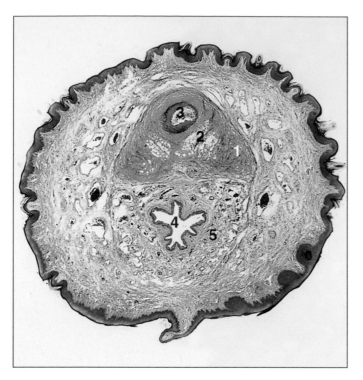

Figure 12.18 Penis (cat). (1) Tunica albuginea. (2) Corpus cavernosum penis. (3) Os penis. (4) Lumen of the urethra. (5) Corpus spongiosum urethrae. Masson's trichrome ×25.

(baculum) and contributes to the rigidity of the penis. Small keratinised epidermal spines are present on the glans penis of the tomcat (**Figure 12.19**) and are sometimes seen in the stallion and billy goat.

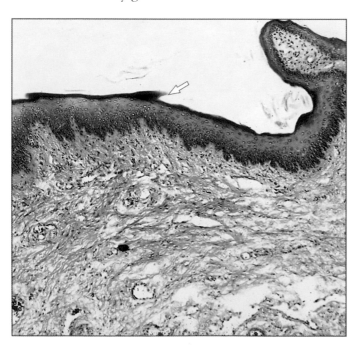

Figure 12.19 Penis (cat). A penile spine is arrowed. Masson's trichrome ×125.

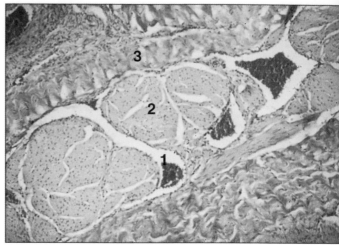

Figure 12.20 Penis (horse). (1) Vascular space. (2) Smooth muscle. (3) Connective tissue. H&E ×100.

The prepuce is a double reflection of skin that covers the distal, free portion of the penis. It is composed of an external and an internal layer. The external layer is typical skin, and the internal layer is the continuation of the external layer at the preputial opening. Hair, sebaceous, and sweat glands occur over a variable distance from the external layer to the preputial opening. This area is extremely sensitive, with many nerve endings (**Figures 12.20** and **12.21**).

Male Accessory Genital Glands

The male accessory genital glands include the prostate gland, the glands of the ampulla (*see* 'Epididymis and Ductus Deferens', above), the vesicular glands, and the

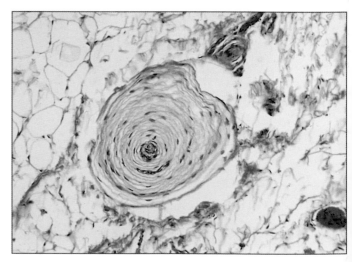

Figure 12.21 Penis (bull). A Pacinian corpuscle lies in the lamina propria of the penile urethra. H&E ×200.

Table 12.1 Male Accessory Genital Glands, Species Variation

Species	Prostate Compact	Prostate Disseminate	Ampulla	Bulbourethral	Seminal vesicles
Stallion	+	−	+	+	+
Bull	+	+	+	+	+
Ram	−	+	+	+	+
Billy goat	−	+	+	+	+
Boar	+	+	−	+	+
Dog	+	+	+	−	−
Tomcat	+	+	−	+	−

+ Gland is present; − gland is absent.

bulbourethral glands (**Table 12.1**). These glands are androgen dependent and exhibit pronounced regressive change after castration, generally resulting finally in loss of function.

Prostate Gland

The prostate gland is androgen-dependent and exhibits profound regressive changes after castration, with a reduction in the secretory epithelium, deposition of fat, and an increase in the connective tissue stroma. The prostate is a compound tubuloalveolar gland and consists in a disseminate internal portion and a compact external portion. In the boar and ruminants, the prostate gland consists mostly of a disseminate portion in the form of a glandular layer in the submucosa of the proximal (prostatic) urethra. In carnivores, the disseminate portion is represented only by scattered glands. The disseminate portion is not present in the stallion. The compact prostate (well developed in the stallion and carnivore and absent in the ram and billy goat) presents a thick fibromuscular capsule extending into the gland to form the supporting framework. The epithelium is pseudostratified with tall columnar secretory cells and small basal reserve cells. In the dog, the secretion is serous (**Figures 12.22** and **12.23**); in other animals, it is seromucous.

Glands of the Ampulla

Ampulla is described in 'Epididymis and ductus deferens', above.

Seminal Vesicles

The seminal vesicles are compound tubular or tubuloalveolar glands with a fibromuscular capsule and elastic fibres in the connective tissue stroma. They are absent in carnivores and are true vesicles in the stallion (**Figure 12.24**). In the boar and ruminants, they are compact with a lobulated surface. The secretory epithelium is tall columnar or

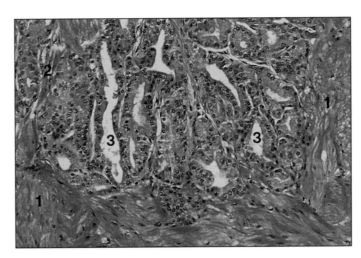

Figure 12.22 Prostate (dog). (1) Fibromuscular capsule. (2) Fibromuscular trabeculae. (3) Lumen of the secretory acinus lined by simple columnar epithelium. H&E ×100.

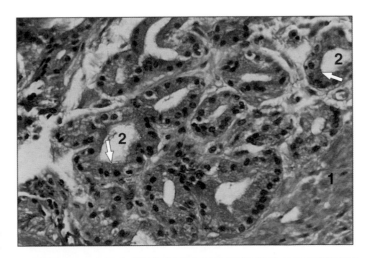

Figure 12.23 Prostate (dog). (1) Fibromuscular trabecula. (2) Lumen of the secretory acinus lined by tall columnar secretory epithelium (arrowed). H&E ×250.

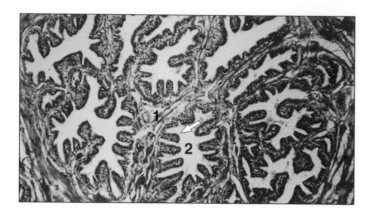

Figure 12.24 Seminal vesicle (stallion). (1) Connective tissue. (2) Lumen is lined by a tall columnar secretory epithelium (arrowed). H&E ×100.

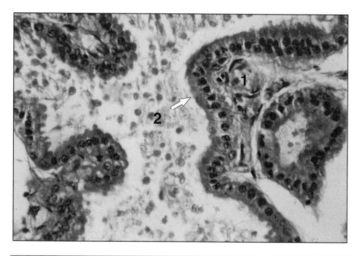

Figure 12.25 Seminal vesicle (stallion). (1) Connective tissue of the capsule. (2) Lumen lined by tall columnar secretory epithelium (arrowed). H&E ×160.

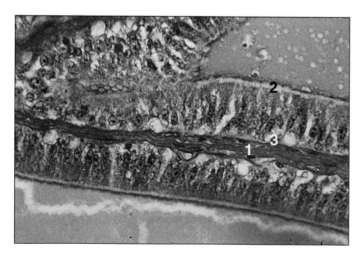

Figure 12.26 Seminal vesicle (bull). (1) Trabecular connective tissue. (2) Pseudostratified columnar secretory epithelium with apical blebs. (3) Basal replacement cells. H&E ×250.

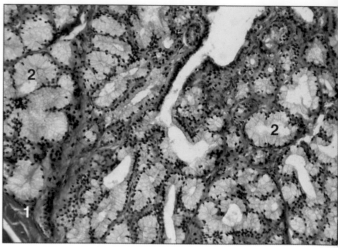

Figure 12.27 Bulbourethral gland (boar). (1) Striated muscle capsule (2) Mucus-secreting secretory epithelium lines the acini. Light green, haematoxylin ×100.

pseudostratified with surface blebs (large flaccid vesicles) giving a ragged appearance to the cells (**Figures 12.25** and **12.26**). The secretion is rich in fructose and is a source of energy for spermatozoa.

Bulbourethral Glands

Bulbourethral glands are compound tubular or tubuloalveolar and mucus secreting. They are present in all domestic mammals except dog. A connective tissue capsule containing a variable amount of striated muscle fibres sends fine trabeculae into the gland (**Figures 12.27** and **12.28**). The glandular epithelium is pseudostratified columnar. The secretion neutralises the vaginal environment and provides lubrication and in pigs contributes to the occlusion of the cervix to prevent sperm loss.

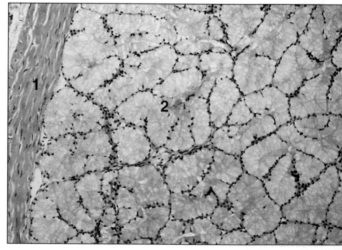

Figure 12.28 Bulbourethral gland (goat). (1) Striated muscle capsule. (2) Mucus-secreting secretory epithelium lines the acini. H&E ×100.

CLINICAL CORRELATES

Developmental problems are important in the pathology of the male reproductive tract (not surprisingly given the complex embryology of this system), but inflammatory, degenerative, and neoplastic conditions are also recognised. These problems can be interrelated (e.g. retained testicles, which can be located at any level along the path of descent, are more susceptible to the development of neoplasia than are normally descended scrotal testicles). These internally retained testicles also undergo degenerative changes.

Bilateral cryptorchidism leads to infertility since normal spermatogenesis requires the testis to be several degrees cooler than core body temperature. Cryptorchidism is important in the horse (**Figure 12.29**) because hormone production by the internally retained testicle maintains stallion-like behaviours. Thus, an animal with a retained testicle may appear to be gelded, but it is more dangerous to handle. Blood testing can identify such animals.

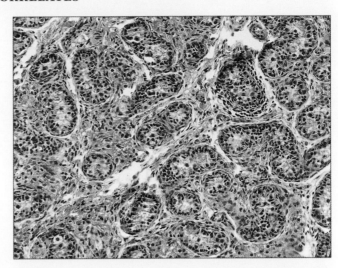

Figure 12.29 Cryptorchid (retained) testicle from a horse. The tubules are atrophic with the absence of spermatogenesis. Fibrous tissue and interstitial cells lie between the tubules. H&E ×125.

Three major types of primary testicular neoplasia are identified. These are the sustentacular cell tumour (**Figure 12.30**), common in the dog, which can be associated with a feminisation syndrome that results from oestrogen production by the tumour cells, the interstitial cell tumour, and the seminoma (**Figure 12.31**). A minority of sustentacular cell tumours and seminomas metastasize, but spread is very rare with interstitial cell tumours.

The pattern of accessory sex organs varies between different species. In the dog and cat, the prostate gland is of primary importance. This gland is very sensitive to hormonal influence, and prostatic hyperplasia is a common finding in older entire male dogs. Enlargement of the prostate, which may result from benign hyperplasia, prostatitis, or neoplasia, can present with difficulty in defecation, as this gland lies beneath the rectum.

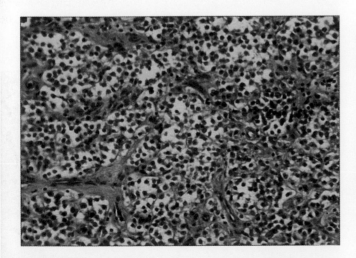

Figure 12.30 Sustentacular (Sertoli) cell tumour in a dog. In this intratubular example, the seminiferous tubules are packed with and distorted by pale-staining neoplastic sustentacular cells with faintly vacuolated cytoplasm instead of normal spermatogenetic cells with an orderly pattern of maturation. H&E ×200.

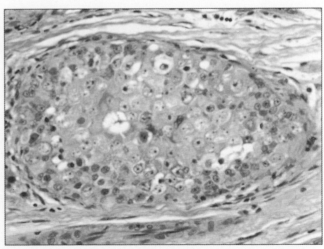

Figure 12.31 Seminoma in a 7-year-old dog. This intratubular example of a seminoma is composed of large polyhedral tumour cells with discrete cell borders and large, round, often vesicular nuclei. Seminomas do not produce hormones. H&E ×400.

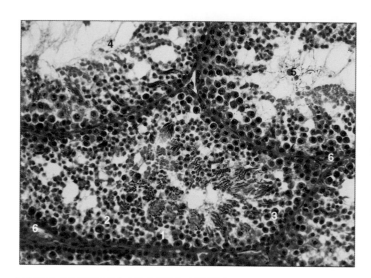

Figure 12.32 Testis (bird). (1) Spermatogonia. (2 and 3) Primary spermatocyte. (4) Spermatozoa. (5) Lumen of the seminiferous tubule. (6) Interstitial tissue with specific interstitial cells. H&E ×200.

Avian Male Reproductive System

The avian testis has a fibrous tunica albuginea but no mediastinum or supporting connective tissue trabeculae. The seminiferous tubules form an anastomosing network, the interstices of which are occupied by blood vessels, fibroblasts, and specific interstitial cells. The seminiferous epithelium is multilayered in the mature adult bird as in the mammal (**Figure 12.32**). The head of the spermatozoon is long, narrow, and gently curved.

The ductus deferens traverses the wall of the urodeum (part of the cloaca; **Figure 12.33**) and is continued as the

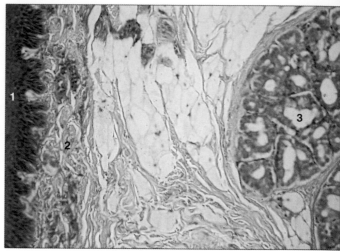

Figure 12.34 Ejaculatory duct (cock). (1) Stratified squamous epithelium. (2) Vascular lamina propria. (3) Simple tubular glands. H&E ×160.

ejaculatory duct (**Figure 12.34**). Deep crypts are present with short tubular glands lying in the lamina propria of the urodeum (**Figure 12.35**). Lymphocytes are present in the subepithelial connective tissue.

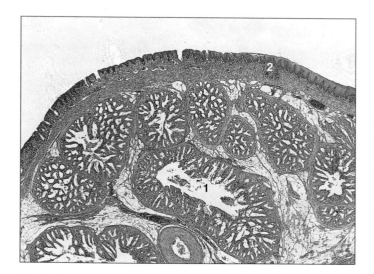

Figure 12.33 Urodeum (domestic fowl cockerel). (1) Simple tubular glands in the wall of the urodeum. (2) Lining epithelium of the urodeum. H&E ×62.5.

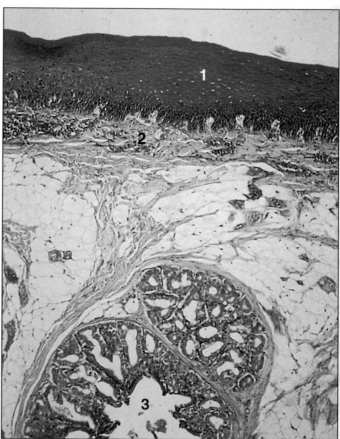

Figure 12.35 Ejaculatory duct (cock). (1) Stratified squamous epithelium. (2) Vascular erectile tissue. (3) Simple tubular gland. H&E ×150.

Reptilian, Amphibian, and Fish Male Reproductive System

The histology of the testis and associated tubular post-testicular reproductive system of fish, amphibians, and reptiles is sufficiently similar to that of mammals and birds (**Figures 12.36–12.38**) that only the differences need to be discussed here. Parallel to the situation that is observed in some mammals and birds that exhibit seasonal waxing and waning of spermatogenesis, the testes of many fish, most amphibians, and many reptiles undergo seasonally induced testicular atrophy and recrudescence. The spermatozoa of these lower vertebrates are morphologically similar to those of higher vertebrates. During the

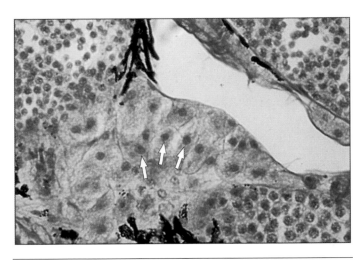

Figure 12.38 A section of the testis of a panther chameleon (*Chamaeleo pardalis*). A nest of large interstitial cells containing finely granular cytoplasm can be seen in the lower third of this image (arrows). H&E ×200.

quiescent portions of this cycle, the seminiferous tubules are devoid of active spermatogenesis. They are lined only by sustentacular cells and a few primary spermatogonia (**Figure 12.39**). The interstitial cells may or may not follow a parallel increase and decrease in activity, size or number. The epididymis and ductus deferents of reptiles are histologically similar to those of mammals.

Internal fertilisation of ova is a rare exception in amphibian reproduction. Spermatozoa are usually discharged over the eggs as they are released by the female during amplexus, the term given to the 'nuptial clasp'. In many caudate amphibians, the female is positioned during courtship over a packet consisting of sperm and accessory gelatinous secretions, called a spermatophore, that is released by a male. This structure is then taken up into the female's cloaca, where the sperm are released and

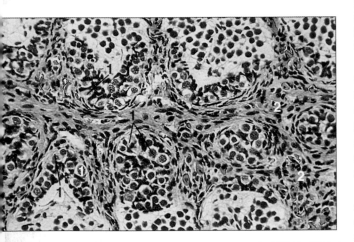

Figure 12.36 Section of the testis of an aquatic salamander (*Amphiuma tridactylum*). The paler staining sustentacular cells (1) are readily distinguished from the darker spermatogonia and spermatocytes. The interstitial cells (2) are contained within small lobules bound within concentric fibrovascular connective tissue stroma. H&E ×125.

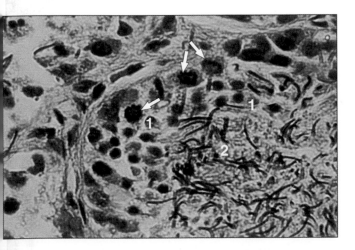

Figure 12.37 Section of a seminiferous tubule of a mature green iguana (*Iguana iguana*). Three spermatogonia are shown in mitosis (arrows). Primary spermatocytes (1) and many tailed spermatozoa (2) lie within the centre of the tubule. H&E ×400.

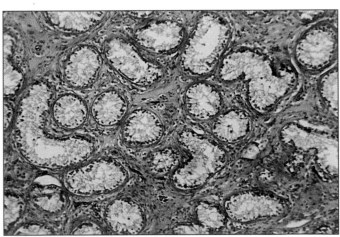

Figure 12.39 Inactive testis of a green sea turtle (*Chelonia mydas*). The seminiferous tubules are lined only by sustentacular cells and a few spermatogonia. H&E ×100.

fertilise ripe eggs. In a few species, females deposit their eggs directly upon a spermatophore. In caecilians, which are among the few amphibians in which internal fertilisation occurs, the sperm are directed into the female's caudal genital tract with an eversible portion of the male's cloacal wall, called a phallodeum, which serves as an intromittant organ. One anuran (*Ascaphus truei*) possesses an intromittant organ that is actually a tail-like extension of the cloaca in which vascular erectile elements are located. The epithelium covering the cranial portion of this structure is mucous; distally, it is keratinised and bears horny spines near its orifice. In the few other species of amphibians in which internal fertilisation occurs, the spermatozoa are introduced into the female by direct cloacal apposition without the intervention of an intromittant organ, or by the uptake of spermatophores that are deposited onto a substrate. Once deposited into a female's body, spermatozoa are stored for a variably prolonged period in branched pouch-like spermathecae that are located in the wall of the roof of the cloaca. These crypt-like structures are lined by cuboidal to low columnar epithelial cells with eosinophilic granular cytoplasm.

Reptiles utilise internal fertilisation. Male chelonians possess a single erectile penis that is often heavily pigmented and is covered by lightly keratinised and mucous epithelium. Male snakes and lizards possess paired erectile intromittant organs, called hemipenes, that contain fibrovascular tissue that, when involuted, invaginates. When erect, the outer surface is covered with keratinised epithelium. Additional flamboyant spines, flounces, and other sexual adornments are species specific. The hemipenes receive lubrication from mucous glands that are located within the lumen of the hemipenial sheath. In some reptiles, modified sebaceous glands are associated with the hemipenes and their sheaths (**Figure 12.40**).

Male crocodilians possess a relatively small erectile penis that superficially resembles that of higher vertebrates; it even has a glans-like swelling at its distal end. The surface is covered with lightly keratinised epithelium without horny projections.

Male tuataras lack an erectile intromittant organ. Internal fertilisation is accomplished by cloacal apposition during which semen is transferred to the female and enters the proctodeum portion of the cloaca.

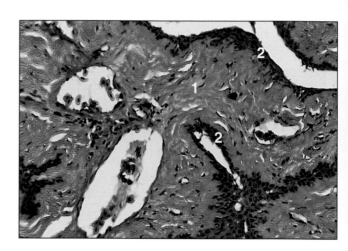

Figure 12.40 Inverted (detumescent) hemipenis of a Carolina anole lizard (*Anolis carolinensis*). The organ is composed of spongy fibrovascular tissues containing many arterial and venous channels (1), and stratified lightly cornified squamous epithelium (2). Lobules of glandular tissue, similar to sebaceous glands, lie at the base of each hemipenis and furnish lubrication and scent-rich secretions to the erectile intromittant organ. H&E ×50.

CLINICAL CORRELATES

True hermaphroditism occurs occasionally in reptiles but is usually only discovered at necropsy. The condition can involve one gonad of each sex, one or more ovotestes, or separate masses of ovarian and testicular tissue.

A Bidder's organ is normally present in male bufonid toads. This interesting structure is a cap-like mass of ovarian tissue attached to the cranial pole of the testis (**Figure 12.41**). It is unclear why androgenic and ovarian hormone secretion does not each suppress their opposites.

As in domestic animals, the reptilian or amphibian testis can be the site of inflammatory disorders, atrophy and fibrosis (**Figure 12.42**), and neoplastic disease (**Figure 12.43**). Malignant tumours are more prevalent than benign variants.

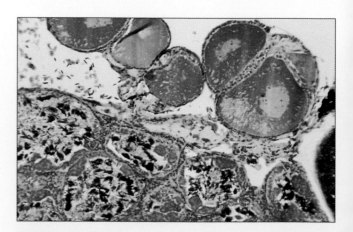

Figure 12.41 Bidder's organ in a western toad (*Bufo boreas*). Although similar to hermaphroditic gonadal tissue, Bidder's organ is a normal constituent found in some bufonid amphibians and consists of a cap-like assemblage of pigmented ovarian tissue, including yolked follicle-like structures attached to the cranial pole of each testis. The presence of ovarian tissue does not suppress spermatogenesis. H&E ×50.

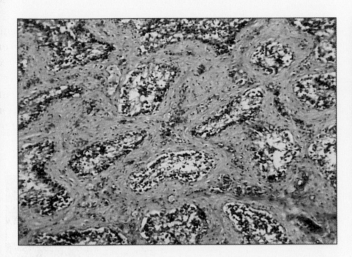

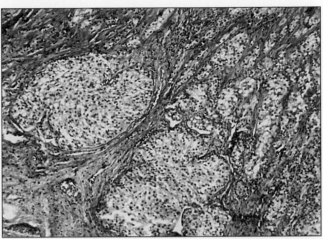

Figure 12.42 Testicular fibrosis in a blood python (*Python curtus*). The seminiferous tubules are atrophic and widely separated from each other by wide bands of mature fibrocollagenous connective tissue. H&E ×50.

Figure 12.43 Sustentacular (Sertoli) cell tumour from a snake. Note large nests of pale-staining sustentacular cells. H&E ×100.

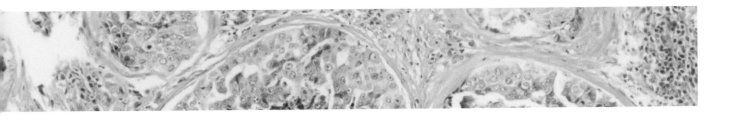

13

FEMALE REPRODUCTIVE SYSTEM

The female reproductive system consists of the ovaries, uterine tubes (oviducts), uterus, usually bicornuate, cervix, vagina, vestibule, and associated glands. Moreover, the external genitalia comprise the vulva, labia, clitoris, and external urethral orifice.

Ovaries

Cortex and Medulla

The ovaries have an exocrine function, the production of the oocyte or female gamete, and an endocrine function, the production of female sex hormones. Each ovary is divided into an outer parenchymatous zone (cortex) and an inner vascular zone (medulla) (**Figure 13.1**), except in the mare, where these structures are reversed. The cortex of each ovary is covered by a simple squamous or cuboidal epithelium that is continuous with the mesothelium of the visceral peritoneum (the germinal epithelium), except at the

hilus where blood vessels and nerves enter and exit from the organ. Beneath the epithelium is a layer of dense irregular connective tissue, the tunica albuginea. Deep to that is the cortical stroma, loose connective tissue surrounding ovarian follicles in various stages of development. The ovarian stroma consists of spindle-shaped cells arranged in whorls surrounding the ovarian follicles. In carnivores, particularly the cat, numerous specific interstitial cells are present in the stroma. They are small, round, epithelioid cells with a round nucleus, and they stain poorly (**Figure 13.2**).

The medulla is highly vascularised and consists of fibroelastic connective tissue and some smooth muscle (**Figure 13.3**). Channels, lined with (densely staining) cuboidal epithelium and called the rete ovarii, are conspicuous components of the medulla in carnivores and ruminants (**Figure 13.4**). They are derived from the mesonephric tubules during embryogenesis.

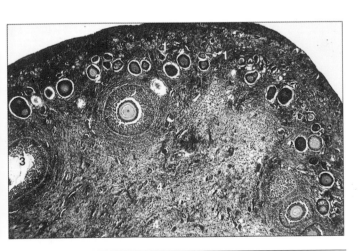

Figure 13.1 Ovary (cat). (1) Parenchymatous zone with primary follicles (2) and secondary follicles (3). (4) Vascular zone. H&E ×7.5.

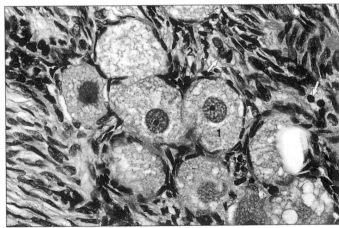

Figure 13.2 Ovary. Parenchymatous zone (cat). (1) Primary follicles; the large oocyte is surrounded by flattened nurse cells. (2) Cellular ovarian stroma. The specific interstitial cells are arrowed. H&E ×250.

DOI: 10.1201/9781003333807-13

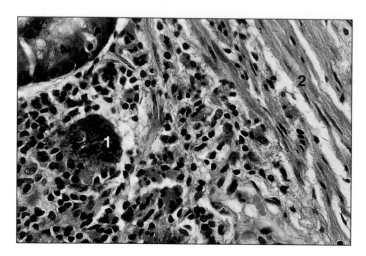

Figure 13.3 Ovary. Vascular zone (bitch). (1) The ovarian stroma is very vascular. (2) Smooth muscle cells. H&E ×200.

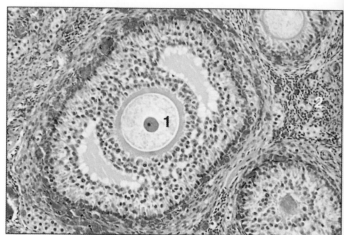

Figure 13.5 Ovary. Parenchymatous zone (sheep). (1) The oocyte is surrounded by the granulosa layer. (2) Ovarian stroma. H&E ×125.

Development of the Follicles

Primordial follicles are the least developed and most numerous follicles of the ovary, lying just below the tunica albuginea. Each consists of a primary oocyte surrounded by a layer of simple squamous follicular cells (*see* **Figure 13.2**). The primary oocytes originate in the late embryonic/early postnatal ovary from oogonia and are surrounded by a flattened layer of follicular cells. Further development is arrested until puberty when a regular cycle of events results in the passage of one or more follicles into the lumen of the uterine tube during reproductive cycles. A number of follicles begin to grow, and the primary oocyte enlarges from 60 μm to between 100 and 120 μm in diameter. Concomitantly, an acidophilic, translucent membrane, the zona pellucida, forms around the oocyte (**Figure 13.5**), and the follicular cells become cuboidal. These changes

lead the primordial follicle to develop into the primary follicle. The follicular cells proliferate and stratify and begin to accumulate fluid in the intercellular spaces. The follicle now comes under the influence of follicle-stimulating hormone (FSH) from the pituitary gland and continues to grow to become a secondary follicle, with a C-shaped, fluid-filled space, or antrum (**Figure 13.6**). Its cells are now called the granulosa layer. The follicle secretes oestradiol, the female sex hormone that prepares the endometrium to receive the fertilised oocyte. The ovarian stroma condenses around the developing follicle to form an inner layer, the theca interna, which is cellular and vascular, and an outer layer, the theca externa, which is composed of fibrous connective tissue.

The follicle continues to increase in size, moving to the surface of the ovary to become a tertiary (mature or Graafian) follicle. The oocyte is surrounded by an accumulation of granulosa cells that protrude into the antrum, the cumulus

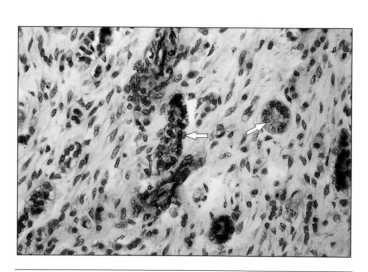

Figure 13.4 Ovary. Vascular zone (sheep). Tubules of the reti ovarii are arrowed. H&E ×200.

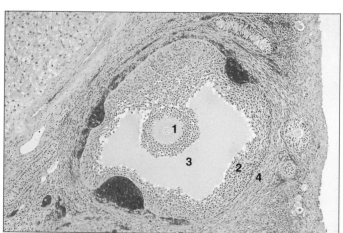

Figure 13.6 Ovary. Secondary follicle (bitch). (1) The oocyte is surrounded by the amorphous zona pellucida. (2) The granulosa cells are stratified. (3) Fluid accumulates to begin antrum formation. (4) Ovarian stroma concentrates around the follicle to form the theca. H&E ×62.5.

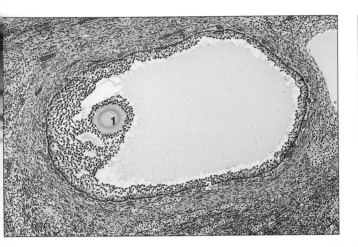

Figure 13.7 Ovary. Tertiary follicle (cow). (1) The oocyte surrounded by the zona pellucida is embedded in a mound of granulosa cells, the cumulus oophorous. (2) Theca or capsule of ovarian stroma. H&E ×62.5.

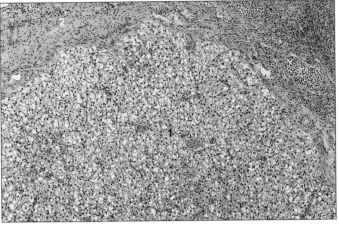

Figure 13.9 Ovary. Parenchymatous zone (sheep). (1) The pale vacuolated cells are part of the corpus luteum. (2) Ovarian stroma. H&E ×62.5.

oophorus (**Figures 13.7** and **13.8**). The single layer of granulosa cells that immediately surround the zona pellucida forms the corona radiata. Ordinarily, a mature vesicular follicle contains a single oocyte. The follicles of certain animals (carnivores, sows, and ewes) may, however, contain up to six oocytes. Maximum size is reached just before ovulation.

Ovulation

At ovulation, the primary oocyte divides into secondary oocytes, one cell retaining most of the cytoplasm; completion of meiosis occurs at fertilisation. The follicle breaks open at the stigma, releasing the oocyte surrounded by the zona pellucida, corona radiata, and cells of the cumulus oophorus, which passes into the uterine tubes. The follicle collapses

and the blood from torn capillaries fills the antrum becoming the corpus hemorrhagicum. Granulosa cells, together with those of the theca interna, hypertrophy, expand into the cavity to form the granulosa lutein (large) and theca lutein (small) cells of the corpus luteum (**Figures 13.9** and **13.10**). They are arranged in long cords separated by vascular connective tissue. The luteal cells secrete progesterone which, with the oestradiol, prepares the uterus for possible conception. With regression of the corpus luteum, stromal elements move in and replace the dead cells with collagen to form a scar: the corpus albicans (**Figures 13.11** and **13.12**). Although many primordial follicles begin the process outlined above, few become mature. The majority undergo a degenerative regression. They are called atretic follicles and can be recognised by the irregular outline of the follicle and the separation of the granulosa cells (**Figure 13.13**). Cells of the theca interna hypertrophy and the zona pellucida becomes swollen. Eventually, the entire follicle is resorbed.

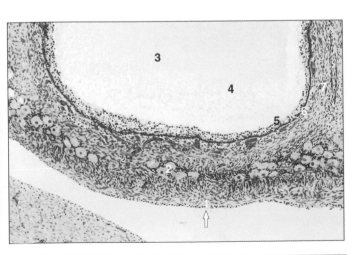

Figure 13.8 Ovary. Parenchymatous zone (cat). The cuboidal epithelium is arrowed. (1) Tunica albuginea is a thin layer of connective tissue. (2) Primary follicles. (3) Vescicular (tertiary) follicle with a large fluid-filled antrum (4) lined by granulosa cells (5) and encapsulated by the vascular theca interna (6) and the fibrous theca externa (7). Gomori's trichrome ×62.5.

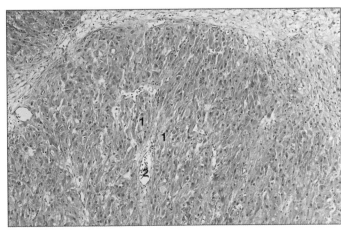

Figure 13.10 Ovary. Parenchymatous zone (sow). (1) Cords of large densely stained luteal cells are separated by (2) blood vessels. H&E ×125.

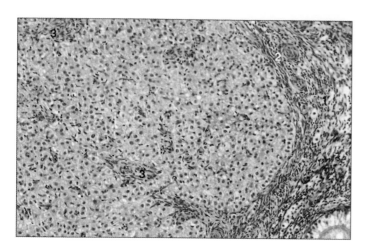

Figure 13.11 Ovary. Parenchymatous zone (sheep). (1) Ovarian stroma. (2) Dense white connective tissue. (3) Active fibroblasts replacing degenerating luteal cells. H&E ×125.

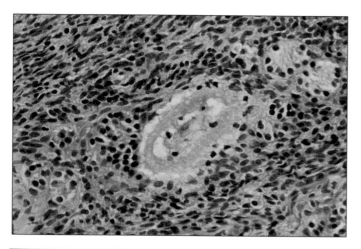

Figure 13.12 Sheep. Parenchymatous zone (sheep). The central pale staining area is dense white connective tissue (scar tissue) and represents a later stage in the development of a corpus albicans. H&E ×250.

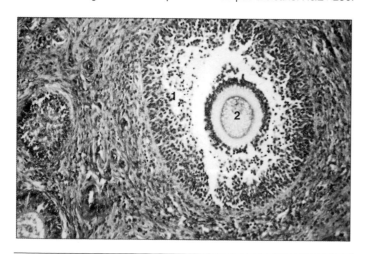

Figure 13.13 Ovary. Parenchymatous zone (bitch). Atretic follicle. (1) The granulosa cells are separating; spaces are present between them. (2) The oocyte is no longer in contact with the lining granulosa cells. H&E ×125.

Uterine Tubes (Oviducts)

The uterine tubes are long flexuous musculo-membranous tubes. They consist of a funnel-shaped, thin-walled expanded cranial end, the infundibulum, which is close to the ovary, a middle region, which is the longest segment, the ampulla, and a caudal narrow part or isthmus that opens into the ipsilateral horn of the uterus. The epithelium is a simple or pseudostratified columnar with secretory and ciliated cells. The lamina propria-submucosa is composed of loose connective tissue with many plasma cells and projects longitudinal folds, thicker and shorter as approaching the isthmus. In some breeds of sheep (blackface and crosses), pigment cells can be present. The muscularis is mainly a layer of circular smooth muscle, thickening at the junction with the uterine horn, with an outer layer of longitudinal muscle. The serosa is loose vascular connective tissue with prominent blood vessels (**Figures 13.14** and **13.15**).

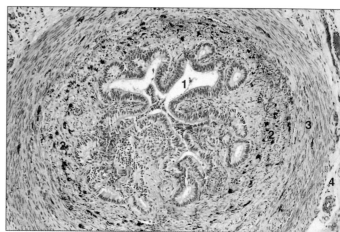

Figure 13.14 Transverse section (TS) uterine tube (sheep). (1) The lumen is lined by a pseudostratified epithelium thrown into deep folds. (2) The lamina propria has deposits of dark brown pigment. (3) The muscularis is, for the most part, circular smooth muscle fibres. (4) The serosa is loose vascular connective tissue with prominent blood vessels. H&E ×62.5.

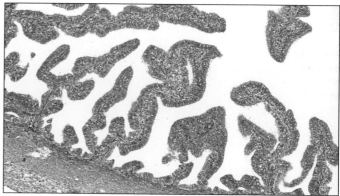

Figure 13.15 Uterine tube (pig). The lining epithelium is pseudostratified columnar with ciliated and nonciliated cells. The mucosa projects into the lumen, forming villi-shaped folds. H&E ×100.

Uterus

The uterus is bicornuate, with right and left horns (cornua), a body (corpus), and a neck (cervix) (**Figures 13.16–13.23**, all illustrations in transverse section [TS]). The uterine wall in the cornua and corpus has three layers: an inner endometrium (mucosa-submucosa), a middle myometrium (muscularis), and an outer perimetrium (serosa).

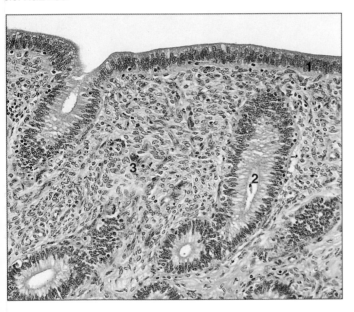

Figure 13.16 Uterus (cat). (1) Endometrium with simple straight tubular glands. (2) Myometrium with circular and longitudinal smooth muscle. H&E ×50.

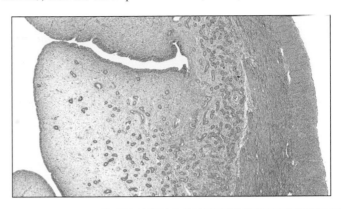

Figure 13.17 Uterus (pig). Endometrium with a simple columnar epithelium, connective tissue and numerous glands, myometrium (muscular layer), and external perimetrium. H&E ×50.

Figure 13.18 Uterus (bitch). (1) The lining epithelium is simple columnar. (2) The endometrial glands are lined by tall columnar epithelial cells with a basal nucleus. (3) The lamina propria is very cellular with many small blood vessels. H&E ×125.

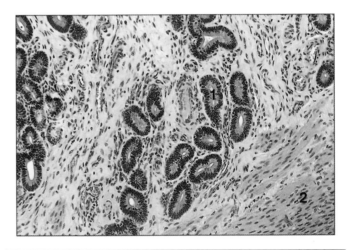

Figure 13.19 Uterus (cow). (1) The endometrial glands are markedly coiled. (2) Thick layer of myometrium. H&E ×100.

Figure 13.20 Uterus (cow). The myometrium is split by the stratum vasculare (arrowed). H&E ×100.

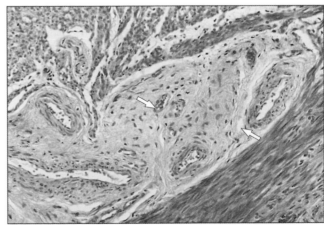

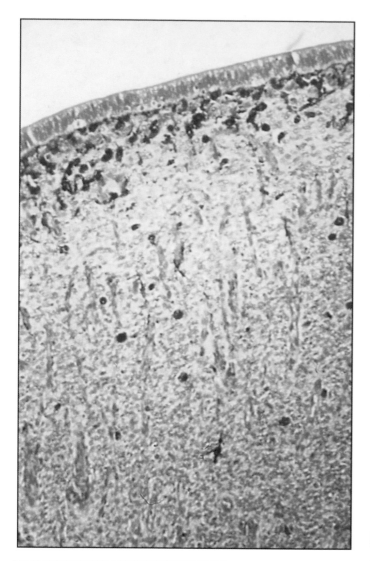

Figure 13.21 Uterus (sheep). Only part of a caruncle is present; the connective tissue is very cellular and well vascularised. H&E ×200.

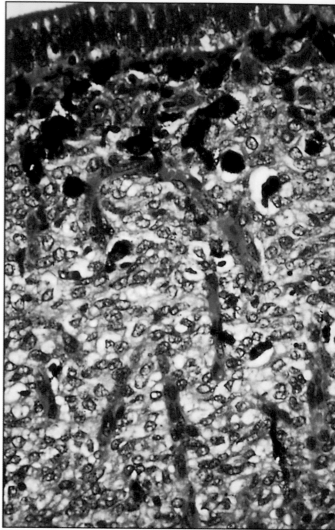

Figure 13.22 Uterus. Caruncle (sheep). Local deposits of pigment (melanin) are present in the connective tissue of some breeds of sheep. H&E ×400.

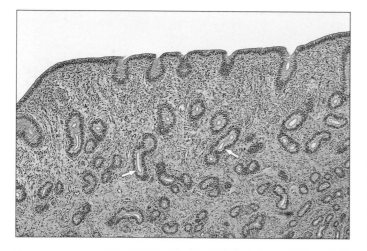

Figure 13.23 Uterus. Intercaruncular area (sheep). The endometrial glands (arrowed) are confined to the intercaruncular area. H&E ×100.

Endometrium

The epithelial lining of the endometrium is simple cuboidal or columnar in the mare and carnivores, and pseudostratified in the sow and ruminants. In ruminants, ciliated cells may be present. The lamina propria-submucosa is a deep layer of highly vascularised loose connective tissue, with simple branched tubular endometrial glands that open into the lumen of the uterus. These glands are straight in carnivores and coiled in the mare, sow, and ruminants. Caruncles are areas of nonglandular connective tissue and project into the lumen of the uterus in ruminants. They are larger when the animal is pregnant.

Myometrium

The myometrium (muscularis) is composed of a deep inner layer of circular smooth muscle and a less clearly defined outer layer of longitudinal smooth muscle. The stratum vasculare is a layer of connective tissue carrying large blood vessels to the uterus that divides the circular muscle into two layers in the cow, separates the inner and outer layers in carnivores, and forms an indistinct layer in the sow.

Perimetrium

This is the outer loose connective tissue layer covered by the peritoneal mesothelium.

Cervix Uteri

The mucosa appears as a series of longitudinal folds, which may become subdivided into secondary and tertiary folds (**Figure 13.24**). The epithelium is a tall columnar with goblet cells and a few ciliated cells (but areas of stratified squamous in the bitch and sow). Simple tubular glands appear in small ruminants and sow. The lamina propria is dense connective tissue, showing considerable variation according to the physiological status of the animal. The muscularis consists of an inner circular and outer longitudinal layer of smooth muscle. The serosa is loose connective tissue (**Figure 13.25**).

Vagina

The vaginal wall consists of four layers: mucosa, submucosa, muscularis, and adventitia or serosa.

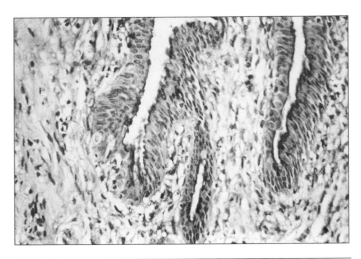

Figure 13.25 Cervix (sheep). The mucus is stained deep pink and is present in the lumen and in the luminal cytoplasm of the columnar cells. H/periodic acid-Schiff (PAS) ×250.

The mucosa is lined with stratified squamous epithelium in all species except the cow, in which the cranial vagina is stratified columnar with goblet cells and forms low longitudinal folds and a few circular folds. The mucosa rests on a papillated lamina propria-submucosa (**Figures 13.26–13.28**). The subepithelial connective tissue is very cellular, with a vascular and fibrous deeper layer. Elastic fibres are present and lymphatic nodules are more frequent in the caudal vagina.

The muscularis consists of inner circular and outer longitudinal smooth muscle. In the bitch, queen, and sow, a thin layer of longitudinal muscle occurs internal to the circular layer.

The serosa (cranially) and adventitia (caudally) are composed of loose connective tissue rich in blood vessels and nerves.

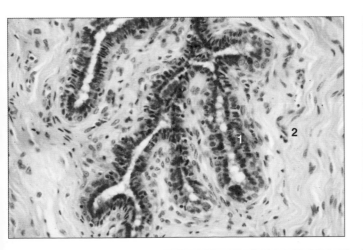

Figure 13.24 Cervix (sheep). The mucosa has deep folds. The epithelium (1) is tall columnar mucus secreting, and the lamina propria (2) is dense connective tissue. H&E ×125.

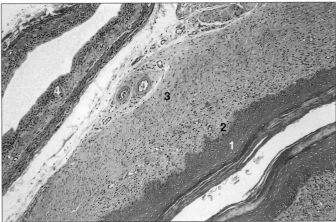

Figure 13.26 Vagina and urethra (bitch). (1) Stratified squamous keratinised epithelium lining the vagina. (2) Papillated lamina propria. (3) Muscularis. (4) Urothelium of the urethra. H&E ×62.5.

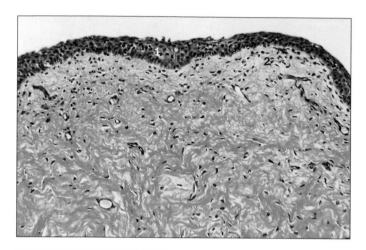

Figure 13.27 Vagina (sheep). (1) Stratified squamous nonkeratinised epithelium. (2) Lamina propria. Masson's trichrome ×125.

During the breeding season, the vaginal epithelium undergoes cyclic changes in carnivores. During anoestrus, the epithelium comprises only two or three layers. There is an oestrogen-determined proliferative stage in pro-oestrus, and the epithelium becomes thicker and keratinises. During oestrus, keratinisation reaches the maximum. With the development of the corpus luteum and secretion of progesterone, the surface cells desquamate and polymorphs migrate into and through the epithelium. These changes are clearly defined in the bitch and queen, and vaginal smears are used to indicate the stage of the cycle, with a high degree of accuracy (**Figures 13.29–13.32**). This has an important practical application in determining the exact time of mating. The vaginal changes are not well defined in the other domestic species. During metoestrus-dioestrus, the number of epithelial cell layers and keratinisation decreases.

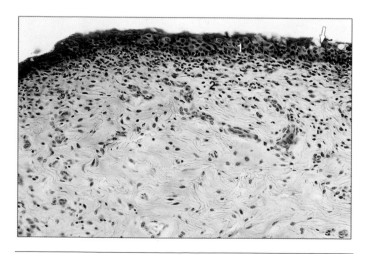

Figure 13.28 Vagina (cow). (1) Pseudostratified epithelium with mucus-secreting cells (arrowed). (2) Cellular lamina propria. Masson's trichrome ×125.

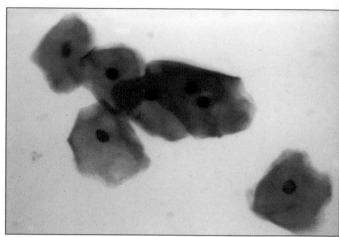

Figure 13.29 Vaginal smear. Bitch in anoestrus. This vaginal smear consists of typical nucleated epithelial cells from an anoestrus bitch. Papanicolaou ×400.

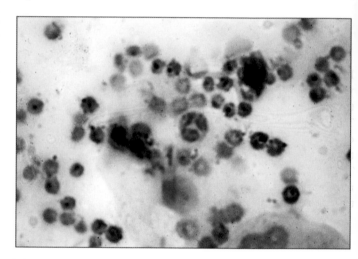

Figure 13.30 Vaginal smear. Bitch in pro-oestrus. Nucleated surface epithelial cells and erythrocytes are present. Papanicolaou ×400.

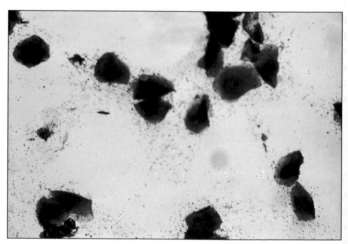

Figure 13.31 Vaginal smear. Bitch in oestrus. This is the typical 'dirty smear' of the oestrus phase of the cycle, with anucleate squames predominating. Papanicolaou ×400.

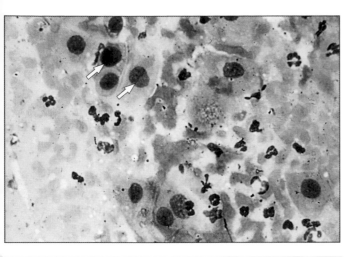

Figure 13.32 Vaginal smear. Bitch in metoestrus. Pale blue staining surface cells (arrowed) mingled with polymorphonuclear leukocytes. Papanicolaou ×400.

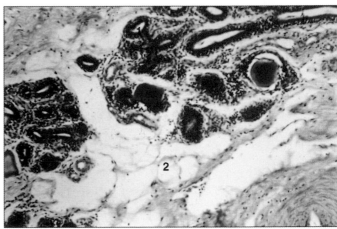

Figure 13.33 Nonlactating mammary gland (cow). (1) Only the densely staining remnants of the duct system are present. (2) Interlobular adipose connective tissue. H&E ×62.5.

Vestibule

In the vestibule, the mucosal folds are less than in the vagina. The vestibule is lined with stratified squamous epithelium. Cyclic changes are less evident than in the vagina. The lamina propria-submucosa is characterised by a plexus of large veins and vestibular bulbs (erectile tissue) and by the major and minor mucus-secreting vestibular glands. The muscularis is a continuation of the vaginal smooth muscle. External to this is the striated muscle: the constrictor vestibuli.

Vulva, Labia, and Clitoris

The stratified squamous epithelium of the vulva is continuous with the skin at the labia; sebaceous and sweat glands are present in the lamina propria. A dense vascular plexus is responsible for the congestion during oestrus, especially in bitch and sow. The clitoris, like the penis, consists of erectile tissue (corpus cavernosum clitoridis), a glans, and a prepuce and presents numerous sensory nerve endings.

Mammary Glands

The mammary glands are the distinguishing feature of the mammals and are modified sweat glands that secrete milk. They are compound tubuloalveolar glands enclosed in a fibroelastic capsule extending into the substance of the gland and dividing it into lobes and lobules. In the nonlactating gland, the lobules consist of a duct system surrounded by loose connective tissue, separated from the adjoining lobules by fatty interlobular tissue

(**Figure 13.33**). During pregnancy, the duct system expands at the expense of this tissue, and it is reduced to thin strands of vascular connective tissue as parturition approaches.

The terminal parts of the ducts expand to form secretory saccules (alveoli) lined with cuboidal/low columnar epithelium (**Figures 13.34–13.36**). Myoepithelial cells, responsible for milk let-down, lie between the secretory cells and the basement membrane. Plasma cells are numerous in the stroma immediately postpartum, and immunoglobulin is passed via the saccules into the colostrum (first milk), where it provides a passive immunity to the neonate. The saccules open onto the interlobular ducts that empty into the lactiferous ducts, which open into the lactiferous sinus at the base of the teat. Initially, the epithelium is simple cuboidal and in the larger ducts and sinuses is

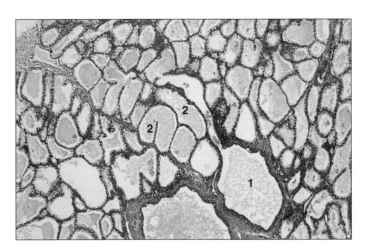

Figure 13.34 Lactating mammary gland (cow). (1) The duct system expands and terminates in (2) the secretory alveoli in the active gland. The connective tissue is reduced to a supporting role. H&E ×62.5.

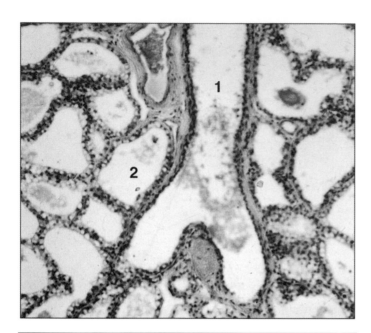

Figure 13.35 Lactating mammary gland (cow). (1) Large excretory lobular duct. (2) Secretory alveoli. H&E ×250.

bistratified cuboidal to columnar with some smooth muscle cells present in the wall.

The papilla or teat contains the teat sinus, which opens onto the teat surface by the papillary duct (or ducts depending on the species: 1 for the cow; 2–3 for the mare and sow; 4–7 for the queen; 7–16 for the bitch; **Figures 13.37** and **13.38**). The epithelium is bistratified cuboidal, changing to keratinised stratified squamous at the external orifice. The lamina propria has abundant elastic fibres. The muscularis consists of inner and outer layers of smooth muscle with a middle circular layer and condenses to form a sphincter in the cow, sow, and bitch. In the cow, complex epithelial

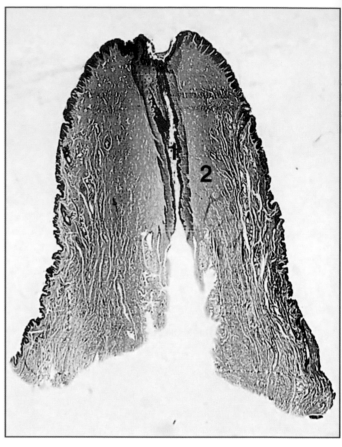

Figure 13.37 Teat (sheep). (1) The teat canal is lined by stratified squamous epithelium. (2) The connective tissue of the lamina propria is continuous with that of the skin. Masson's trichrome ×20.

folds are present in the upper part of the papillary duct. In the cow and sow, the skin covering the teat is hairless and nonglandular; in the mare, bitch, and queen, abundant sebaceous glands and fine hairs are present.

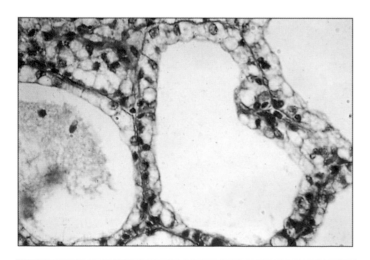

Figure 13.36 Lactating mammary gland (cow). The secretory alveoli are lined by a cuboidal epithelium. Loss of fat in processing causes the empty appearance of the cells. H&E ×250.

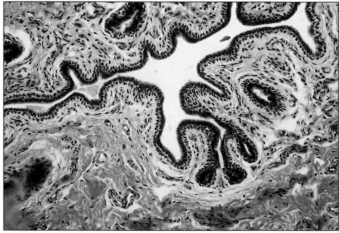

Figure 13.38 Teat canal (cow). The columnar epithelium lining the canal is folded. H&E ×100.

Placentation

The placenta is formed during pregnancy and allows physiological exchange between the mother and the fetus, which also have an important endocrine role. This organ includes fetal and maternal components. The fetal membranes are the amnion, the yolk sac, the allantois, and the chorion, and these are derived from the three basic germ layers (ectoderm, mesoderm, and endoderm) in a variety of combinations. This is illustrated diagrammatically in **Figures 13.39–13.43**, and a classification is given in **Appendix Table 3**.

The histological classification of the placenta is based on the concept that placental exchange during pregnancy is dependent on the approximation of the fetal and the maternal capillary bed: the interhaemal barriers. Fetal and maternal blood may be separated by as many six layers. The fetal layers

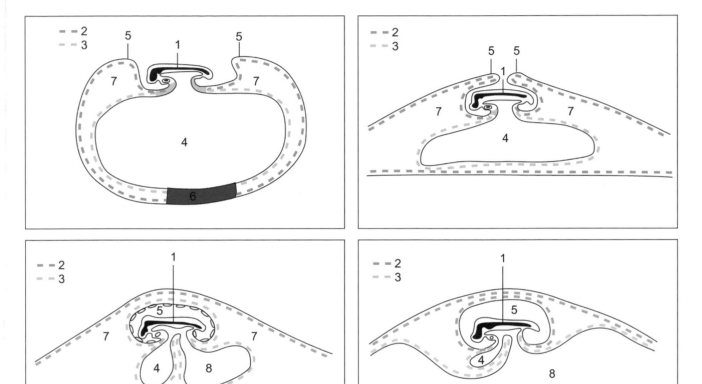

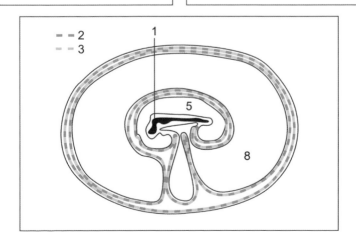

Figure 13.39–Figure 13.43 **Figures 13.39–13.41** illustrate the common early stages of development of the extra-embryonic membranes. **Figure 13.42** illustrates the cow, sheep, and pig, and **Figure 13.43** illustrates the mare and carnivore. (1) Embryo. (2) Somatopleure (ectoderm + somatic mesoderm) forms the lateral and ventral walls of the fetus. (3) Splanchnopleure (endoderm + splanchnic mesoderm) forms the digestive tube. (4) Yolk sac. (5) Amniotic folds. (6) Unsplit mesoderm. (7) Extra-embryonic coelom. (8) Allantois.

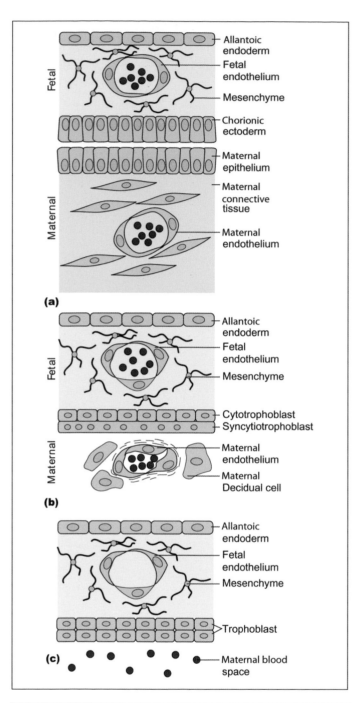

Figure 13.44 Histological classification of the placenta. (a) Epitheliochorial placenta, sow and mare; (b) endotheliochorial placenta, dog, and cat; and (c) hemochorial placenta, rodents, primates, and man.

the apposition occurs over the extensive chorionic sac: a diffuse placenta.

Ruminants have a synepitheliochorial placenta where binucleate chorionic epithelial cells migrate and fuse with endometrial epithelial cells forming syncytia. The apposition of maternal and fetal tissues occurs at the caruncle–cotyledon interface giving rise to a cotyledonary placenta. In the carnivores, there is considerable destruction of the maternal tissues in the central band of the gestational sac and the chorionic epithelium apposed to the endothelium of the maternal blood vessels. This is a zonary endotheliochorial placenta. In rodents, lagomorphs, and primates, the maternal tissue is destroyed over a limited discoidal area, and the trophoblast is bathed by maternal blood. It is a hemochorial placenta. Some maternal tissue is lost at parturition; therefore, the placenta is deciduate (**Figure 13.44**).

Sow

Sows have a diffuse, folded, epitheliochorial placenta, and a gestation period between 113 and 115 days.

In the early placenta, the chorioallantois is folded and simply apposed to the endometrial folds. Both epithelia are columnar, and the capillary bed is subepithelial. As the pregnancy advances, the membranes increase in volume, the epithelial surfaces become interlocked, the epithelium is reduced in height, the folds become more complex, and the capillaries on both sides in the immediate subepithelial tissue often project between the epithelial cells where they are known as intraepithelial capillaries (**Figures 13.45** and **13.46**). Specialised areas, the areolae, exist for the absorption of uterine milk containing uteroferrin, the primary source of iron for the fetus. The endometrial glands secrete into a space between the chorion and the endometrium, and the trophoblast cells are tall columnar absorptive (**Figures 13.47** and **13.48**).

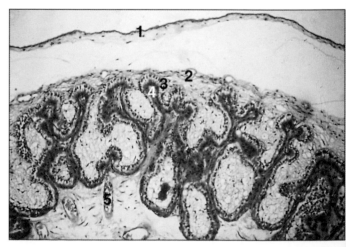

Figure 13.45 Epitheliochorial placenta (sow). (1) Allantoic endoderm lining the allantoic sac. (2) Mesenchyme. (3) Trophectoderm. (4) Maternal epithelium. (5) Maternal blood vessels in the endometrium. The simple folding of the early pig placenta is shown here. H&E ×100.

are the allantoic capillary endothelium, the chorioallantoic mesenchyme, and the chorionic epithelium (trophoblast), and the maternal layers are the uterine epithelium, the endometrial connective tissue, and the uterine capillary endothelium.

The type of placenta where all six layers persist is the epitheliochorial placenta. There is no destruction of maternal tissue and no loss at parturition; therefore, it is nondeciduate. This type of placentation is seen in the sow and mare, where

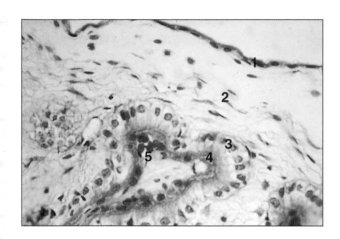

Figure 13.46 Epitheliochorial placenta (sow). (1) Allantoic endoderm. (2) Mesenchyme. (3) Trophectoderm. (4) Maternal epithelium. (5) Maternal blood vessels in the endometrium. H&E ×200.

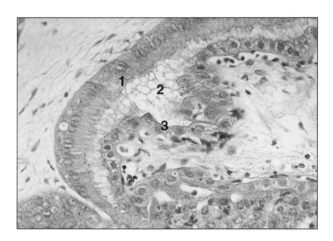

Figure 13.47 Epitheliochorial placenta. Areola (sow). (1) Tall columnar absorptive trophectodermal cells. (2) Endometrial gland secretion. (3) Maternal epithelium. H&E ×200.

Mare

Mares have a diffuse, villous, epitheliochorial placenta and a gestation period of 335–345 days.

The membranes develop slowly in the mare, and a choriovitelline placenta is established in early pregnancy; simple folding very similar to that in the sow is seen. Short simple villi now appear over the chorionic sac with the development of the chorioallantois (**Figure 13.49**). As pregnancy advances, the villi branch repeatedly forms complex tufts embedded in the corresponding uterine crypts; these are called microplacentomes.

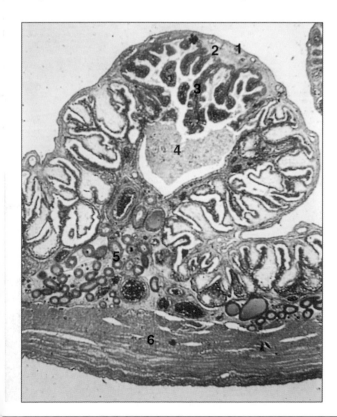

Figure 13.48 Epitheliochorial placenta. Areola of late gestation (sow). (1) Allantoic endoderm. (2) Mesenchyme. (3) Fetal villi covered by absorptive trophectoderm. (4) Endometrial gland secretion. (5) Endometrial glands. (6) Myometrium. H&E ×50.

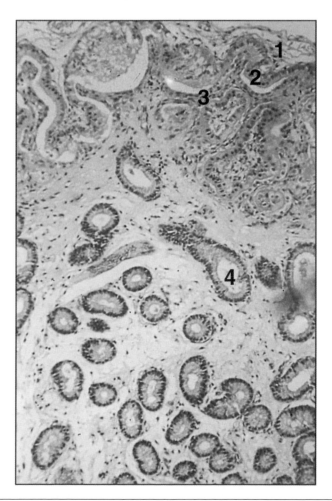

Figure 13.49 Epitheliochorial placenta (mare). In this early gestation example of the mare placenta, short simple villi indent the maternal tissue. (1) Mesenchyme. (2) Trophectoderm. (3) Maternal epithelium. (4) Endometrial glands. H&E ×100.

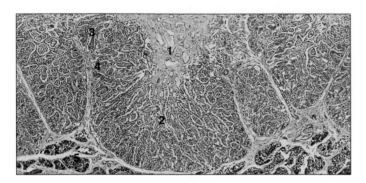

Figure 13.50 Epitheliochorial placenta (mare). (1) Main stem fetal villus. (2) Microplacentome. (3) Absorptive trophectoderm, the areola. (4) Uterine milk. (5) Endometrial glands. H&E ×50.

The chorionic epithelium between them is highly absorptive, and the endometrial glands secrete into the space to form an areola (**Figures 13.50** and **13.51**).

Endometrial cups are peculiar to the mare and are only present from weeks 5–20 of gestation. The endometrial

Figure 13.51 Epitheliochorial placenta. Mare in late gestation. (1) Main stem fetal villus. (2) Absorptive trophectoderm. (3) Interlocking fetal/maternal tissue of the microplacentome. H&E ×62.5.

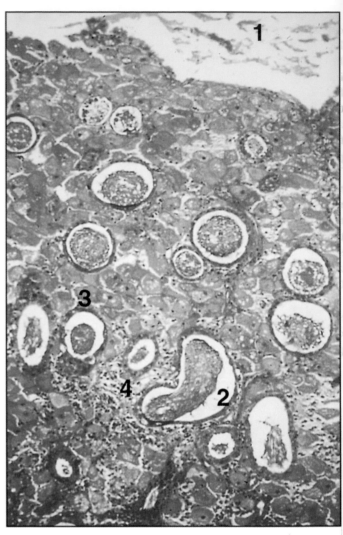

Figure 13.52 Epitheliochorial placenta. Endometrial cup (mare). Only the endometrium is present. (1) Area of eroded maternal epithelium. (2) Distended endometrial glands. (3) Decidual cells. (4) Lymphocytes. H&E ×62.5.

epithelium disappears over small areas, and the endometrial glands hypertrophy and decidual cells (of fetal origin) migrate into the maternal tissue (**Figure 13.52**). A brown coagulum accumulates between the fetal and the maternal tissue, the richest known source of gonadotrophic hormone. This is responsible for the second wave of follicles and corpora lutea seen in the mare after the loss of the corpus luteum of pregnancy. The endometrial cups degenerate when they detach from the uterine mucosa and are surrounded by chorionic folds forming allantochorionic pouches.

Ruminants

Ruminants have a synepitheliochorial, villous, cotyledonary placenta and a gestation period of 278–290 days in the cow and 145–155 days in small ruminants.

Apposition to the endometrium begins with the development of focal areas of short simple villi, called cotyledons, on the chorionic surface opposite the caruncle. Folding of the caruncle creates crypts, and the villi project into these to form a complex arrangement of interlocking fetal/maternal tissue: the placentome. In the cow, the epithelium is cuboidal or even columnar through pregnancy, and binucleate giant cells are present in the trophoblast from day 16 of gestation; these binucleate cells secrete placental lactogens, progesterone, and prostaglandin. The caruncular crypts are lined with cuboidal or squamous epithelium within which some binucleated or trinucleated giant cells can be observed. Maternal blood is released at the tips of the septae and absorbed by the trophoblast: arcade haemorrhages. The uterine glands secrete onto the uterine surface in the intercaruncular zone, and areas of absorptive trophoblast form areolae (**Figures 13.53–13.55**).

The placentome development in small ruminants is similar to the cow. Large syncytia can be observed in the caruncular epithelium together with isolated migrated trophoblastic giant cells (**Figures 13.56–13.58**).

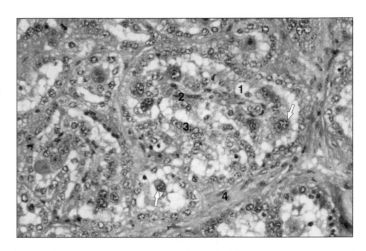

Figure 13.54 Synepitheliochorial placenta (cow). (1) Mesenchyme. (2) Trophectoderm. (3) Maternal epithelium. (4) Maternal connective tissue. The binucleate giant cells are arrowed. H&E ×125.

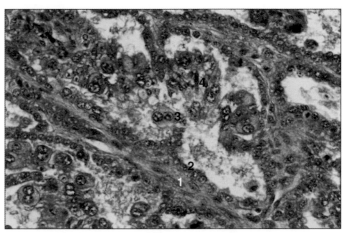

Figure 13.55 Synepitheliochorial placenta (cow). (1) Dark strands of maternal connective tissue. (2) Maternal epithelium. (3) Trophectoderm with binucleate giant cells. (4) Mesenchyme. H&E ×250.

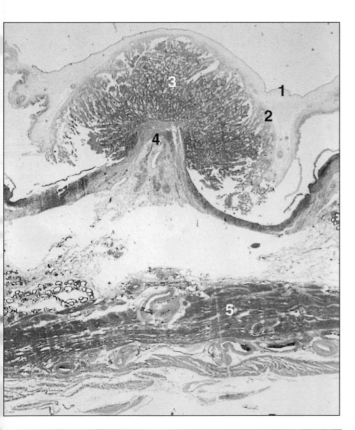

Figure 13.53 Synepitheliochorial placenta (cow). This is an early bovine placentome; note the convex appearance. (1) Endoderm lining the allantois. (2) Chorioallantoic membrane. (3) Interlocked fetal/maternal tissue. (4) Endometrium at the caruncle. (5) Myometrium. H&E ×17.5.

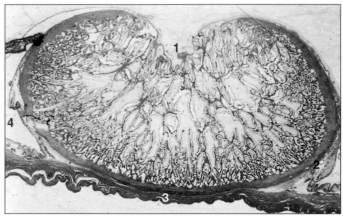

Figure 13.56 Synepitheliochorial placenta. Placentome (sheep). Note the concave appearance of the placentome. (1) The chorioallantoic membrane projecting into the caruncle. (2) Endometrium of the caruncle surrounding the chorioallantois. (3) Myometrium. (4) Intercaruncular area. H&E ×7.5.

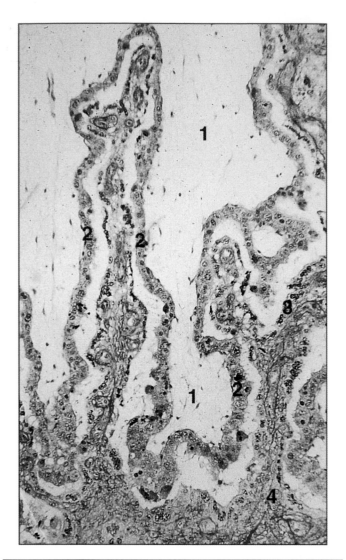

Figure 13.57 Synepitheliochorial placenta. Placentome (sheep). (1) Mesenchyme of the chorioallantois. (2) Trophectoderm with giant cells. (3) Maternal epithelium. (4) Maternal connective tissue. H&E ×62.5.

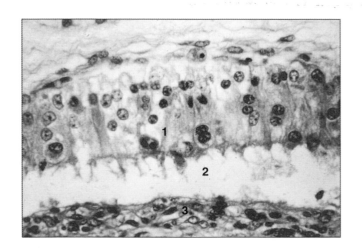

Figure 13.58 Synepitheliochorial placenta. Intercaruncular area (sheep). (1) Absorptive trophectoderm with giant cells. (2) Uterine milk. (3) Endometrium, note the apparent absence of epithelium. H&E ×400.

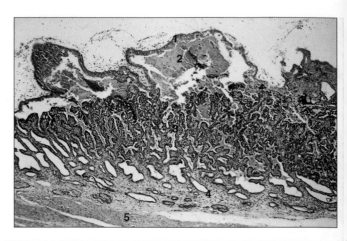

Figure 13.59 Endotheliochorial placenta. Cat placental band. (1) Absorptive trophectoderm. (2) Haematoma, pools of maternal blood. (3) Interlocking zone of fetal/maternal tissue. (4) Endometrium. (5) Myometrium. H&E ×20.

Carnivores

Carnivores have an endotheliochorial, zonary, labyrinthine placenta, and the following gestation periods: bitch, 58–65 days, and queen, 63–67 days.

The trophoblast in the central zone of the choriovitelline placenta becomes stratified with the presence of syncytial cells, the syncytiotrophoblast, and discrete cells, the cytotrophoblast. Cords of trophoblast block the mouths of the uterine glands and secrete lytic enzymes, destroying the maternal tissue. With the development of the more invasive chorioallantois, the maternal tissue is further eroded, and the trophoblast is apposed to the endothelium of the maternal blood vessels (**Figures 13.59** and **13.60**). The zone of association between the allantochorion and the uterine blood vessels is called labyrinth because of the initial difficulty in distinguishing fetal and maternal components.

Decidual cells, probably fetal in origin, are also present. Haematomas are present on the margin of the zonary band

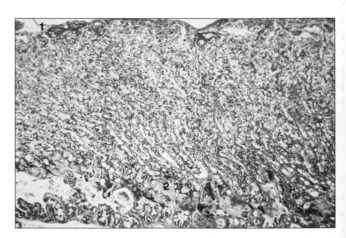

Figure 13.60 Endotheliochorial placenta. Placental band in the bitch. The lamellae are more regular than in the cat. (1) Chorioallantoic membrane. (2) Junctional zone marks the deep penetration of fetal tissue. H&E ×25.

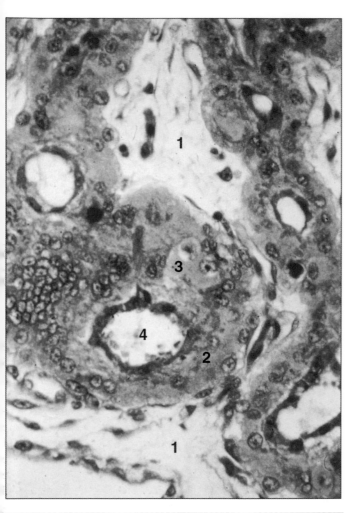

Figure 13.61 Endotheliochorial placenta (bitch). (1) Chorioallantoic mesenchyme. (2) Trophectoderm. (3) Decidual cell. (4) Maternal blood vessels lined by thickened endothelium. H&E ×400.

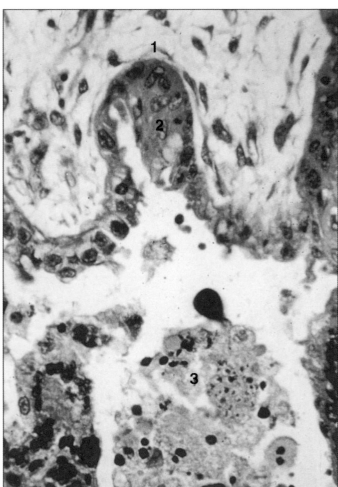

Figure 13.62 Endotheliochorial placenta (bitch). This represents the junctional zone, the limit of the fetal invasion. (1) Chorioallantoic mesenchyme. (2) Trophectoderm. (3) Maternal tissue debris. H&E ×400.

(bitch) and centrally dispersed (queen). These are pools of maternal blood surrounded by absorptive trophoblast, where maternal erythrocytes have been destroyed; these are the haemophagous zones. A green deposit of uteroverdin occurs in the bitch and a brown deposit in the queen. Outside the zonary band, there is simple apposition of chorion and endometrium with no loss of maternal tissue at parturition, as opposed to the zonary band where damage is considerable (**Figures 13.61** and **13.62**).

Umbilical Cord

The umbilical cord is the communications link between the fetus and the placenta. The umbilical arteries leave the fetus and carry deoxygenated blood to the placenta. The umbilical veins carry nutrient blood to the fetus (these commonly fuse to form a single vein). The umbilical vesicle is the remnant of the yolk sac, and the small canal is the lumen of the allantoic duct (**Figure 13.63**).

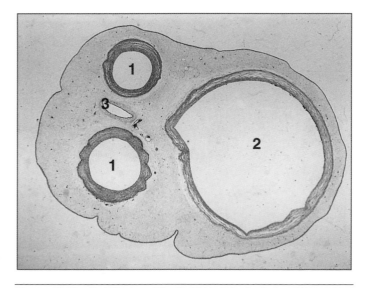

Figure 13.63 Umbilical cord (foal). (1) Umbilical arteries. (2) Umbilical vein. (3) Allantoic duct. H&E ×7.5.

CLINICAL CORRELATES

MAMMARY GLANDS

Inflammation of the mammary glands, termed mastitis (**Figure 13.64**), can affect any mammal and is most common in the lactating mammary gland, where the moist, nutrient-rich secretions provide an ideal environment for the growth of microorganisms.

A variety of mammary tumours are recognised, and several classification schemes exist that divide the tumours according to the cell types and patterns of growth present. For example, a mixed mammary tumour (**Figure 13.65**) is derived from more than a single germ layer and includes epithelium, myoepithelium, cartilage, and possibly bone. The anaplastic carcinoma is a highly recurrent and metastatic malignant tumour composed by round to polygonal cells with abundant cytoplasm and a round to oval nuclei (sometimes multinucleated) with one large or multiple nucleoli with the presence of necrotic areas and an abundant collagenous stroma with an important inflammatory reaction (**Figure 13.66**). In the bitch, early ovariohysterectomy is known to reduce significantly the incidence of future mammary tumour development. True behavioural malignancy is recognised in a minority of canine mammary tumours but is present in a higher proportion of mammary tumours in the cat (**Figures 13.67** and **13.68**).

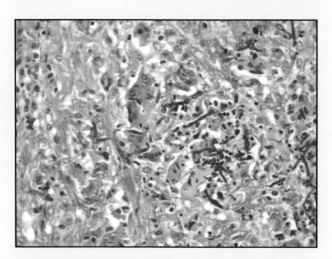

Figure 13.64 Mastitis (goat). Fungal hyphae (dark red) within a mixed inflammatory cell infiltrate is observed. PAS ×250.

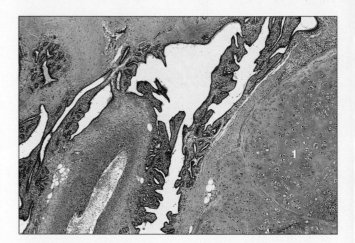

Figure 13.65 Mixed mammary tumour from an elderly bitch. In addition to glandular epithelial and stromal components, cartilage (1) is formed. The glandular lumina are dilated, and the epithelium is arranged in a single layer. This tumour is benign. H&E ×62.5.

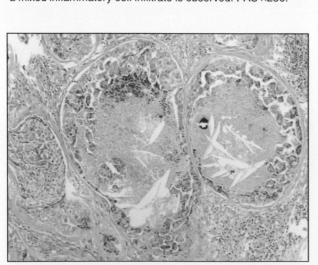

Figure 13.66 Anaplastic mammary tumour (bitch). Pleomorphic tumoural cells with abundant cytoplasm and a round to oval nuclei (some of them multinucleated) with one large or multiple nucleoli. Areas of necrosis and an abundant collagenous stroma with an important inflammatory reaction. H&E ×100.

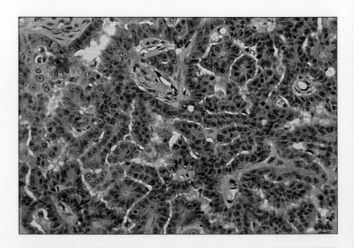

Figure 13.67 Feline mammary papillary cystadenocarcinoma. The cells in this malignant tumour form multiple frond-like papillae that extend into cystic spaces. H&E ×125.

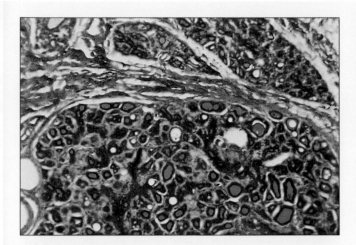

Figure 13.68 Feline ductal adenocarcinoma. The thin-walled ductal structures are filled with eosinophilic proteinaceous secretion and are divided by fibrocollagenous connective tissue. Invasion into surrounding tissue and lymphatic metastasis is common with this type of tumour. H&E ×62.5.

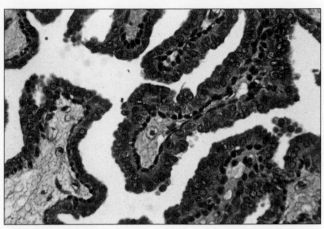

Figure 13.69 Mammary adenoma from an adult female rabbit. This high-power view shows frond-like papillary projections with fibrous cores overlain by well-differentiated columnar epithelium. H&E ×250.

Rats of both sexes have extensive mammary-type glandular tissue and may present with masses of this origin anywhere from the neck to the inguinal region. Most are benign fibroadenomas. Mammary tumours are also recognised in the rabbit (**Figure 13.69**).

FEMALE REPRODUCTIVE SYSTEM

Developmental disorders of the genital system occur in all species of domestic animals but are uncommon. These disorders are caused by abnormalities of genetic origin or by aberrant hormonal influences. Often, precise mechanisms have not been defined, but specific syndromes, such as freemartinism in cattle, are recognised. A freemartin is a genetically female calf twinned with a male. If, as is common, anastomoses form between the placental circulations, then factors passed between the twins lead to abnormalities in the female reproductive system, including inhibition of ovarian development or testis-like differentiation within the ovary and the absence of parts of the tubular tract. Effects on the male twin are minimal.

A spectrum of cystic changes is recognised in the mammalian ovary. Cystic ovarian disease in cows is important as a cause of reproductive failure and hence economic loss. Tumours that arise from tissues that are specifically ovarian can be divided into three broad categories: tumours of the surface coelomic epithelium, those of the gonadal stroma, and those of the germ cells. Tumours of the surface epithelium are significant only in the bitch as papillary and cystic adenomas and rarely papillary adenocarcinomas. Tumours that arise from the gonadal stroma include granulosa (**Figure 13.70**) and thecal (**Figure 13.71**) cell tumours. These

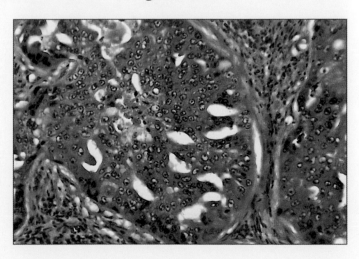

Figure 13.70 Canine ovarian granulosa cell tumour. This is the most common gonadostromal tumour in all species. Histological appearances vary. This section shows a lobular mass with clefting between the tumour cells. H&E ×125.

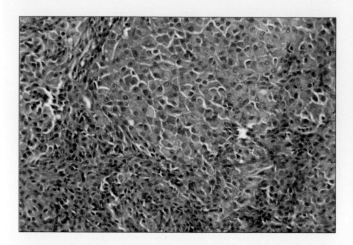

Figure 13.71 Canine ovarian thecoma in a 12-year-old Springer Spaniel bitch. The tumour cells are large and polyhedral, contain finely granular eosinophilic cytoplasm, and are arranged in solid lobules separated from each other by a fine fibrovascular stroma. H&E ×125.

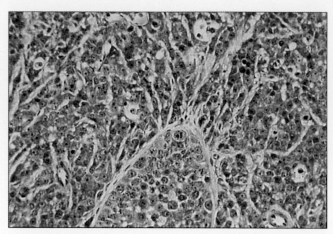

Figure 13.72 Canine ovarian dysgerminoma. The tumour is composed of lobular sheets of large cells with central nuclei and prominent nucleoli that resemble the testicular seminoma. Giant cells may be present. H&E ×125.

neoplasms can produce hormones and are rarely malignant in any species. Germ cell tumours include dysgerminomas (**Figure 13.72**) and teratomas (**Figure 13.73**). The dysgerminoma is morphologically similar to primordial germ cells and resembles its testicular homologue, the seminoma. In teratomas, totipotential germ cells have undergone somatic differentiation, and a variety of tissues of different germ lines are present within the tumour.

Obstructive and inflammatory conditions are recognised in the uterine (fallopian) tubes. A range of inflammatory, hyperplastic or, cystic changes (**Figure 13.74**) that are under hormonal influence can affect the uterus itself. Hyperplastic change, which is usually focal within the uterus, does not appear to be preneoplastic in domestic animal species but is an important precancerous indicator in humans. Uterine neoplasia is uncommon in most domestic

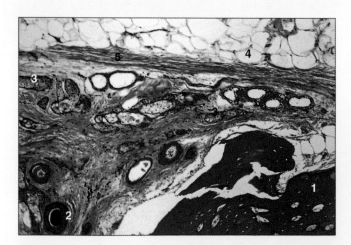

Figure 13.73 Teratoma from a 2-year-old bitch. Several germ lines are represented. Within this section, (1) dense bone, (2) well-developed hair follicles and (3) sebaceous glands, and (4) adipose and (5) collagenous connective tissue can be seen. H&E ×62.5.

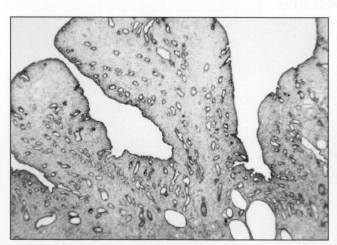

Figure 13.74 Cystic endometrial hyperplasia in a 13-year-old cat. Localised papillary outgrowths and cyst formation are present within the uterine lining. Hydrometra or mucometra may develop concurrently. Progestagen administration is the most common cause of this condition. H&E ×44.

species, although obviously many female domestic animals are neutered. In the rabbit, however, uterine adenocarcinoma occurs in a large percentage of adult females (**Figure 13.75**).

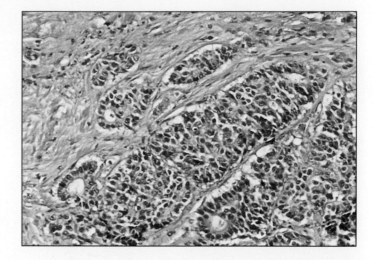

Figure 13.75 Uterine adenocarcinoma in a domestic rabbit. The myometrium is infiltrated by intersecting cords of neoplastic acini and distorted duct-like structures with hollow lumens filled with pink staining proteinaceous fluid. H&E ×44.

Avian Female Reproductive System

Ovary

In the majority of birds, only the left ovary and oviduct are functional. The ovary consists of an outer cortex that envelops a vascular medulla. The stroma is composed of very vascular loose connective tissue with sinusoidal blood vessels and follicles of varying sizes, postovulatory and atretic follicles, thecal gland cells, and interstitial cells. The main structural components of the cortex are solid, stalked ovarian follicles that never develop an antrum. Each follicle consists of a growing yolk-laden oocyte with a rounded nucleus surrounded by several layers: a simple cuboidal surface epithelium, theca externa, theca interna, and granulosa cells. At ovulation, the oocyte and granulosa cells are released, forming the theca cells a transient corpus luteum (**Figures 13.76–13.79**).

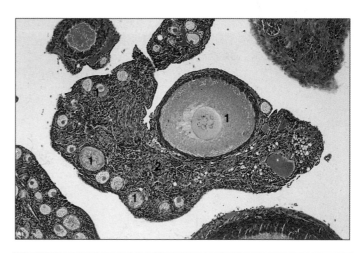

Figure 13.77 Ovary (bird). (1) The range of follicle sizes depends on the size of the ovum. (2) Vascular stroma. H&E ×50.

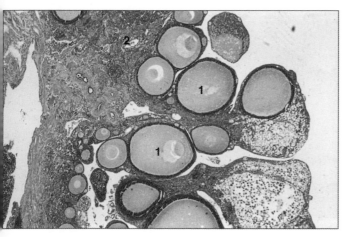

Figure 13.76 Ovary (bird). (1) The follicles consist of the megalethic ovum surrounded by a single layer of granulosa cells. (2) Vascular ovarian stroma. H&E ×25.

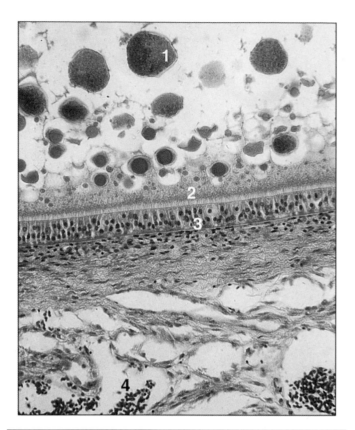

Figure 13.78 Ovary (bird). (1) Yolk granules. (2) Plasma membrane of the egg. (3) Granulosa cells. (4) Vascular ovarian stroma. H&E ×200.

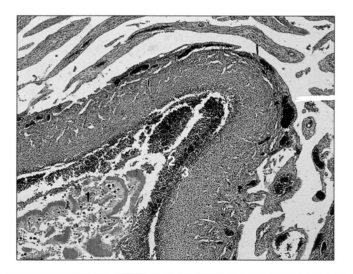

Figure 13.79 Ovary. Atretic follicle (bird). (1) Remnants of the yolk-filled ovum. (2) Disrupted granulosa layer. (3) Theca formed from the ovarian stroma. H&E ×100.

Oviduct

In the domestic fowl, the functional left oviduct consists of five regions: infundibulum, magnum, isthmus, uterus or shell gland, and vagina. The wall of the oviduct consists of a mucosa made up of columnar to the pseudostratified epithelium and a glandular lamina propria rich in lymphocytes and plasma

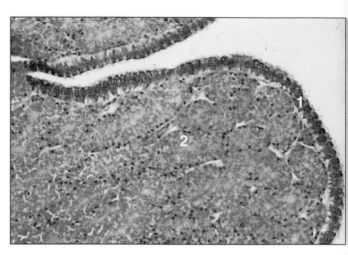

Figure 13.80 Oviduct. Magnum (bird). (1) The lining epithelium is pseudostratified columnar ciliated. (2) The lamina propria is filled with simple tubular glands lined by columnar cells packed with eosinophilic granules. H&E ×100.

cells. Longitudinal folds in the mucosa extend spirally down the length of the oviduct but vary in height and thickness. The muscularis is a smooth muscle with inner circular and outer longitudinal layers increasing gradually in thickness. Loose connective tissue covered by mesothelium forms the serosa.

The infundibulum engulfs the shed oocyte and, after fertilisation, lays down the continuous and extravitelline membranes. The albumen is produced by the next and longest part of the duct: the magnum. The mucosal glands of the magnum are lined with columnar cells packed with eosinophilic granules before the arrival of an egg and depleted after its passage (**Figures 13.80** and **13.81**). The shell membranes are formed in the next short, narrow region, the isthmus. Here, the mucosal glands are lined with poorly staining vacuolated cells (**Figure 13.82**) that do not exhibit such marked secretory phases as those seen in the magnum. The

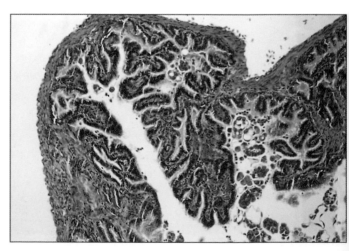

Figure 13.81 Oviduct. Magnum (bird). After the passage of the egg, the glands are empty and the deep mucosal folds project into the lumen. H&E ×100.

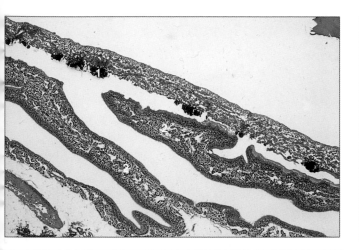

Figure 13.82 Oviduct. Shell gland/uterus (bird). (1) Shell membranes, the dark blue areas, are sites of calcification. (2) The long mucosal folds lie parallel to the developing shell. H&E ×62.5.

Figure 13.84 Eggshell with the membranes removed. (1) Gaseous pore. (2) Cuticle. (3) Multiple layers of calcite containing organic matrix. (4) Mammillary layer. Scanning electron micrograph ×1600.

mucosal folds of the isthmus are elongated and lie in leaf-shaped folds. Once received by the 'uterus' or shell gland (**Figure 13.83**), the egg remains in this region for about 20 h, during which calcification of the shell and formation of the cuticle take place (**Figures 13.84** and **13.85**). Watery fluid is also added to the albumen. The uterine mucosa forms flat, leaf-shaped, longitudinal folds. The epithelium is a pseudostratified columnar epithelium with basal and apical cells. The basal cells have a restricted apical surface; the apical cells are ciliated. The tubular coiled glands of the uterus are lined with cells that contain pale staining granules both before and during the phase of shell formation but which are subsequently depleted.

The vagina is short and narrow and has a well-developed muscularis. Short, simple tubular glands, the sperm host glands, are found near the junction of the vagina with the uterus and lie within the mucosal tissue (**Figure 13.86**).

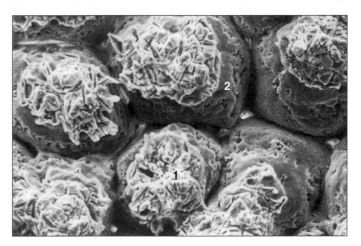

Figure 13.85 Mammillary layer of the eggshell. This is the organic component. The individual structural units are the mammillae arranged as (1) mammillary caps and (2) mammillary cones. Scanning electron micrograph ×3200.

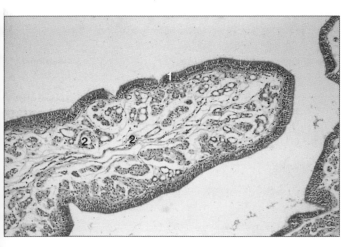

Figure 13.83 Oviduct. Shell gland/uterus (bird). (1) The lining epithelium is pseudostratified columnar with two distinct rows of nuclei. (2) The mucosal glands in the long folds appear empty. H&E ×62.5.

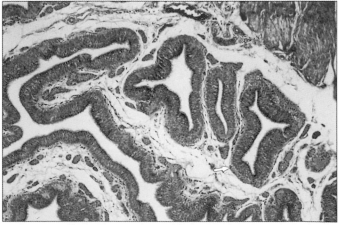

Figure 13.86 Vagina (bird). A pseudostratified columnar epithelium lines the vagina; small tubular glands (sperm-host glands) are found in the lamina propria (arrowed). H&E ×62.5.

As their name suggests, their function is to store sperm after insemination. The mucosal folds are long and slender at this point and bear short secondary folds.

The surface is lined with pseudostratified columnar epithelium and mucous cells. The vagina opens into the cloacal urodeum.

Reptilian, Amphibian, and Fish Female Reproductive System

Paired ovaries are typical in fish, amphibians, and reptiles, and the histology of oogenesis is similar to that in mammals (**Figures 13.87** and **13.88**). The ovaries are composed of germinative, stromal, vascular, and nervous tissue. The ova begin their development as oogonia that are mitotically derived from successive generations of oocytes. Diploid primary oocytes then undergo meiotic division to become primary oocytes and primary polar bodies that are discarded. Another reduction division yields haploid ova and secondary polar bodies that are also discarded.

The ovum is encircled by a cell membrane, a variably narrow zona pellucida, and a layer of follicle cells. During vitellogenesis, the yolk is added. This process occurs after a variable period of time after ovulation. Although the ova of most amphibians are uninuclear, some species are known to produce multinucleated ova. However, before fertilisation, all of the nuclei, except one, become inactivated. Reptile ova typically have a uninuclear ovum. However, binucleated ova are produced occasionally and, after being fertilised, may yield twin embryos.

After ovulation, the corpora lutea, then the corpora albicans and, finally, the corpora atretans replace the ovulated follicles.

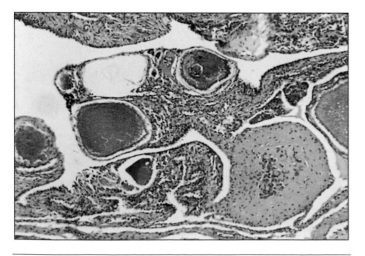

Figure 13.87 Ovary and proximal oviduct (fimbrium) of a desert tortoise (*Xerobates agassizi*). H&E ×25.

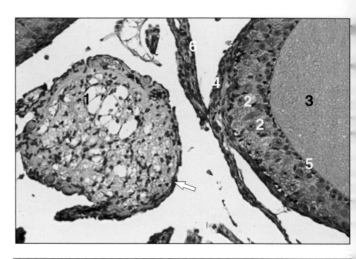

Figure 13.88 Corpus luteum (arrowed) of a leopard gecko (*Eublepharis macularius*). (1) Granulosa lutein cells. (2) Primordial follicles. (3) Oocyte cytoplasm. (4) Tunica albuginea. (5) Theca externa. (6) Stroma. H&E ×62.5.

The appearance of reptilian oviducts varies depending on the species and whether the female is egg laying (oviparous) or live bearing (oviviviparous or viviparous). However, they are readily recognisable. The histological features of the tubular oviduct change with each segment as it courses distally from the infundibulum. It is lined by ciliated, often mucus-secreting, glandular columnar epithelium for at least part of its length. The thin walls of the oviducts contain alveolar or tubuloalveolar glands that secrete albumin and shell substrate onto the yolked egg as it descends. The cuboidal to columnar epithelial cells comprising these glands tend to be characterised by their distinctive eosinophilic cytoplasmic granularity (**Figures 13.89** and **13.90**). The caudal oviduct of many reptiles also contains straight or branched crypt-like depressions that are surrounded by cuboidal epithelium with eosinophilic granular cytoplasm (**Figure 13.91**). In many species, these glandular crypts serve as spermathecae in which spermatozoa are nourished and stored for prolonged periods of time. The oviducts of viviparous and oviviviparous reptiles are thick-walled, muscular and vascular, and contain glands with secretions that help nourish the developing embryo(s). They often exhibit marked plaiting, which facilitates their distension during gravidity or pregnancy. These modified oviducts are called the 'uterus' by some authorities.

The embryos of some viviparous reptiles develop a primitive vascular placenta. In the lizard (*Xantusia vigilis*) it is disc shaped; in others, it is more diffuse. Although it was long thought that embryonic development did not occur in shelled eggs until they were deposited and exposed to atmospheric oxygen, it has been demonstrated that significant embryonic development can occur before egg deposition in some species.

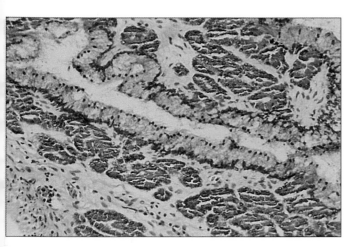

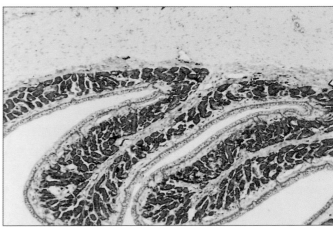

Figure 13.89 and Figure 13.90   Shell gland portion of the oviduct of a Pacific pond turtle (*Clemmys marmorata*). The luminal lining epithelium is pseudostratified columnar and contains numerous goblet cells. Subjacent to the mucosa are lobules of shell-secreting glands with characteristic eosinophilic granule-containing cells. H&E ×125.

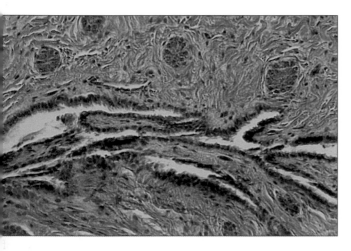

Figure 13.91   Caudal oviduct from a sexually mature female green iguana (*Iguana iguana*). This portion of the oviduct is characterised by numerous shallow crypts that connect with small lobules of glandular epithelial cells. These crypts are thought to be sites of sperm storage and nourishment that sustain spermatozoa for prolonged periods that may exceed 4–6 years in some reptiles. H&E ×62.5.

CLINICAL CORRELATES

Ovarian and oviductal (or 'uterine') inflammation and infections are less common in fish, amphibians, and reptiles than in mammals. Rupture of yolked eggs with subsequent leakage of yolk into the coelomic cavity is, however, a relatively frequent reproductive disorder in oviparous reptiles. Once lipid has gained entry into the vascular system, it is disseminated widely in the form of yolk-lipid emboli throughout the body (**Figures 13.92 and 13.93**).

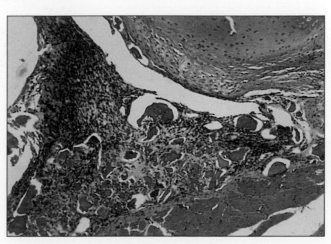

Figure 13.92   Egg-yolk serocoelomitis from an iguana. The intense exudative reaction to yolk characterised by numerous histiocytic macrophages with engulfed yolk lipid. H&E ×62.5.

Ovarian neoplasms have been observed in fish and amphibians but are less prevalent than in mammals, birds, and reptiles. As with mammals, relatively common reptilian ovarian neoplasms are granulosa cell tumours, luteal tumours, and thecal cell tumours. Dysgerminoma and teratoma (**Figure 13.94**) occur less often. The teratoma is distinguished by usually having tissue from all three germ layers present: hair (or scales), bone, cartilage, teeth, muscle (of all three types), glandular tissue, nerve and brain-like structures, and so on.

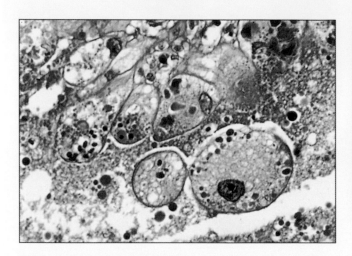

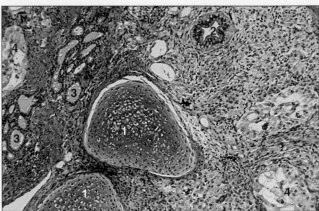

Figure 13.93 Yolk serocoelomitis in a green iguana (*Iguana iguana*). The smooth muscle tunic of the viscus organ is covered with yolk, and an inflammatory exudate is comprised of macrophages with engulfed lipid. H&E ×125.

Figure 13.94 Ovarian teratoma from a green iguana (*Iguana iguana*). This section illustrates (1) two masses of cartilage, (2) a duct-like structure, (3) an aggregate of thyroid-like follicles filled with pink staining protein resembling colloid, and (4) nervous tissue. H&E ×62.5.

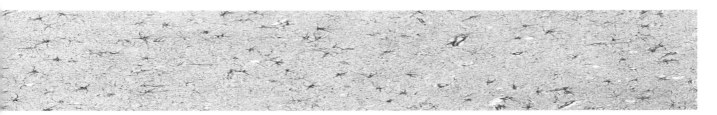

14

NERVOUS SYSTEM

The nervous system is subdivided into the central nervous system (CNS), which comprises the brain and spinal cord, and the peripheral nervous system (PNS), which covers all other nervous tissue and acts to interconnect all the tissues of the body with the CNS, including cranial and spinal nerves and associated nerve roots and ganglia. The nervous system is derived from a specialised region of surface ectoderm along the dorsal midline of the flat embryonic disc. The ectodermal cells thicken to become neural ectoderm. During flexion, the neural folds form and the lips fuse to become the neural tube. This is incorporated into the developing embryo to form the brain and spinal cord. A separate population of neural ectodermal cells remains separate from the neural tube. This is the neural crest and forms the PNS.

The nervous tissue parenchyma consists of neurons, the basic cellular unit of the nervous system, and supportive cells call neuroglia. Neurons are generally incapable of mitosis, although there is some evidence of this possibility, e.g., in some olfactory neurons and neuronal stem cells. The neuron comprises a large cell body, also known as perikaryon or soma, with a single nucleus and a prominent nucleolus. Basophilic granules (Nissl's granules) are present in the cytoplasm and represent the rough endoplasmic reticulum (**Figure 14.1**). The function of the neuron is to receive and transmit impulses. Neurons can show different sizes and shapes depending on their function. Elongated processes extend from the cell body: the dendrites and axons. The dendrites are branching processes at the receiving end, and the axon is the single, long process for onward transmission of the impulse. Information passed along a chain of neurons is transmitted from axon to dendrite at a specialised junction: the synapse. A synapse comprises a presynaptic element, a synaptic cleft, and a postsynaptic element. Depending on the type of synapse, the presynaptic element can be constituted by a terminal bulb of the axon (terminal bouton) or by swellings along the axon

called preterminal bulbs (boutons en passage), where each bouton could be a synaptic site. Those elements contain synaptic vesicles with neurotransmitters, mitochondria, and vesicles of the smooth endoplasmic reticulum. The synaptic cleft is a narrow space (about 20 nm) between the presynaptic and postsynaptic elements. The postsynaptic element is constituted by a thickened postsynaptic membrane containing neurotransmitter receptors.

Neurons can be classified as unipolar, bipolar, or multipolar, depending on the number of neuronal processes displayed in the cell body. Multipolar neurons are the most common type. The single axon arises from one pole at a granule-free, pale-stained area of the neuron body: the axon hillock. Dendrites arise from the opposite pole and branch extensively. Examples of these neurons are found in the CNS, in the cerebral and cerebellar cortex, and in

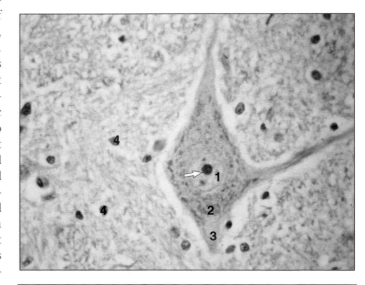

Figure 14.1 Motor neuron. Trigeminal nerve (dog). (1) Nucleus and nucleolus (arrow). (2) Basophilic granules (Nissl's granules). (3) Axon hillock. (4) Neuroglial cells. H&E ×200.

DOI: 10.1201/9781003333807-14

Figure 14.2 Multipolar neuron. Cerebellar cortex (dog). (1) Cell body. (2) Branching processes. Golgi's silver ×125.

the spinal cord (**Figure 14.2**). Bipolar neurons have one axon and one main dendrite, with little branching, and can be found in the retina and the olfactory region. Unipolar neurons show a single axon that bifurcates into central and peripheral branches and can be found in the sensory ganglia located in the roots of cranial and spinal nerves. In mammals, unipolar neurons are often called pseudounipolar as they originate as bipolar and become unipolar during development.

The axon is a cylindrical process, relatively long, that originates from the axon hillock, although, in some neurons, it can emerge as a branch from a dendrite. Axons are covered by a myelin sheath, the neurilemma, forming together the nerve fibre. The neurilemmal sheath is composed of individual cells that form a continuous investment from the origin of the axon almost to the peripheral termination (**Figure 14.3**). These cells produce a lamellar system of membranes with a high lipidic content. In the

PNS, the neurilemmal sheath is produced by neurolemmocytes or Schwann cells, and in the CNS by neuroglial cells: oligodendrocytes. The degree of the investment varies; heavily invested fibres are known as myelinated and minimally invested fibres as nonmyelinated. The axon has no sheath at the peripheral nerve endings; these are naked fibres. The individual sheaths divide the nerve fibre into segments. Where adjoining cells meet, a node (of Ranvier) is formed so that each internodal segment represents an investing neurilemmal cell. The high lipidic content of the sheath means that lipid stains are best used to demonstrate the myelinated fibres (**Figure 14.3**).

Peripheral Nervous System

The PNS is derived from the neural crest and is composed of nerves and ganglia. Nerves are composed of nerve fibres, and ganglia are localised groups of nerve cells. The individual fibres are gathered into bundles or fascicles by a connective tissue sheath, the epineurium, and form an anatomical nerve. The epineurium extends into the nerve and subdivides the fascicles with a vascular connective tissue investment: the perineurium. This, in turn, surrounds each nerve fibre with fine vascular connective tissue: the endoneurium (**Figures 14.4–14.7**). In the dorsal root ganglion, the neurons are peripherally arranged, with the nerve fibres in the centre of the ganglion (**Figure 14.8**). The large neuron bodies are surrounded and supported by satellite cells (**Figure 14.9**). In the sympathetic ganglion, the neurons are evenly distributed, the cell bodies are smaller, and the nucleus is eccentrically placed in the cell (**Figure 14.10**). Parasympathetic ganglia are small groups of neurons within the terminal tissue (**Figure 14.11**).

Figure 14.3 Longitudinal section (LS) of a peripheral nerve (cat). (1) Axon. (2) Neurilemmal sheath. (3) Node. Osmic acid ×200.

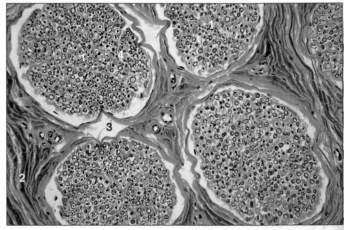

Figure 14.4 Transverse section (TS) of a peripheral nerve (dog). (1) Bundles of axons cut in transverse section, fascicles. (2) Epineurium (3) Perineurium. (4) Endoneurium. Masson's trichrome ×25.

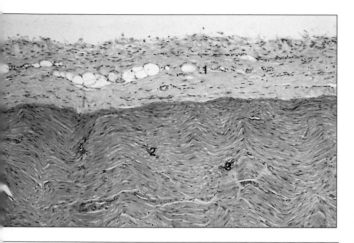

Figure 14.5 LS peripheral nerve (dog). (1) Epineurium with fat cells. (2) Nuclei of the neurilemmal cells. (3) Axons cut longitudinally; note the wavy appearance. H&E ×62.5.

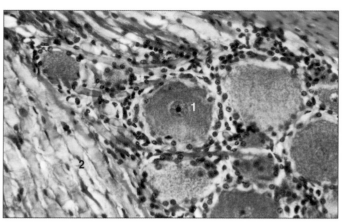

Figure 14.8 Dorsal root ganglion (cat). (1) Neuron bodies. (2) Nerve fibres and supporting neuroglial cells. Masson's trichrome ×160.

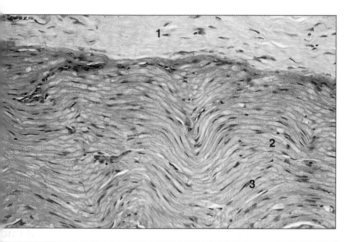

Figure 14.6 LS peripheral nerve (dog). (1) Epineurium. (2) Axons. (3) Neurilemmal cell nuclei. H&E ×125.

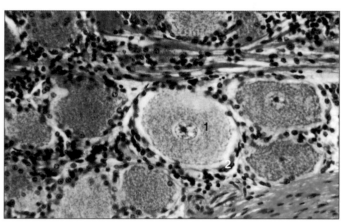

Figure 14.9 Dorsal root ganglion (cat). (1) Neuron body. (2) Satellite cells; these are supporting neuroglial cells. Masson's trichrome ×160.

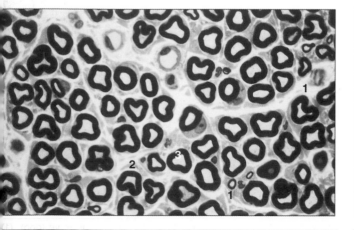

Figure 14.7 TS laryngeal nerve (horse). (1) Perineurium with blood vessels. (2) Endoneurium. (3) Myelinated axons. Toluidine blue ×62.5.

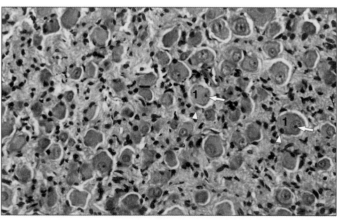

Figure 14.10 Sympathetic ganglion (dog). (1) Neuron body with eccentric nucleus (arrow) and satellite cells (arrowhead). H&E ×125.

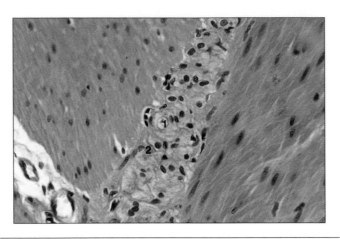

Figure 14.11 Parasympathetic ganglion. Stomach (pig). (1) Neuron body. (2) Nerve fibres and supporting neuroglial cells. H&E ×400.

Central Nervous System

The CNS consists of the brain and spinal cord, together with the optic nerve and the retina which originate as an extension of the brain in the embryo. The brain is divided into the brainstem, cerebellum, and cerebrum. The brain and spinal cord and the roots of peripheral nerves are enveloped in connective tissue membranes, the meninges. The outer dense membrane is the dura mater, also called pachymeninx. The spinal dura mater is surrounded by an epidural space. The arachnoid (middle) and pia mater (inner) are also called leptomeninges (**Figure 14.12**). The subarachnoid space separates the arachnoid from the pia mater, and it is filled with cerebrospinal fluid (CSF). The arachnoid is a fine cobweb-like membrane composed of outer layers of fibrocytes and inner loosely arranged fibrocytes and bundles of collagen fibres. Arachnoid trabeculae are thin strands that establish continuity with the pia mater. Arachnoid villi are small projections that penetrate into the dura mater acting as one-way valves for drainage of CSF. The pia matter is a highly vascularised layer covering the spinal cord and brain extending into the fissures and sulci and is separated from the ependyma only by the basal lamina.

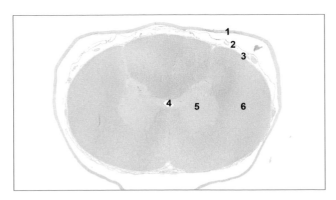

Figure 14.12 TS of the spinal cord (dog). (1) Dura mater. (2) Arachnoid. (3) Pia mater. (4) Spinal canal lined by ependyma. (5) Grey matter with dorsal and ventral horns. (6) White matter. H&E ×15.

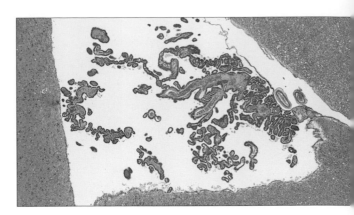

Figure 14.13 Choroid plexus in a lateral ventricle (dog). H&E ×50.

The central fluid-filled canal and the ventricles of the brain (lateral ventricles, third and fourth ventricles, and the aqueduct) are lined with a columnar epithelium: the ependyma. The choroid plexus is a secretory tissue found in each of the brain ventricles and has the main function of producing CSF. Choroid plexus consists of simple cuboidal epithelial cells surrounding a core of fenestrated capillaries and connective tissue (**Figures 14.13** and **14.14**).

In the CNS, collections of functionally related neurons are called nuclei and collections of nerve fibres are called tracts. The fibres have an investing sheath cell: the oligodendrocyte. The presence of lipid gives a white glistening appearance, and the fibres are called white matter. By contrast, the cell bodies appear grey and form grey matter (**Figure 14.15**).

Neuroglial cells are the supporting cells of the CNS, taking the place of the connective tissue of other systems. Oligodendrocytes surround the axon, provide the myelin sheath, and act as supporting satellite cells. Astrocytes are stellate cells with long processes. In fibrous astrocytes, the processes are thin with minimal cytoplasm, whereas in protoplasmic astrocytes, they are thicker (**Figures 14.16–14.19**). The cell processes extend to the blood vessels in the pia mater, acting as anchors and transferring nutrients. The blood

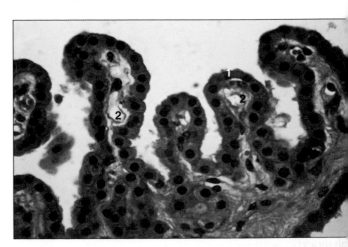

Figure 14.14 Choroid plexus in the fourth ventricle (cat). (1) Ependyma. (2) Capillaries. H&E ×400.

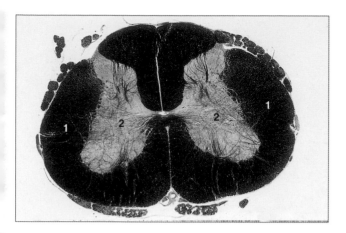

Figure 14.15 TS of the spinal cord (cat). (1) Black stained areas are the lipid-rich white matter. (2) Pink stained areas are the grey matter. Osmic acid ×7.5.

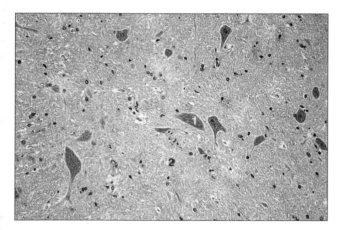

Figure 14.16 Spinal cord (dog). (1) Neuron. (2) Nuclei of protoplasmic astrocytes. H&E ×125.

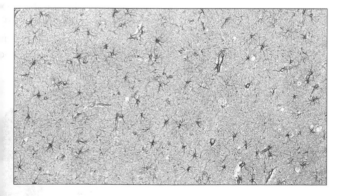

Figure 14.17 Cerebral cortex (dog). Immunohistochemical staining of astrocytes (brown colour). Immunohistochemistry (IHC) (GFAP-Glial fibrillary acid protein) ×75.

vessels are lined by a continuous endothelium. There is a well-developed basal lamina that, with the protoplasmic astrocyte processes, forms the blood–brain barrier. Microglia are the phagocytic, scavenging cells of the CNS (**Figure 14.20**).

As a general rule, the spinal cord has internal grey matter (richer in neuronal cell bodies) and external white

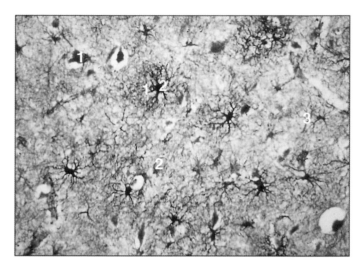

Figure 14.18 Spinal cord (dog). (1) Neuron. (2) Nuclei of protoplasmic astrocytes. (3) Fibrous astrocytes. H&E ×125.

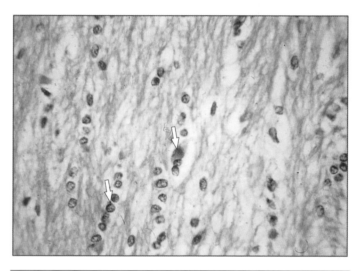

Figure 14.19 Corpus callosum (dog). Oligodendrocyte nuclei in orderly columns (arrowed) providing the neurilemmal sheath in the central nervous system. H&E ×400.

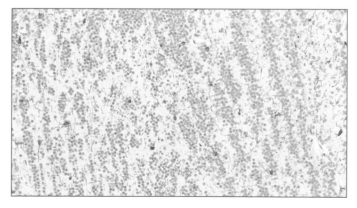

Figure 14.20 Microglial cells in the olfactory bulb (forebrain – central nervous system, mouse). IHC staining using Iba1 antibody, a cell marker of microglia ×100.

matter (richer in neuronal processes), while the brain has external grey matter and internal white matter. In most species, the brain can be divided in three distinct areas:

1. The forebrain, which consists of the cerebrum, the thalamus, and the hypothalamus. The cerebrum is the major part of the forebrain, and the two (left and right) hemispheres are separated by the corpus callosum. The outer layer (cortex) can be divided into the neocortex and the small allocortex, the latter containing the olfactory system and the hippocampus.
2. The midbrain, short in length acting as a pathway for fibres running from the hindbrain and forebrain.
3. The hindbrain, which consists in the cerebellum, the pons, and the medulla oblongata, the latter connecting with the spinal cord.

The cerebral and cerebellar cortices are highly specialised folded areas of outer grey matter with layers of neurons of various sizes and supporting cells, as well as a central core of white matter (**Figures 14.21–14.29**). The cerebellar surface shows a number of folia, separated by ridges. The outer layer is composed of grey matter (cerebellar cortex) divided into three layers: the superficial molecular layer, composed predominantly of neuropil (the space between neuronal and glial cell bodies that is comprised of dendrites, axons, synapses, glial cell processes, and microvasculature), the piriform cell (or Purkinje cell) layer, a thin layer of large neurons that send axons into the molecular layer; and the granular cell layer, composed of mainly granule cells (small neurons with heterochromatic nuclei) adjacent to the white matter.

In many mammals, the cerebral surface has well-developed gyri (ridges) demarcated by sulci (grooves). The outer layer, as in the cerebellum, is composed of grey matter (cerebral cortex). In mammals, most of the cerebral cortex is

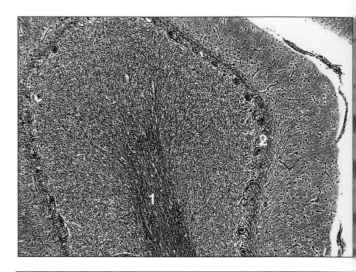

Figure 14.22 Cerebellum (cat). Each folium has (1) a central core of white matter, the nerve fibres, and supporting cells and (2) a surface cellular layer of grey matter. Cajal's silver ×62.5.

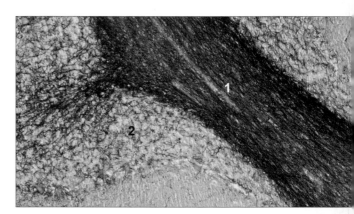

Figure 14.23 Cerebellum (dog). White matter with abundant nerve fibres (1) and grey matter with abundant neuronal cell bodies and supporting cells (2). Silver staining ×200.

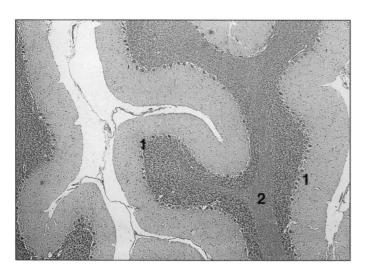

Figure 14.21 Cerebellum (cat). (1) Outer cellular layer of grey matter and (2) inner fibrous layer of white matter. H&E ×25.

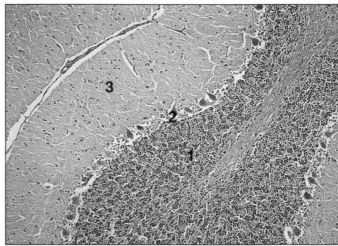

Figure 14.24 Cerebellum (cat). (1) Inner granular layer of small neurons. (2) Middle Purkinje layer of large neurons. (3) Outer molecular layer with few neurons and many branching processes. H&E ×100.

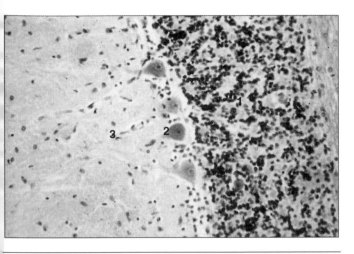

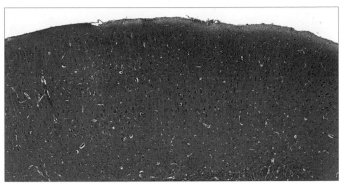

Figure 14.28 Cerebral cortex (mouse). The neocortex layers are difficult to identify, but different sizes and shapes of neurons can be observed in different areas. Note that the mouse does not show gyri and sulci in the cerebral cortex. H&E ×150.

Figure 14.25 Cerebellum (sheep). (1) Granular layer. (2) Purkinje cell layer. (3) Molecular layer. H&E ×250.

called neocortex and can be divided into six layers of cells that vary in thickness depending on the region and function. These layers are (from superficial to deep):

  i. Molecular layer, containing few scattered neurons.
 ii. External granular layer, containing small pyramidal neurons and abundant stellate neurons.
 iii. External pyramidal layer, containing medium-sized pyramidal neurons.
 iv. Internal granular layer, containing different types of stellate and pyramidal cells.
  v. Internal pyramidal layer, containing large pyramidal neurons.
 vi. Fusiform (or multiform layer), containing many spindle-shaped and multiform neurons.

The cerebral cortex is adjacent to the cerebral white matter containing abundant nerve fibres.

Nerve endings and receptors are specialised terminal parts of axons or dendrites. They include motor end plates, pressure receptors (such as Pacinian corpuscles), and taste buds (**Figures 14.30–14.32**).

Nerve tissue is so characteristic that it is readily identified, even in invertebrates.

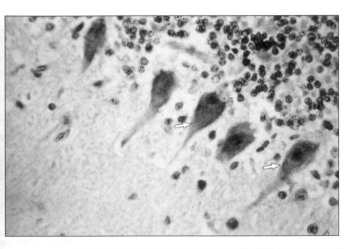

Figure 14.26 Cerebellum (dog). Large Purkinje cells are arrowed. H&E ×400.

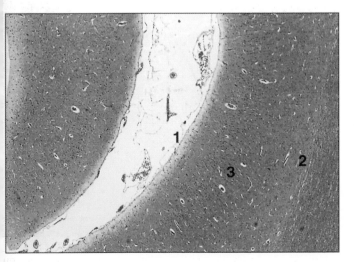

Figure 14.27 Cerebral cortex (cat). (1) Meninges. (2) White matter. (3) Grey matter. H&E ×100.

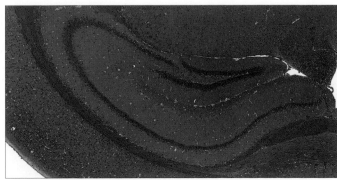

Figure 14.29 Hippocampus (mouse). Allocortex area within the cerebral cortex. H&E ×100.

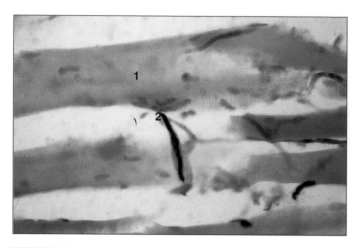

Figure 14.30 Motor end plate in striated muscle (dog). (1) Striated muscle fibre. (2) Terminal nerve fibre. Cajal's silver ×400.

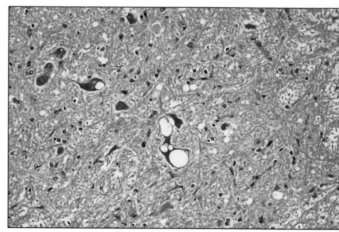

Figure 14.33 Scrapie (sheep). This is a section of the midbrain from a sheep with scrapie. The characteristic large intraneuronal vacuoles and diffuse vacuolar change in the neuropil can be seen. No stainable material is present in the vacuoles. H&E ×125.

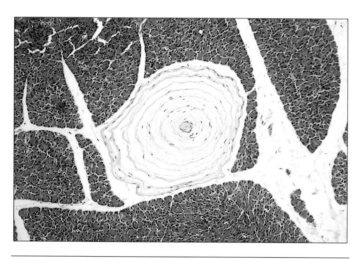

Figure 14.31 Lamellar (Pacinian) corpuscle in a serous gland (dog). H&E ×20.

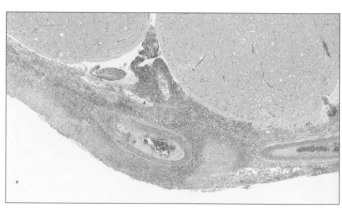

Figure 14.34 Severe meningitis observed in a cat with nervous clinical signs due to a bacterial infection. Accumulation of abundant inflammatory cells within the meninges and to a lesser extent in the brain. H&E ×50.

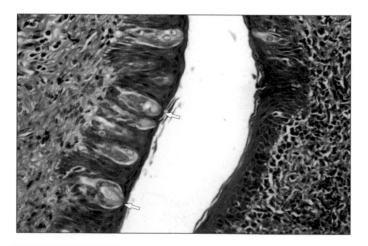

Figure 14.32 Circumvallate papilla. Tongue (cow). Taste buds are arrowed. Masson's trichrome ×250.

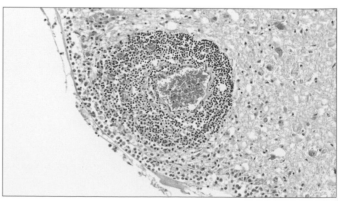

Figure 14.35 Perivascular cuffing or presence of inflammatory cells around blood vessels in the brain of a cat. H&E ×200.

CLINICAL CORRELATES

The CNS is affected by a variety of congenital disorders that may occur with relatively high frequency, possibly because of the susceptibility of this complex system to teratogenic insult. Inherited CNS disease is also recognised, and a heritable basis is suspected for conditions such as idiopathic epilepsy.

Inflammatory, neoplastic, metabolic, and parasitic disorders are all recognised in the CNS, with certain specific entities appearing in particular species. For example, feline parvovirus, which is trophic for dividing cells, attacks the external germinal layer of the cerebellum in kittens when infection occurs before or shortly after birth. Primary CNS neoplasms are not uncommon in the dog and cat, and tumours that arise from almost every type of cell within the nervous system have been recognised. The functional distinction between benign and malignant neoplasms within the CNS is not as significant as in other systems, as any space-occupying lesion, even the most biologically benign, has serious consequences.

Transmissible spongiform encephalopathies (TSE) are a group of diseases that may be induced by a very particular infectious agent: prions, not containing any type of genetic material. The prions can enter via the oral route and proliferate in lymphoid tissues and in the gastrointestinal tract, but no effects are observed until it reaches the target areas of the CNS. The sheep TSE, scrapie (**Figure 14.33**), has been known for a very long time, and other TSEs have been described in cattle (BSE), deer (CWD), etc. Scrapie produces no significant gross pathology within the CNS, but microscopically, spongiosis (presence of large vacuolae in the neuronal soma or neuropil) can be observed in specific neuron nuclei.

Inflammation of the meninges (meningitis) or the brain (encephalitis) or both simultaneously (meningoencephalitis) are frequent processes in animals induced by infection by bacteria, viruses, parasites, or fungi. The usual clinical signs include fever, neck pain and rigidity, depression and may lead to partial paralysis, agitation, and loss of consciousness. The inflammation is presented microscopically as an influx of inflammatory cells in the area of brain and meninges affected, forming and many cases perivascular accumulations (cuffing) of inflammatory cells around blood vessels (**Figures 14.34** and **14.35**).

Viruses can also induce neuronal loss and death and associated inflammatory response in the brain. Pestiviruses may induce a syndrome called "congenital tremor" in neonatal piglets. It is a disease in newborn piglets showing sporadic body contractions, lack of coordination, and muscle tremors when walking. Neuronal degeneration is observed in several areas of the brain (**Figure 14.36**). The brain can also be affected by storage disease. For example, Lafora disease, which has a genetic component in dogs and humans, is produced by excessive storage of polyglucosan in neurons (Lafora bodies) (**Figure 14.37**). The clinical course is characterised by a progressive myoclonus epilepsy (recurrent seizures).

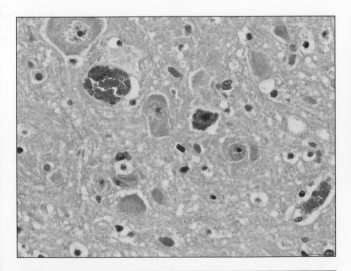

Figure 14.36 Neuronal degeneration in a piglet with congenital tremor. H&E ×400.

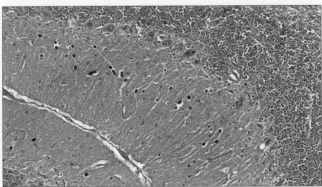

Figure 14.37 Lafora bodies (inclusions of dark stained polyglucosan stored in neurons) in the cerebellum cortex (molecular layer) in an affected dog. H&E ×200.

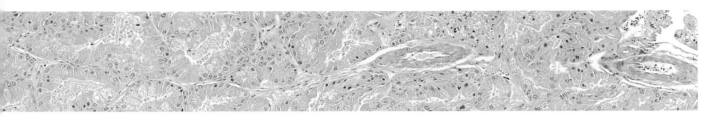

15

SPECIAL SENSES

Eye

The eye is the organ of vision and comprises the eyeball or globe of the eye and the optic nerve. It is protected by the eyelids and lacrimal apparatus.

Eyelids

The eyelids are movable folds of skin in front of the eyeball protecting and lubricating the surface of the eye. The outer surface is covered with stratified squamous epithelium, tactile hairs, sebaceous glands (glands of Zeiss), and sweat glands (glands of Möll) (**Figure 15.1**). The inner surface is lined with the palpebral conjunctiva, a thin transparent mucous membrane. The bulbar conjunctiva is continuous with the surface of the cornea at the limbus. The epithelium

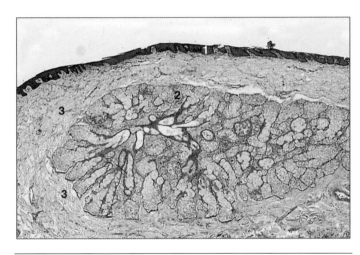

Figure 15.2 Eyelid (horse). (1) The conjunctiva, stratified columnar epithelium with mucus-secreting cells. (2) Sebaceous glands in the lamina propria. (3) Tarsal plate. H&E ×25.

may be stratified cuboidal to columnar, with goblet cells. Between the dermis of the skin and the lamina propria of the conjunctiva is a plate of dense connective tissue with elastic fibres: the tarsus or tarsal plate, which is surrounded by the tarsal (Meibomian) glands (**Figure 15.2**), large multilobar sebaceous glands. The nictitating membrane (third eyelid) is situated at the medial angle of the eye. It is a semicircular fold of conjunctiva enclosing a plate of cartilage (hyaline in ruminants and dogs; elastic in the horse, pig, and cat; **Figures 15.3–15.7**).

Lacrimal Apparatus

The lacrimal glands are compound tubuloalveolar or tubuloacinar: serous in the cat, seromucous in the horse, ruminant, dog, and pig (in the horse and ruminant, they

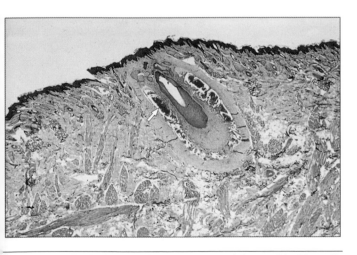

Figure 15.1 Eyelid (horse). Outer skin layer of the eyelid with tactile hair (arrowed). H&E ×7.5.

DOI: 10.1201/9781003333807-15

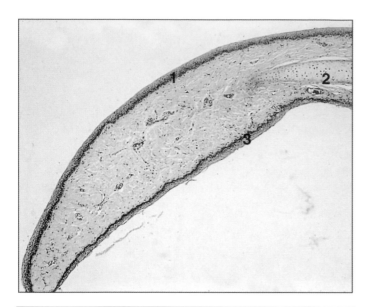

Figure 15.3 Nictitating membrane (membrana nictitans; horse). (1) Anterior surface is covered by conjunctiva. (2) Elastic cartilage plate in a central core of connective tissue. (3) Posterior plate covered by conjunctiva. H&E ×7.5.

are predominantly serous, and in the pig, predominantly mucous). Lymph nodules are seen in the lamina propria (**Figure 15.7**). The pig and ox also have a deep gland (Harder gland) of the nictitating membrane. Lacrimal fluid accumulates in the lacrimal sac and drains through the nasolacrimal duct. Both structures are lined with conjunctival epithelium.

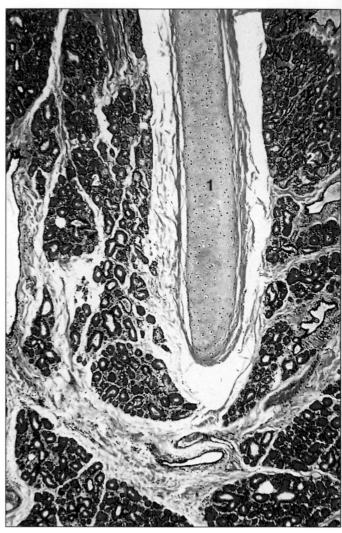

Figure 15.5 Nictitating membrane (dog). (1) Hyaline cartilage. (2) Seromucus-secreting glands in the lamina propria. H&E ×50.

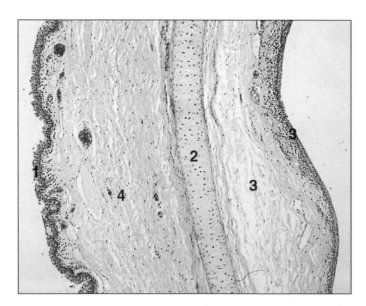

Figure 15.4 Nictitating membrane (horse). (1) Anterior conjunctival surface. (2) Elastic cartilage plate. (3) Posterior conjunctival surface. (4) Lamina propria. H&E ×50.

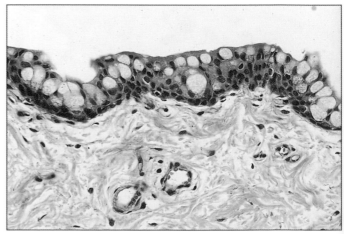

Figure 15.6 Nictitating membrane (pig). The mucus-secreting cells in the outer epithelium are stained pale pink. H/ periodic acid-Schiff (PAS) ×200.

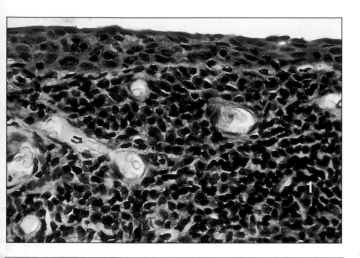

Figure 15.7 Nictitating membrane, posterior surface (dog). (1) Lymphoid cells in the lamina propria. H&E ×400.

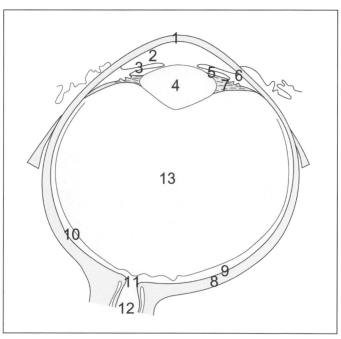

Figure 15.9 Developing eye in a 35-day cat embryo. (1) The lens fibres almost fill the lens vesicle. (2) Loose vascular mesenchyme. (3) Retina. H&E ×25.

Development of the Eye

The eyes are recognisable in the early embryo as lateral diverticulae of the diencephalon. As the diverticulum approaches the surface ectoderm, it invaginates to form the optic cup. The inner layer becomes the light-sensitive retina; the outer layer becomes the retinal pigment, pigmented ciliary, and anterior iris epithelium. The surface ectoderm overlying the optic cup thickens and invaginates to form the lens. The reconstituted ectoderm and the local mesenchyme form the cornea; the surrounding mesoderm provides the connective tissue, blood vessels, and ocular muscles (**Figures 15.8** and **15.9**).

Structure of the Eyeball

The eyeball (**Figure 15.10**) is spheroid and composed of a lens and a wall that is divided into three layers: an outer fibrous tunic (includes the sclera and cornea); a middle

vascular tunic (includes the choroid, ciliary body, and iris); and an inner neuroepithelial tunic, which consists of a 10-layered photosensitive retina and a bilayered nonphotosensitive portion that covers the ciliary body and posterior surface of the iris.

The eye contains two fluid-filled compartments: the anterior and posterior compartments. The anterior compartment is located between the cornea and the vitreous body and is divided into two chambers (anterior and posterior)

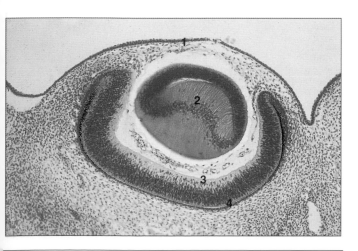

Figure 15.8 Developing eye in a 35-day cat embryo. (1) Surface ectoderm. (2) Developing lens. The optic cup has a thick inner layer (3) and a thin outer layer (4) with the pigment cells. H&E ×7.5.

Figure 15.10 Diagram of the eye. (1) Cornea. (2) Anterior chamber. (3) Iris. (4) Lens. (5) Posterior chamber. (6) Ciliary body. (7) Ciliary processes. (8) Sclera. (9) Choroid. (10) Retina. (11) Optic papilla. (12) Optic nerve. (13) Vitreous humour.

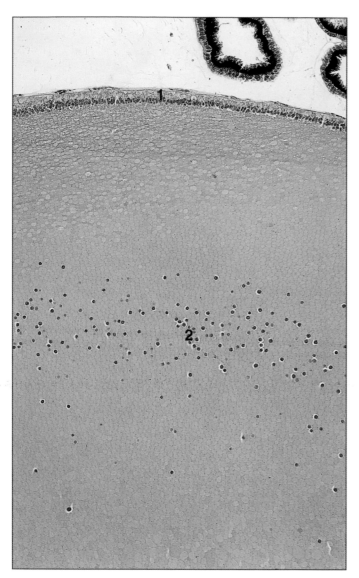

Figure 15.11 Lens (dog). (1) Anterior epithelium. (2) Nuclei of the lens fibres. H&E ×62.5.

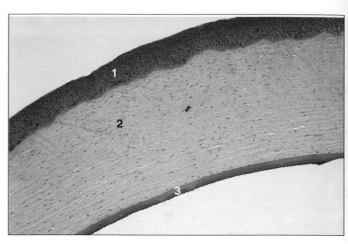

Figure 15.12 Cornea (horse). (1) Stratified squamous epithelium of the anterior surface. (2) Corneal stroma. (3) Simple cuboidal endothelium of the posterior surface resting on a thick basement membrane (Descemet's). H&E ×62.5.

sclera. The avascular cornea is transparent, covered with nonkeratinised stratified squamous epithelium (anterior epithelium) resting on a basement membrane (Bowman's). The underlying stroma, the corneal stroma or substantia propria, consists of thin lamellae of collagenous fibres and flattened fibrocytes running parallel to the corneal surface in a mucoid ground substance rich in sulphated glycosaminoglycans (**Figures 15.12-15.14**). The caudal limiting membrane (Descemet's) separates the connective tissue of the substantia propria from the simple squamous corneal endothelium (also called posterior epithelium; **Figure 15.12**).

The opaque sclera consists of interlacing bundles of collagen fibres with a few elastic fibres. Melanocytes may also be present. The area of the sclera adjacent to the choroid is a layer of delicately pigmented connective tissue known as the lamina fusca (**Figure 15.15**). The optic nerve passes through the sclera at the lamina cribrosa (**Figure 15.16**).

filled with aqueous humour. The anterior chamber is bordered by the cornea, iris, and lens; the posterior chamber is located between the iris, lens, zonular fibres, and ciliary processes. The posterior compartment is located between the lens and the retina and is filled with the vitreous body (**Figure 15.10**). The lens is a biconvex transparent body composed of epithelial cells within a homogeneous outer capsule. The anterior epithelial cells are cuboidal at the pole and become elongated, prismatic, and arranged in meridional rows at the equator where they form lens fibres, which lack of nucleus and organelles (**Figure 15.11**). The aqueous humour is elaborated by the ciliary epithelium, capillaries of the ciliary processes and fibrocytes, circulates continuously and is drained via the iridocorneal angle.

The fibrous tunic is divided into a transparent anterior segment, the cornea, and an opaque posterior segment, the

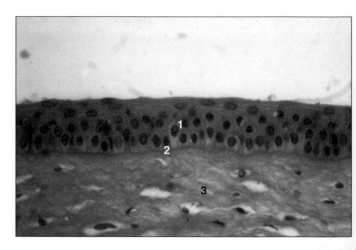

Figure 15.13 Cornea. Anterior surface (dog). (1) Stratified squamous epithelium of the anterior surface. (2) Basement membrane (Bowman's). (3) Corneal stroma. H&E ×400.

Figure 15.14 Cornea. Anterior surface (dog). (1) Corneal stroma. (2) Bowman's membrane. (3) Surface epithelium. H&E ×62.5.

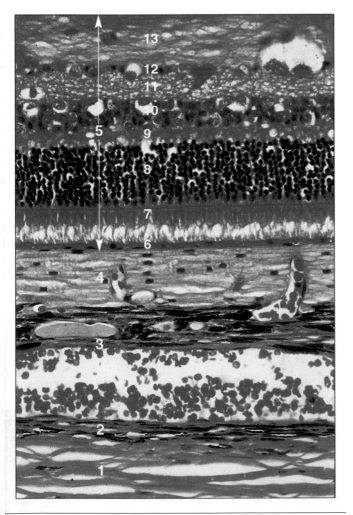

Figure 15.15 Eye. Tapetum fundus (dog). (1) Sclera composed of dense white fibrous tissue. (2) Lamina fusca. (3) Choroid with capillaries. (4) Tapetum cellulosum. (5) The photosensitive retinal layers are (6) pigment layer, (7) rods and cones, (8) outer nuclear layer, (9) outer plexiform layer, (10) inner nuclear layer, (11) inner plexiform, (12) ganglion cells, and (13) optic nerve fibre layer. H&E ×160.

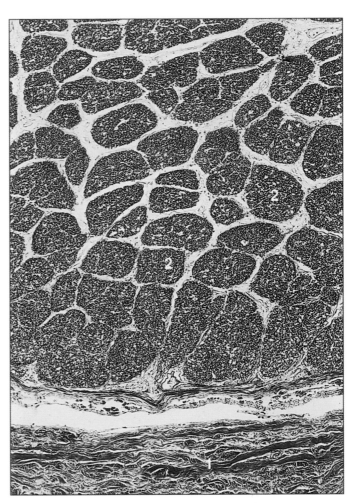

Figure 15.16 Optic disc (horse). (1) Scleral connective tissue. (2) Optic nerve bundles. H&E ×25.

The choroid is highly vascular and has numerous melanocytes. The ciliary body is an anterior continuation of the choroid that extends from the *ora serrata* to the base of the iris. The inner surface is a continuation of a non-light-sensitive retina: the pars ciliaris retinae (a two-layer epithelium, an outer simple cuboidal pigmented layer, and an inner simple cuboidal to columnar nonpigmented layer). The loose connective tissue of the stroma contains the ciliary muscle, bundles of smooth muscle fibres. An inner nonpigmented choriocapillary layer contains a capillary network that supplies the retina (**Figures 15.17–15.20**). The choroid of some species has an iridescent reflecting tissue layer: the *tapetum lucidum* (dog, **Figure 15.15**; cat, **Figure 15.19**). This gives their eyes the property of shining in the dark and allows incident light two opportunities to stimulate the retinal receptors. It is located between the choriocapillary and vascular layers of the choroid in the dorsal portion of the eye. In ungulates and the horse, a fibrous area of the choroid composed by parallel collagen fibres with few

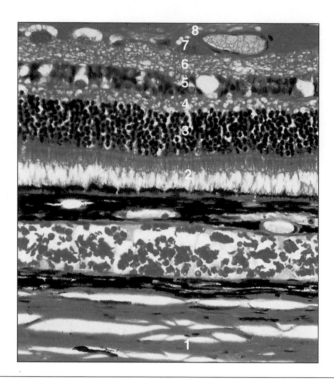

Figure 15.17 Eye. Nontapetum fundus (dog). The retinal layers are (1) sclera, underneath the pigment layer, (2) rods and cones, (3) outer nuclear layer, (4) outer plexiform layer, (5) inner nuclear layer, (6) inner plexiform layer, (7) ganglion cells, and (8) optic nerve fibre layer. H&E ×160.

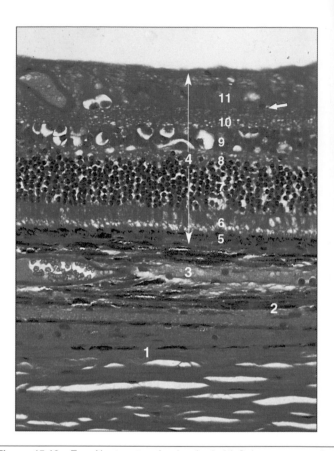

Figure 15.18 Eye. Nontapetum fundus (cat). (1) Sclera composed of dense white fibrous tissue. (2) Lamina fusca. (3) Choroid with capillaries. (4) The photosensitive retinal layers are (5) pigment layer, (6) rods and cones, (7) outer nuclear layer, (8) outer plexiform layer, (9) inner nuclear layer, (10) inner plexiform layer, and (11) optic nerve fibre layer. Ganglion cells are arrowed. H&E ×160.

Figure 15.19 Eye. Tapetum fundus (cat). (1) Sclera. (2) Choriocapillaris. (3) Nonpigmented tapetum cellulosum. (4) The photosensitive retinal layers are (5) pigment layer, (6) rods and cones, (7) nuclei of the rods and cones, (8) outer plexiform layer, (9) inner nuclear layer, (10) inner plexiform layer, and (11) optic nerve fibre layer. H&E ×160.

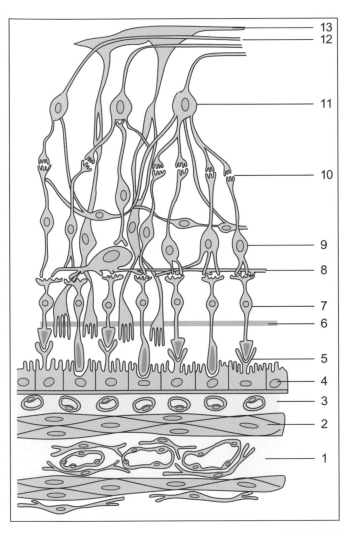

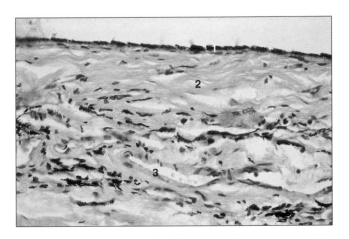

Figure 15.21 Eye. Tapetum fibrosum (horse). (1) Pigment layer of the retina. (2) Compact layer of fibrous connective tissue, the tapetum fibrosum. (3) Choroid with blood vessels and some pigment cells. H&E ×100.

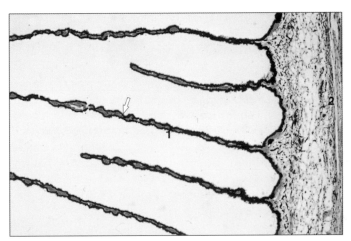

Figure 15.20 Eye. Layers of the retina. (1) Vessel layer of the choroid. (2) Connective tissue/tapetum lucidum. (3) Choriocapillaris. (4) Pigmented epithelium. (5) Photoreceptors, layers of rods and cones. (6) External limiting membrane. (7) Outer nuclear layer. (8) Outer plexiform layer. (9) Inner nuclear layer. (10) Inner plexiform layer. (11) Ganglion cell layer. (12) Optic nerve fibre layer. (13) Internal limiting membrane.

Figure 15.22 Ciliary processes (dog). (1) The long ciliary processes extend from the ciliary body, a localised expansion of the vascular coat. The epithelium is arrowed. (2) Ciliary muscle. H&E ×62.5.

fibrocytes forms the *tapetum fibrosum* (**Figure 15.21**). In carnivores, several layers of flat polygonal cells form the *tapetum cellulosum* (**Figure 15.19**).

The ciliary processes form a circle of radial folds surrounding the lens. This epithelial basal layer consists of pigmented cuboidal cells and the surface layer of nonpigmented cuboidal or columnar cells (**Figure 15.22**).

The iris, the most anterior part of the vascular tunic, is a muscular diaphragm rostral to the lens and pierced centrally by an opening: the pupil. The base of the iris is attached to the ciliary body. The connective tissue stroma supports many blood vessels and contains pigment cells and is lined on its anterior surface by a discontinuous flattened mesothelium. The epithelium of the posterior surface is continuous with ciliary epithelium: the *pars iridica retinae*. Both layers of epithelial cells are pigmented (**Figures 15.23** and **15.24**).

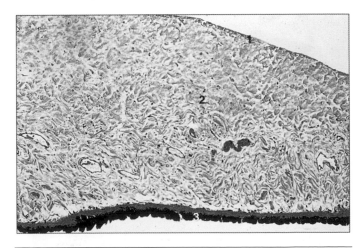

Figure 15.23 Iris (dog). (1) The anterior surface is covered by a discontinuous flattened mesothelium. (2) The core of the iris is vascular connective tissue. (3) The posterior surface epithelium is two layers of cells and part of the retina. The pigment is present. H&E ×62.5.

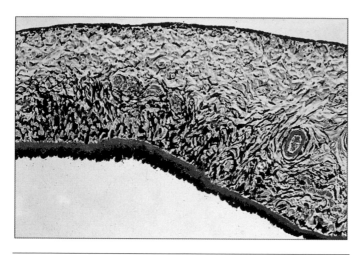

Figure 15.24 Iris (dog). Compare this with **15.23** and note the pigment; this eye will be much darker. H&E ×62.5.

Two muscles of neuroepithelial origin regulate the aperture of the pupil: the sphincter muscle (at the pupillary margin) and the dilator muscle (constituting the deep or anterior layer of the iris epithelium).

The retina is the innermost layer of the wall of the eye. The photosensitive portion lines the inner surface of the eye posteriorly from the *ora serrata*, the point of transition between the photosensitive and nonphotosensitive areas of the retina. The latter consists of two layers of non-light-sensitive epithelium forming the *pars iridica retinae* and the *pars ciliaris retinae*. From the choroid to the cavity of vitreous humour, the 10 layers (**Figure 15.20**) of the photosensitive area are as follows:

- Pigmented epithelium (the outermost layer, a simple cuboidal epithelium derived from the inner layer of the embryonic cup).
- Layer of rods and cones (modified dendrites that act as photoreceptors).
- External limiting membrane, formed by zonulae adherents between processes of radial glial (Müller) cells and rods and cones.

- Outer nuclear layer (nuclei in the cell bodies of the rods and cones).
- Outer plexiform layer (connecting the rod and cone axon terminal with the dendrites of the bipolar neurons and processes of horizontal cells).
- Inner nuclear layer of bipolar neurons relaying the impulses received at layers 2–8. Horizontal and amacrine cells are also present.
- Inner plexiform layer (linking bipolar neurons with ganglion cells and amacrine cells with ganglion cells).
- Ganglion cell layer (contains the perikarya of the ganglion cells).
- Optic nerve fibre layer (the axonal processes of the ganglion cells converge at the optic disc and become myelinated to form the optic nerve; this sieve-like part of the sclera is the lamina cribrosa).
- Internal limiting membrane, constituted by a layer of flattened radial glial (Müller) cells processes and a basal lamina.

Layers 2–10 are derived from the outer layer of the embryonic optic cup.

Lacrimal and Harderian Glands

Lacrimal glands are compound tubuloacinar or tubuloalveolar glands. They produce serous secretion in cats and seromucous secretion in dogs and ungulates. Like other glands, their acinar cells may contain lipid inclusions and be surrounded by myoepithelial cells. The secretions flow through secretory ducts lined by a simple or stratified cuboidal epithelium.

Harderian glands are pigmented lacrimal glands situated posterior to the ocular globes. They can be found in mammals, birds, reptiles, and amphibians and are well developed in rodents (**Figure 15.25a,b**). This gland can also serve as a source of pheromones and lipids involved in thermoregulation and as an active site of immune responses to pathogens.

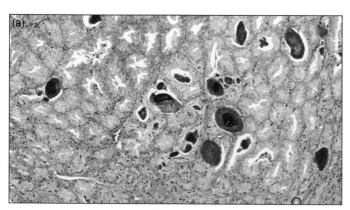
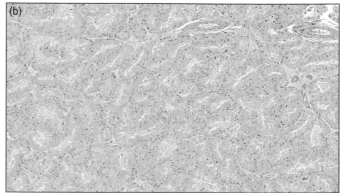

Figure 15.25 Harderian gland (Syrian golden hamster; *Mesocricetus auratus*). (a) Pigmented (female) and (b) Nonpigmented (male). H&E ×100.

CLINICAL CORRELATES

Numerous inherited ocular abnormalities are recognised in purebred dogs. In some cases, control or monitoring schemes exist. Such defects may affect structures of the globe or associated tissues, such as the eyelids. Some dog breeds have exaggerated palpebral fissure shapes, which can predispose them to entropion and may require surgical correction. Dermoids, foci of cutaneous-type differentiation on the cornea or conjunctiva that may produce hair and other adnexal structures, are recognised in all species.

Inflammatory lesions of the eyes are named according to the structures affected and may result from infectious agents, chemical irritation, or trauma. Ocular lesions may be part of a generalised disease syndrome, as in keratitis associated with malignant catarrhal fever (MCF) in cattle (**Figure 15.26**). Migrating parasitic larvae can reach the eye with potentially damaging consequences, and occasional incidents of human visceral larva migrans caused by infections with intermediate stages of ascarid parasites (Toxocara canis) occur. Thelazia spp. of spiuroid worms inhabit the conjunctival sacs and lacrimal glands of horses, cattle and in North America dogs, and cats also.

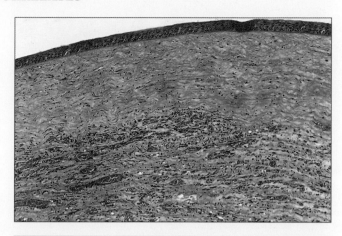

Figure 15.26 Keratitis (cow). This section, illustrating keratitis, or inflammation of the cornea, is from a cow with MCF. The effects of this highly fatal herpes virus infection of ruminants are multisystemic and characterised by vasculitis. The normally avascular cornea shows neovascularisation – the production of a capillary network, fibroplasia, oedema, and infiltration by mononuclear inflammatory cells around the vessels. H&E ×62.5.

Important primary neoplasms of the ocular structures include squamous cell carcinoma, tarsal gland tumours, and melanomas. The eye may also be affected in cases of multicentric lymphosarcoma in several species and may be affected by metastatic tumour spread.

Avian Eye

The avian sclera has a ring of overlapping scleral ossicles enclosed in dense connective tissue (**Figures 15.27** and **15.28**). The avian cornea is similar to the mammalian but has a more prominent basement membrane. The choroid is thick and well vascularised with numerous pigment cells; there is no tapetum. The avian retina has the same number of layers as the mammalian but is avascular. The cones contain oil droplets, thought to enhance colour vision. The *pecten oculi* is a thin, folded, heavily pigmented membrane projecting into the vitreous

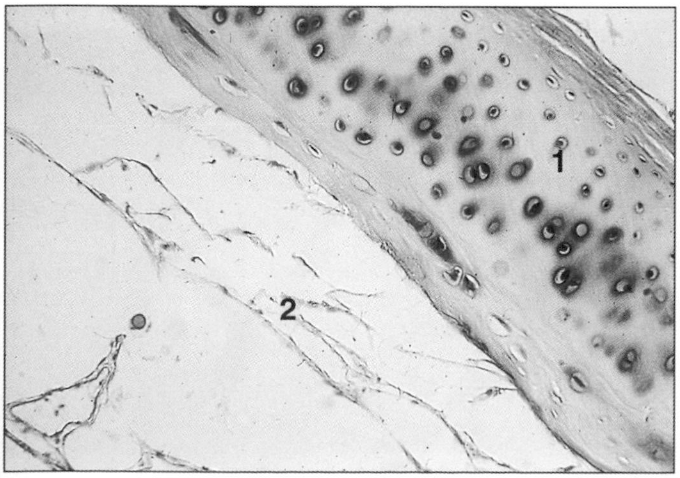

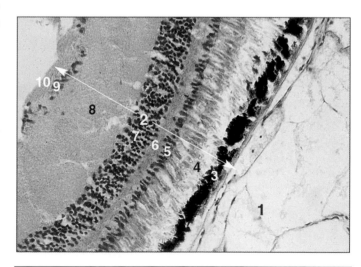

Figure 15.28 Avian eye. (1) Choroid. (2) Retinal layers are (3) pigment layer, (4) rods and cones, (5) outer nuclear layer, (6) outer plexiform layer, (7) inner nuclear layer, (8) inner plexiform layer, (9) ganglion cell layer, and (10) optic nerve fibre layer. H&E ×125.

Figure 15.27 Avian eye. (1) Sclera with hyaline cartilage. (2) Choroid. H&E ×62.5.

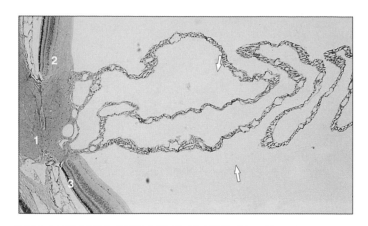

Figure 15.29 Avian eye. The pecten (arrowed) is a heavily folded, vascular pigmented membrane projecting into the vitrous humour from the posteroventral surface of the eye. (1) Optic nerve. (2) Retina. (3) Choroid. H&E ×12.5.

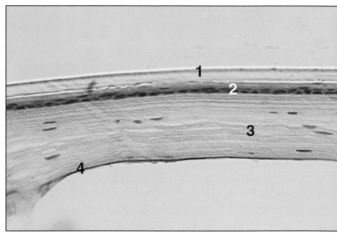

Figure 15.30 Cornea and tertiary spectacle of a regal (ball) python (*Python regius*). The thin keratinised spectacle (1) is contiguous with the periorbital integument, and it is shed and renewed with each moult. (2) Anterior epithelium. (3) Corneal stroma. (4) Descemet's membrane. H&E ×125.

humour from the optic disc (**Figure 15.29**). The lens is composed of a body and an annular pad forming a ring around the equator of the lens. The lens fibres of the annular pad are arranged radially, but in the lens body, the fibres run parallel to the optical axis.

Reptilian, Amphibian, and Fish Eyes

Fish, amphibians, reptiles, and birds are relatively close phylogenetically to each other. Generally, the histological features of their eyes are similar, but there are some notable exceptions.

The pupillary shapes in fish, amphibian, and some reptilian eyes vary considerably between families within a given class.

Some species possess nictitating membranes; others do not. Snakes and some lizards lack moveable eyelids, and their corneas are covered with a tertiary spectacle, an optically clear keratinised epidermal tunic that is derived from the integument. This structure is shed and renewed with each moult of the epidermis (**Figure 15.30**). Like many birds, numerous reptiles have eyes supported in part by scleral ossicles. The crystalline lens in the eyes of fish, amphibians, and reptiles is histologically similar to that in mammals. The posterior segment of the eye, particularly the visual epithelium, varies markedly within each class. The layers of the retina and choroid are similar to those in mammals (**Figures 15.31** and **15.32**). In many reptiles, particularly diurnal species, a vascular and heavily pigmented conus papillaris projects anteriorly into the globe from the head of the optic nerve (**Figure 15.32**).

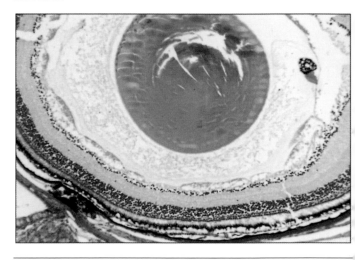

Figure 15.31 Whole mount section of the eye of a small yucca night lizard (*Xantusia vigilis*), a species that lacks moveable eyelids. H&E ×7.5.

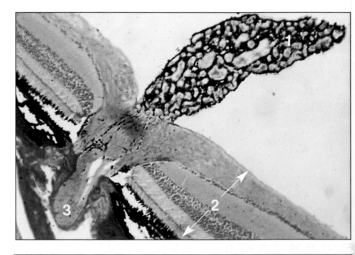

Figure 15.32 Visual epithelium of a panther chameleon (*Chamaeleon pardalis*). (1) Conus papillaris. (2) Retina. (3) Optic nerve. H&E ×20.

CLINICAL CORRELATES

Essentially, all of the clinically significant ophthalmic disorders affecting mammalian eyes can affect the eyes of lower vertebrates. However, some conditions are unique to some lower vertebrates because their eyes are characterised by structures that are lacking in the mammalian eye, for example, inflammation of the tertiary spectacle and the subspectacular space could only occur in an animal whose eyes are lidless and covered with an epidermally derived spectacle (**Figure 15.33**). Some species are more prone to certain ophthalmic disorders induced by their incorrect captive diets. Examples of such lesions are the lipid and calcified corneal opacities of some frogs. Cholesterol deposits occasionally develop in the corneas of some reptiles.

Figure 15.33 Suppurative keratitis in a mountain kingsnake (*Lampropeltis zonata*). The corneal stroma (1) contains numerous heterophilic granulocytic leukocytes. (2) Tertiary spectacle. (3) Cornea. Congo Red ×160.

Ear

The ear is composed of three divisions: external ear, middle ear, and internal (inner) ear.

External Ear

The external ear comprises the auricle (pinna) and the external auditory canal. The auricle consists of a central plate of elastic cartilage covered by the skin rich in sebaceous glands. There are also some sweat glands and a variable amount of hair (**Figure 15.34**). The external auditory canal is a rigid tunnel lined with thin skin; the upper portion is supported by elastic cartilage, the remainder by bone. It contains large coiled sweat glands, some fine hairs, associated sebaceous glands, and ceruminous glands (apocrine sweat glands) that secrete cerumen (ear wax; **Figure 15.35**).

Middle Ear

The external auditory canal ends at the tympanic membrane, which separates the external ear from the middle ear. The external surface of the membrane is covered by

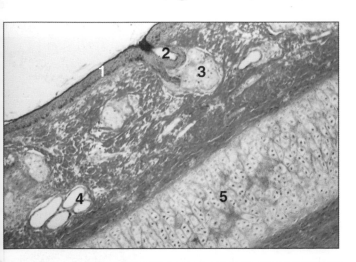

Figure 15.34 External ear canal (dog). (1) Stratified squamous keratinised epithelium lines the canal. (2) Hair follicles. (3) Sebaceous glands. (4) Sweat glands. (5) Elastic cartilage. Masson's trichrome ×50.

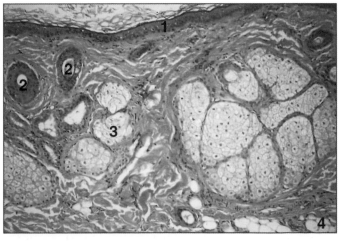

Figure 15.35 Ear canal (dog). (1) Stratified squamous keratinised epithelium. (2) Hair follicles. (3) Sebaceous glands. (4) Sweat glands. H&E ×50.

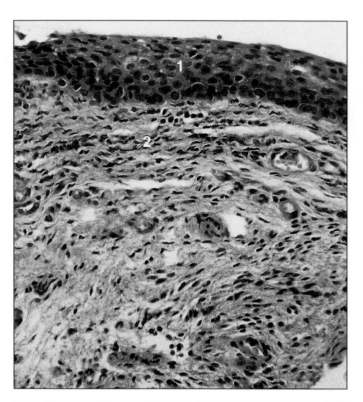

Figure 15.36 Tympanic membrane (goat). (1) External surface covered by stratified squamous epithelium. (2) Dense connective tissue. H&E ×400.

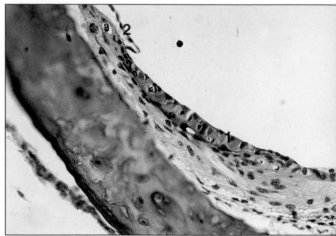

Figure 15.37 Inner ear (cat). (1) Specialised neuroepithelial cells in the vestibule continuous with (2) endothelial cells lining the labyrinth. H&E ×200.

skin, and the internal surface is by a simple squamous epithelium. Between both epithelia there are dense collagen bundles arranged radially, peripherally, and concentrically in the central part (**Figure 15.36**).

The tympanic cavity is lined by simple squamous or cuboidal epithelium. Four auditory ossicles, the malleus, the incus, the lenticular, and the stapes, associated with muscle tissue, form a chain of small bones from the tympanic membrane to the oval window, which separates the middle from the internal ear. Sound is transmitted by the ossicles to the perilymph of the internal ear and generates movement of the delicate basilar membrane of the cochlea. The expanded rostral part of the tympanic cavity forms the auditory (Eustachian) tube and connects the middle ear and the pharynx (*see also* Chapter 8, Guttural pouch in **Figure 8.6**).

Internal Ear

The internal ear consists of the osseous (bony) labyrinth and the membranous labyrinth. The osseous labyrinth is a space within the temporal bone and consists of the vestibule, semicircular canals, and cochlea. The membranous labyrinth is inside the osseous labyrinth and consists of the utricle and saccule (within the vestibule), the

semicircular ducts (within the semicircular canals), and the cochlear duct (within the cochlea). It contains endolymph and is separated from the walls of the osseous labyrinth by perilymph. It adheres to the wall of the osseous labyrinth by means of fine connective tissue strands derived from the connective tissue lamina propria of the lining simple squamous epithelium. This epithelium is replaced at certain points by neuroepithelial cells (**Figure 15.37**). In the semicircular ducts, local expansions of the membranous ampullae house sensory structures: the *cristae ampullaris*. The neuroepithelial sensory hair cells (type I and type II) and supporting (sustentacular) cells of each crista are covered by a gelatinous cupula (**Figure 15.38**).

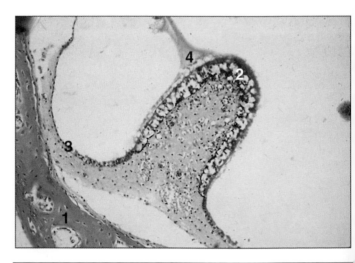

Figure 15.38 Inner ear. Crista (cat). (1) Bone of the osseous labyrinth. (2) Neuroepithelial (hair) cells and supporting cells of the crista continuous with (3) endothelium lining the membraneous labyrinth. (4) Cupula. H&E ×125.

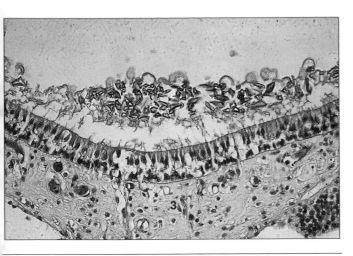

Figure 15.39 Inner ear. Macula (cat). (1) Columnar epithelium of the macula consists of neuroepithelial cells and supporting cells continuous with the lining epithelium of the vestibule. (2) Otoliths at the top. (3) Connective tissue. H&E ×160.

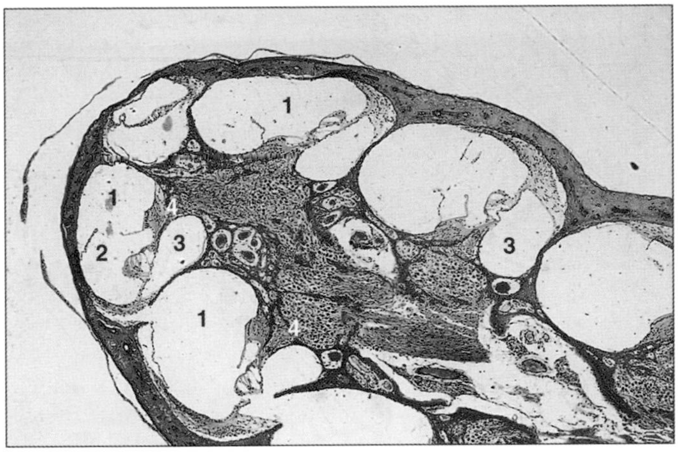

Figure 15.40 Inner ear. Cochlea (cat). (1) Dorsal *scala vestibuli*. (2) Middle cochlear duct. (3) Ventral *scala tympani*. (4) Spiral ganglion. H&E ×7.5.

When the latter is deflected during rotational movements of the head, the sensory cells are stimulated and impulses are sent to the brain.

Both the utricle and saccule are lined in part by maculae, patch-like collections of sensory hair cells (type I and type II) and supporting cells. Maculae are covered with a gelatinous otolithic membrane, the statoconial membrane, in which are embedded calcium carbonate crystals: the otoliths. As the membrane shifts in response to gravity acting upon the otoliths, sensory cells of the maculae are stimulated. They enable the animal to determine the position of its head in space, control of posture, gait, and equilibrium, and to assess linear acceleration and deceleration (**Figure 15.39**). The sensory cells are surrounded by terminals of the vestibular nerve.

The osseous cochlea surrounding the cochlear canal in a spiral around a central pillar of bone, the modiolus, contains, in turn, the spiral lamina, a thin shelf of bone that travels up the modiolus. The canal is divided into three compartments: the dorsal *scala vestibuli* and ventral *scala tympani*, which contain perilymph and are lined with squamous epithelium, and the cochlear duct between them (**Figure 15.40**). The *scala vestibuli* and *scala tympani* are communicated by the helicotrema, a small opening at the cupula of the cochlea. *Scala vestibuli* contacts with the vestibular window and *scala tympani* with the cochlear window. The floor of the cochlear duct is formed from the fibrous basilar membrane and the roof from the vestibular (Reissner's) membrane, lined by a simple squamous epithelium. The acoustically sensitive spiral organ (of Corti) rests on the basilar membrane and is composed of neuroepithelial hair cells and sustentacular cells. The lower surface of the membrane, facing the ventral *scala tympani*, is lined with simple squamous epithelium. The *stria vascularis* is the outer wall of the cochlear duct, which is constituted

by a stratified cuboidal epithelium penetrated by numerous capillaries and participates in the production of endolymph. Overlying the spiral organ and extending from the *spiral limbus* (an elevation of protective tissue above the spiral lamina) is the tectorial membrane, which is produced by interdental cells and rests on the stereocilia of the hair cells. The stereocilia are displaced when the basilar membrane vibrates in response to sound waves passing through the fluid-filled scalas. Nerve terminals form a web around the bases of the hair cells and transmit stimuli to the spiral ganglion of bipolar neurons. The axons form the cochlear division of the eighth cranial nerve, the vestibulocochlear nerve (**Figure 15.41**).

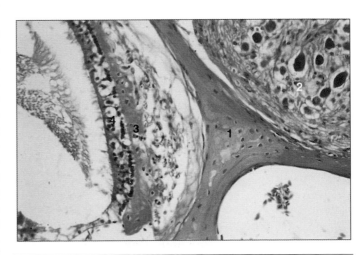

Figure 15.41 Inner ear. Spiral organ (of Corti) (cat). Located on the floor of the cochlear duct. (1) Osseous spiral. (2) Spiral ganglion. (3) Basilar membrane. (4) Sensory and supporting cells. H&E ×160.

CLINICAL CORRELATES

Diseases of the ear in domestic animals are probably not commonly investigated by pathologists, although the three broad categories otitis externa, otitis media, and defects of hearing may be considered. Otitis externa (**Figure 15.42**), or inflammation of the external ear canal, may be a problem restricted to the waxy integument of this site or it may be part of a widespread skin condition. Usually, only chronic cases of this very common condition, in which the severe thickening of the walls of the ear canal may raise clinical suspicion of tumour development, are submitted for histopathological examination.

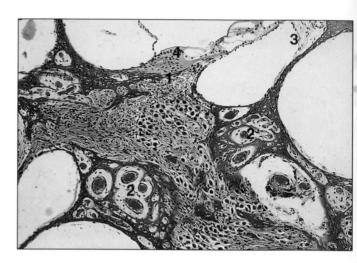

Figure 15.43 Avian inner ear. (1) Spiral ganglion. (2) Osseous labyrinth. (3) Membraneous labyrinth. (4) Neuroepithelial sensory cells. H&E ×125.

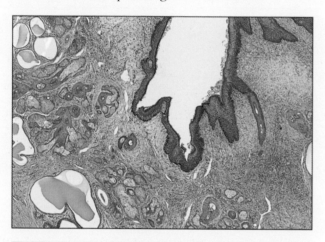

Figure 15.42 Otitis externa (dog). The surface epithelium (centre) is hyperplastic and there is dermal fibrosis. The patchy appearance of the dermis is caused by a mixed, mainly mononuclear, inflammatory infiltrate. The sebaceous glands are hyperplastic, and there is cystic dilatation of the ceruminous glands, some of which are seen filled with eosinophilic cerumen. H&E ×20.

Avian Ear

The external ear lacks a pinna and ceruminous glands. The auditory ossicles are fused to form a cartilaginous rod, the columella, which extends from the tympanic membrane to the oval window. The membranous labyrinth is essentially similar to that of the mammal (**Figure 15.43**). The saccule, however, differs in that it contains two maculae. The cochlear duct also possesses a terminal expansion that is peculiar to birds, the lagena, and is separated from the dorsal *scala vestibuli* by the *tegmentum vasculosum*, a thin connective tissue membrane integrated with a highly folded vascular epithelium.

Reptilian and Amphibian Ears

The external pinna and ceruminous glands are absent. Some terrestrial frogs and toads that utilise audible calls in their courtship and territorial behaviour have acute hearing.

Many amphibians possess external tympanic membranes; others lack them entirely. However, some aquatic amphibians have lateral line systems with which they detect water-transmitted vibratory signals and hydrostatic pressure changes.

Crocodilians are capable of hearing air-transmitted sounds. They have slit-like auditory openings that can be closed when they submerge themselves.

Most lizards have ears with tympanic membranes that are located close to the integumentary surface of the skull. In some species, the tympanic membrane is flush with the skin surface. In others, it lies within a shallow depression or deeper auditory meatus.

Snakes, chelonians, and the tuatara lack external auditory structures. The philosophical question of whether their sense of 'hearing' is limited to substrate-transmitted vibrations is a topic of spirited debate.

The inner ear is responsible for the spatial and postural orientation of the animal within its environment. The morphological and histological features of the amphibian and reptilian inner ear are sufficiently similar to the mammalian and avian ear that further discussion is not warranted here.

Organ of Smell (Olfactory Organ)

The nose, in addition to its role in the respiratory system, also functions as the olfactory organ. In the nasal cavity, brown–yellow olfactory epithelium occurs together with pinkish respiratory epithelium. It is a pseudostratified columnar epithelium with olfactory (sensory) cells, basal cells, and supporting cells. Tubular mucoserous glands (Bowman's) lie in the lamina propria and secrete onto the surface through simple ducts lined with cuboidal cells (**Figures 15.44** and **15.45**).

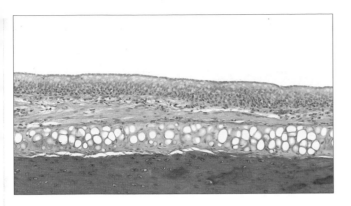

Figure 15.44 Olfactory organ (hamster). The olfactory epithelium is a pseudostratified ciliated columnar with many rows of nuclei; the more superficial are of the sustentacular cells; and the deeper nuclei are of the olfactory nerve cells. Underlying vascular connective tissue and bone (turbinates). H&E ×100.

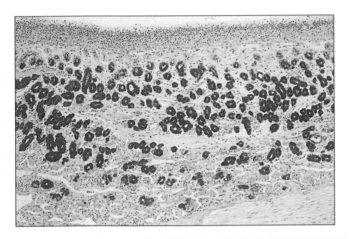

Figure 15.45 Olfactory organ (horse). The mucous cells stain deep blue. Alcian blue ×100.

Other Specialised Sense Organs

Besides sight, hearing, taste, touch, and the perception of pain, heat, and cold, many of the lower vertebrates possess highly specialised organs that augment their awareness of their external environment.

Parietal Eye

In addition to their paired lateral eyes, many lizards and the primitive tuatara (*Sphenodon punctatus*) have a parietal eye. This photosensitive organ consists of a scale-like and cell-poor cornea, a cellular lens, a central chamber filled with clear fluid and a few macrophages, and a cup-shaped pigmented visual epithelium that is analogous to the retina and choroid (**Figure 15.46**). The parietal eye is partially responsible for regulating basking and other thermoregulating behavioural activities, and it can warn of the approach of potential predators whose shadows are detected by the upward-directed eyelike structure.

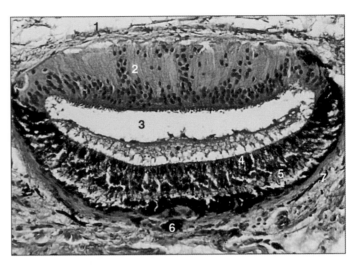

Figure 15.46 Whole mount section of the parietal eye of a green iguana (*Iguana iguana*). At the top of the image is a relatively thick, but avascular, cornea (1). A cellular lens (2), composed of tall columnar cells packed closely and parallel to each other, lies beneath the cornea. The lumen (3) of the central chamber contains clear fluid and is surrounded at the sides and back by heavily pigmented photosensitive retinal cells (4) and unpigmented ganglion cells (5). A giant melanin-packed macrophage (6) can be seen at the bottom of the capsule. The parietal nerve exits the rear of the parietal eye and courses through the parietal foramen. A thin connective tissue capsule (7) envelops the parietal eye, where it is surrounded by calvarial bone. H&E ×125.

Facial and Labial Pit Organs

Rattlesnakes, water moccasins, copperhead snakes, and other pit vipers possess paired facial pit organs (**Figure 15.47**), with which they sense very small differences in the background thermal environment. This ability to discriminate slight temperature variations aids in prey detection both before and after the prey animals have been envenomated.

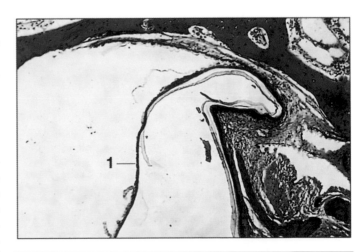

Figure 15.47 Facial pit organ of a western diamondback rattlesnake (*Crotalus atrox*) is surrounded on its inner surfaces by a cup-shaped bony depression. A thin, lightly keratinised diaphragm-like membrane (1) divides the posterior of the pit chambers into two compartments. Within the membrane are embedded numerous dendritic neuron endings. H&E ×20.

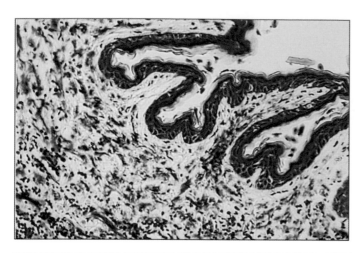

Figure 15.48 Labial pit organ of a reticulated python (*Python reticulatus*). These infrared superficial depressions over much of the external surface of upper lips of some boas and pythons consist of parallel branched or unbranched shallow passages lined by a thin layer of lightly keratinised squamous epithelium. A rich network of dark staining, fine dendritic nerve endings penetrate to just beneath the epithelium. Bodian's silver stain ×125.

Nonvenomous snakes of the family Boidae (boas, pythons, and anacondas that are ambush predators) possess labial pit organs (**Figure 15.48**), which help them locate warm-blooded prey. These branched pit-like depressions are lined with lightly keratinised squamous epithelium through which numerous dendritic sensory neurons penetrate.

Vomeronasal Organ

Most snakes and lizards have well-developed vomeronasal (Jacobson's) organs (**Figure 15.49**), which assist in sampling and discriminating chemosensory stimulatory particles such

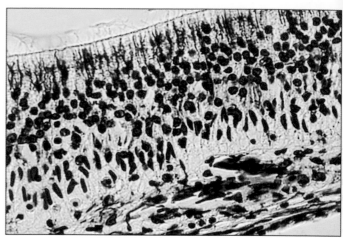

Figure 15.50 Section through the superficial surface of the vomeronasal organ of a green iguana (*Iguana iguana*). Note the myriad number of thin dendritic nerve endings that course between and penetrate the epithelium to the lumenal surface. Bodian's silver stain ×250.

as prey-related scent or pheromones. The vomeronasal organ consists of a mushroom-shaped rounded column with a cartilaginous core. This column is surrounded by a cup-shaped spherical cavity. The luminal surfaces of the column and cavity are covered with ciliated columnar or pseudostratified columnar epithelium through which myriad numbers of tiny dendritic nerves pass between adjacent cell membranes. These nerve endings project out into the lumen (**Figure 15.50**).

Lateral Line Organ

Most fish and many amphibians (particularly aquatic frogs, newts, and salamanders) have lateral line systems (**Figure 15.51**) that are sensitive to slight changes in

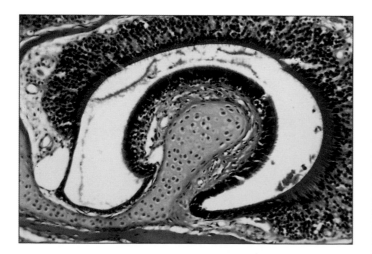

Figure 15.49 Whole mount sagittal section of the vomeronasal (Jacobson's) organ of a small skink (*Scinella lateralis*). The raised mushroom-shaped protruberance is supported by a core of hyaline cartilage and is surrounded by a narrow cavity; the surfaces of which are covered on all sides by ciliated simple columnar epithelium. H&E ×125.

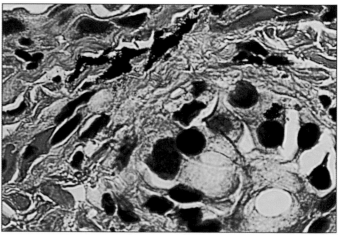

Figure 15.51 Cross-section of the dermis and the pressure-sensitive lateral line of an aquatic salamander (*Amphiuma tridactyla*). The clear central cavity is lined with large, plump columnar glandular secretory cells. H&E ×250.

hydrostatic pressure and water-borne vibration. Tall cuboidal to fully columnar cells form vibration- and pressure-sensitive neuromasts that receive and transmit impulses from the aquatic environment to the animal's central nervous system.

Electric Organ

Some teleost fish, such as the electric eel and electric catfish, and at least one family of elasmo-branch ray, such as the torpedo (*Torpedo* spp.), possess specialised muscles arranged into discrete electric organs (electroplax) that produce and detect powerful electric impulses. Paired electric lobes on the medulla oblongata are the motor centres for the integration of electroplax activity. When these specialised muscular organs suddenly discharge their electrical potential, prey fish and predators are subjected to pulses of high-amperage electrical current that can be incapacitating or fatal. Some electric eels produce pulses of direct current that measure 600 V.

Swim Bladder

Most but not all teleost fish possess a specialised elongated gas- (or oil-) filled organ: the swim bladder. This is a major hydrostatic organ that helps these fish maintain their buoyancy and orientation within a column of water. In some bottom-feeding species, the swim bladder is much reduced or may be absent. In fast swimming fish, it is elongated and, thus, enhances streamlining. The swim bladder is formed from thin sheets of dense fibrocollagenous connective tissue laid at acute angles to each other. This arrangement aids in maintaining its shape and reducing deformation. In some fish, skeletal muscles insert into its outermost surface. In other species, the swim bladder is only attached to the body wall along its dorsal surface. The swim bladder is lined by a thin, much flattened, nonkeratinised squamous epithelium.

Several bacterial, viral, and protozoan diseases are characterised by inflammation of the swim bladder. Thus, when performing a necropsy on a fish, it is important to examine this organ for haemorrhage(s), oedematous thickening, discolouration, or other abnormality.

16

LYMPHATIC SYSTEM

The lymphatic system has a dual function: the lymphatic vessels drain interstitial tissue, returning fluid to the bloodstream; and lymphoid tissue produces phagocytes and immunologically competent cells, for the defence mechanism against invasion by pathogens. During fetal development, the immune system is structured into two principal types of tissues: the diffuse lymphatic tissues, scattered throughout loose connective tissues of digestive, respiratory, and urogenital systems, and the organised lymphatic tissues: lymph nodes and spleen.

Lymphoid tissue consists predominantly of lymphocytes (B and T cells). These and a variable number of plasma cells, macrophages, and other cells are supported by a delicate network of reticular fibres that fill the spaces between the trabeculae. Diffuse lymphatic tissue and lymphatic nodules are the components of most lymphatic organs and also appear in the connective tissue of the digestive, respiratory, urinary, and reproductive organs, among other locations. The former is characterised by a moderate concentration of scattered lymphocytes; the latter comprises an aggregation of mostly small, densely packed lymphocytes.

Thymus

The thymus is a central lymphoid organ that originated from the endoderm, exporting a specialised subpopulation of lymphocytes (T cells) to other sites, such as lymph nodes, bone marrow, spleen, and tonsils. In the embryo, a network of endodermal cells from the pharyngeal pouches is infiltrated by large numbers of lymphocytes (thymocytes) originating in the bone marrow. Whorls of these endodermal cells, often with a keratinised core, are seen in the thymus and are called thymic (Hassall's) corpuscles (**Figure 16.1**). The thymus is enclosed in a fine connective tissue capsule; trabeculae extend into the gland dividing it into lobes and lobules (**Figure 16.2**). A framework

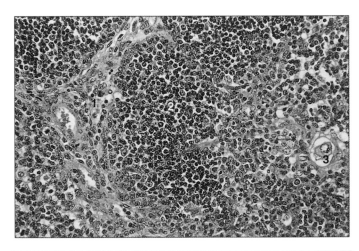

Figure 16.1   Thymus (ox). (1) The supporting epithelial cells derived from the pharyngeal endoderm. (2) Thymocytes. (3) Hassall's corpuscle. H&E ×125.

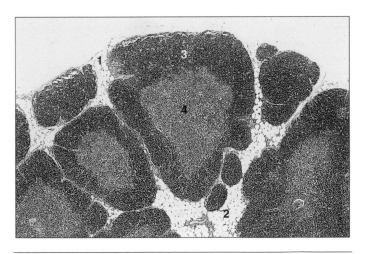

Figure 16.2   Thymus (ox). (1) Fine connective tissue capsule. (2) Fat-laden interlobular connective tissue. Thymic lobule with (3) outer cortex of densely packed thymocytes and (4) inner less cellular medulla. H&E ×62.5.

DOI: 10.1201/9781003333807-16

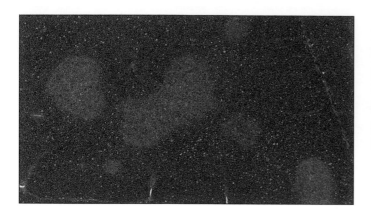

Figure 16.3 Thymus (pig). Various lobules where we can observe the outer cortex with densely packed thymocytes and inner medulla. H&E ×35.

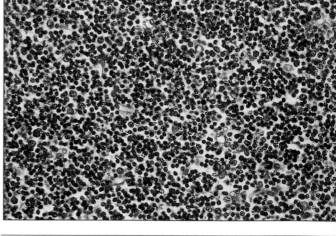

Figure 16.5 Thymus (ox). The small dark blue cells are the densely packed cortical thymocytes. H&E ×250.

of epithelial (reticular) cells supports the thymocytes, separating them from the circulating blood by forming a blood–thymus barrier. The cortex of each lobule is densely populated by small lymphocytes.

The medulla is much less dense, and the epithelial cells are more numerous (**Figures 16.3–16.6**). Spaces in the cortical area are caused by apoptotic lymphocytes phagocytosed by resident macrophages. Fat is present in the capsule and the interlobular connective tissue. The epithelial cells are also regarded as the source of thymic hormone (thymosin) that promotes the maturation of T lymphocytes and hormone-like substances (thymosin and thymopoietin) that induce differentiation of thymocytes. The fetal thymus, after birth, is a very extensive organ, with no fat in the connective tissue capsule. Some regression occurs after birth (**Figures 16.7** and **16.8**).

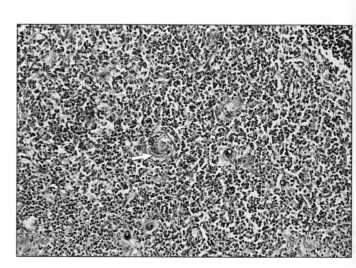

Figure 16.6 Thymus (ox). The central medullary zone is less cellular; a thymic (Hassall's) corpuscle is arrowed. H&E ×125.

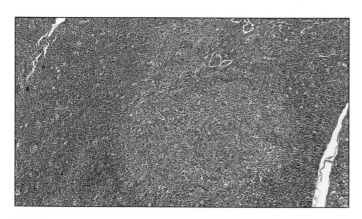

Figure 16.4 Thymus (pig). Cortex with abundant densely packed thymocytes and medulla with epithelial cells and few thymocytes Hassall's corpuscles are seen in the centre of the medulla. H&E ×50.

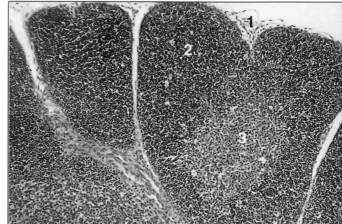

Figure 16.7 Fetal thymus (ox). (1) The connective tissue of the capsule and the supporting trabeculae; there are no fat cells. Thymic lobules with (2) cortex and (3) medulla. H&E ×62.5.

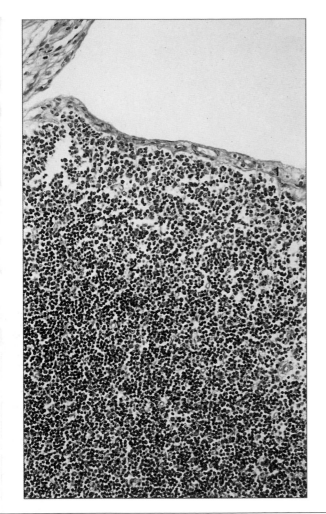

Figure 16.8 Fetal thymus (ox). (1) Connective tissue capsule. (2) Cortical zone of the thymic lobule with loosely arranged thymocytes. The intercellular supporting framework is composed of epithelial cells. H&E ×250.

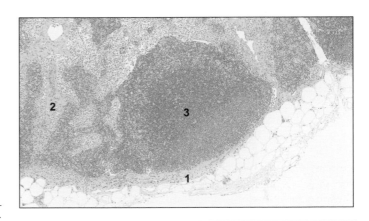

Figure 16.9 Lymph node (dog). (1) Connective tissue capsule. Cortex with (2) lymphatic follicles and (3) paracortex. Masson's trichrome ×50.

Figure 16.10 Lymph node (dog). (1) Connective tissue capsule with some adipose cells. (2) Connective tissue trabeculae. (3) Lymphatic tissue (cortical follicle). H&E ×150.

Lymph Nodes

Lymph nodes are situated along the lymphoid vessel drainage system in the body. Lymph nodes contain diffuse and nodular lymphatic tissue and lymphatic sinuses that are organised into a cortical and medullary region. A capsule of connective tissue extends into fine vascular fibrous trabeculae into the parenchyma of the lymph node. Lymph nodes and the only lymphatic organs with both afferent (flow lymph into the lymph node) and efferent (flow out of the lymph node) lymph vessels and sinuses. Afferent lymphatic vessels enter the capsule and drain into the subcapsular sinus. From there, the lymph drains into a labyrinth of sinuses extending along the trabeculae, eventually emptying into the efferent lymphatics in the hilus. In the cortex, circular aggregations of lymphocytes form follicles or nodules (**Figures 16.9–16.12**). The primary follicle is a solid-packed, evenly distributed mass of cells. The secondary follicle has an outer rim of

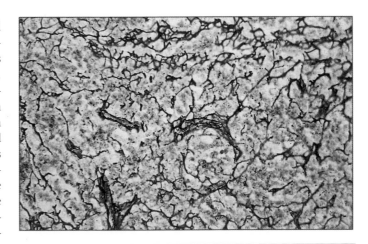

Figure 16.11 Lymph node (cat). The reticular fibres form a black network and support the cells of the lymphatic tissue. Gordon and Sweet ×100.

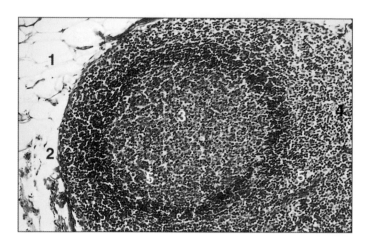

Figure 16.12 Lymph node (dog). (1) Connective tissue capsule with some adipocytes. (2) Subcapsular sinus. (3) Cortical follicle. (4) Parafollicular tissue. (5) Sinusoid. H&E ×125.

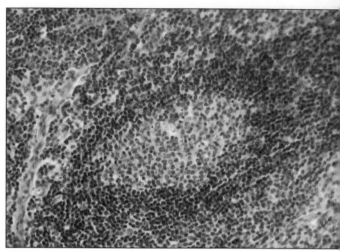

Figure 16.14 Lymph node. Cortex (dog). The germinal centre is com posed of lymphoblasts, reticular cells, and macrophages. H&E ×200.

densely packed small lymphocytes and a pale staining, loosely packed germinal centre, with a mixed population of lymphoblasts, dendritic reticular cells, and macrophages (**Figures 16.10–16.14**). The secondary follicle responds actively to antigen stimulus, and B lymphocytes (originating in the bone marrow) are present, as are macrophages. Thymus-derived T lymphocytes between the follicles form the paracortex (**Figure 16.15**).

The lymph node medulla is less organised, and there is a looser aggregation of cells. The lymphatic tissue extends from the cortical zone as medullary cords (Rosette-like clusters of lymphocytes, plasma cells, and macrophages), separated by a network of sinuses and trabeculae. (**Figures 16.16–16.18**). Sinusoidal spaces are lined with endothelial cells on a reticular framework with

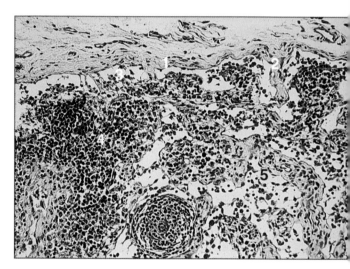

Figure 16.15 Lymph node. Paracortex (pig). (1) Connective tissue capsule. (2) Connective tissue trabeculae. (3) Subcapsular sinus (4) Thymus-derived T lymphocytes. (5) Sinus. Alcian blue ×125.

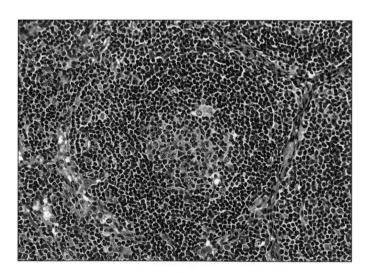

Figure 16.13 Lymph node. Cortex (dog). A single follicle is present. The central zone is the pale staining reactive germinal centre. There is an outer rim of closely packed small lymphocytes. H&E ×200.

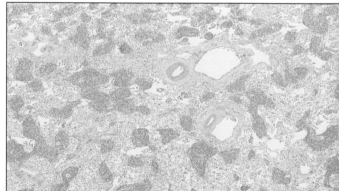

Figure 16.16 Lymph node medulla (dog). Medullary cords full of lym phocytes, plasma cells, and macrophages separated by a network o connective tissue trabeculae. H&E ×50.

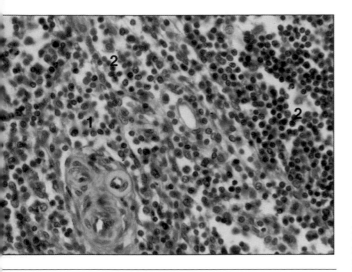

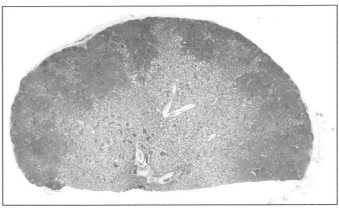

Figure 16.19 Lymph node (goat). Cortex (outer layer) with lymphoid follicles and parafollicular tissue, and medulla in the centre of the lymph node. H&E ×8.

igure **16.17** Lymph node. Medulla (dog). (1) Loose aggregation of ymphoid tissue. (2) Open meshwork of sinuses. H&E ×400.

ttached macrophages. The sinuses are fenestrated and ymphocytes and macrophages have free access.

There are some species differences in the architecure of lymph nodes. Normally, the cortex and paracorex are arranged at the outer layers of the lymph node nd the medulla at the centre. The afferent lymph vessels enetrate the capsule at several different sites, and the nain vasculature and innervation enter the lymph node hrough the hilus together with the efferent lymph vessels **Figure 16.19**). However, the lymph nodes are quite different in the pigs. Most of the lymphoid follicles occupy deep position at the centre of the lymph node, and the eriphery of the node is composed of mostly loose lymphatic tissue (**Figure 16.20**).

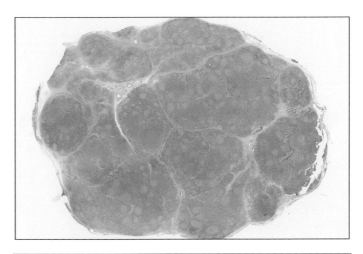

Figure 16.20 Lymph node (pig). Follicles and parafollicular tissue occupying the centre of the lymph node with areas of 'medullary tissue' scattered in the organ. H&E ×15.

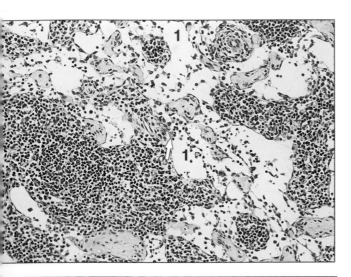

Figure **16.18** Lymph node. Medulla (dog). (1) Sinus lined by macophages. Also, clusters of small lymphocytes (arrowed). Alcian lue ×125.

Immunohistochemistry (IHC) has been widely used to identify different subsets of lymphocytes in the lymphatic tissues, using monoclonal antibodies against surface markers (CDs: clusters of differentiation). Most T cells are CD3 positive and B cells are CD19/20 positive. The cell surface glycoproteins CD4 and CD8 are expressed on exclusive populations of mature T cells (helper and cytotoxic cells respectively) in the parafollicular and deep cortex of lymph nodes. Other cell markers have been used extensively to differentiate different subsets of immune cells, e.g. CD56 for natural killer (NK) cells or CD68 for macrophages in primary and secondary lymphoid organs (**Figures 16.21–16.25**).

Haemal lymph nodes, described only in ruminants, are dark red aggregations of lymphoid tissue, with an unknown function, possibly related to immune responses against blood-borne pathogens. The sinuses are filled with blood instead of lymph (**Figure 16.26**).

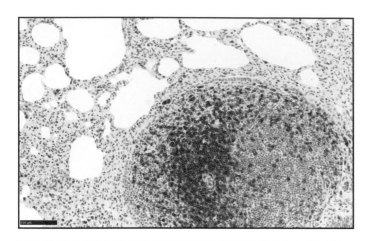

Figure 16.21 Lung (rhesus macaque). IHC with anti-CD3 antibody. T lymphocytes can be observed within the inner part of the lymphoid follicle, adjacent to the blood vessel. IHC (diaminobenzidine, DAB; brown staining) ×100.

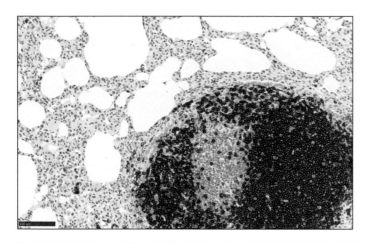

Figure 16.22 Lung (rhesus macaque). IHC with anti-CD20 antibody. B lymphocytes can be observed within the outer part of the lymphoid follicle. IHC (DAB) ×100.

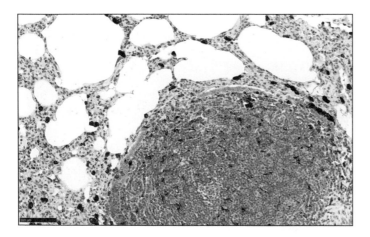

Figure 16.23 Lung (rhesus macaque). IHC with anti-CD68 antibody. Positively marked macrophages can be observed within the alveoli and parenchyma; reticular cells are also observed within the follicle. IHC (DAB) ×100.

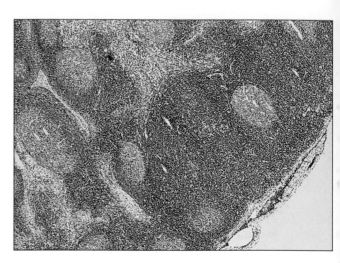

Figure 16.24 Lymph node (cat). IHC with anti-CD4 monoclonal antibody. Positive staining in a population of T cells in the parafollicular and deep cortex of the lymph node. IHC (DAB) ×62.5.

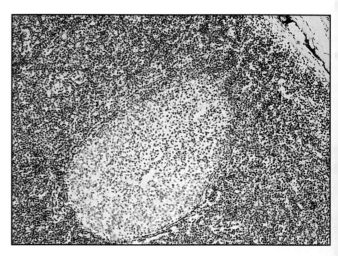

Figure 16.25 Lymph node (cat). IHC with anti-CD4 monoclonal antibody. Positive staining in T cells in the parafollicular and deep cortex of the lymph node. IHC (DAB) ×125.

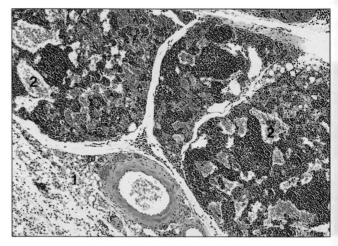

Figure 16.26 Haemal lymph node (ox). (1) Connective tissue capsule (2) Blood-filled sinuses. H&E ×20.

Spleen

The spleen is the largest secondary organ in the lymphatic system. It is usually situated in the cranial part of the abdominal cavity on the left of the stomach (in ruminants on the left lateral wall of the reticulum). It has no afferent lymphatics, and it is involved in filtering the blood and mounting immune responses against pathogens. The capsule consists of smooth muscle, collagen, and elastic fibres with fibrocytes and extends into the parenchyma to form the supporting stroma, dividing the parenchyma into red and white splenic pulp (**Figures 16.27–16.30**).

White pulp contains lymphatic follicles (which may be primary or secondary), together with dense accumulations of T lymphocytes arranged around arteries to form the periarterial lymphatic sheaths. Splenic corpuscles, arterioles with a cuff of T lymphocytes, occupy an eccentric position in the white pulp. The arteriolar branches leave the white pulp and enter the red pulp as straight penicillar arterioles. Some acquire a coat of reticular fibres and become ellipsoids (well developed in the cat) and empty into the splenic sinusoids of the red pulp (**Figure 16.31**). These sinusoids are wide channels lined with endothelial cells with gaps occupied by macrophages. Foreign material is recognised and removed as part of the immune

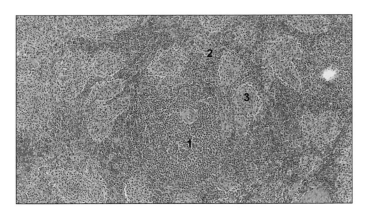

Figure 16.29 Spleen (pig). (1) White pulp. (2) Red pulp with ellipsoids. (3) H&E ×100.

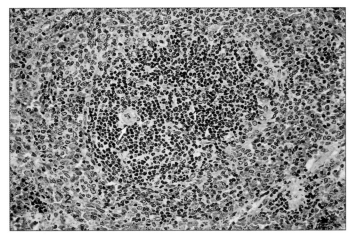

Figure 16.30 Spleen (dog). (1) White splenic corpuscle with an eccentric arteriole (arrowed). (2) Sinusoids of the red pulp filled with erythrocytes. H&E ×200.

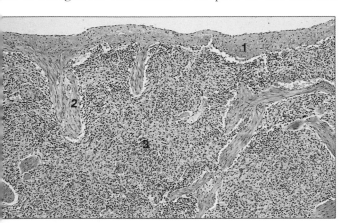

Figure 16.27 Spleen (horse). (1) Fibromuscular capsule. (2) Fibromuscular trabeculae. (3) Splenic pulp. H&E ×125.

Figure 16.28 Spleen (sheep). (1) White pulp. (2) Red pulp. (3) Trabecula. (4) Capsule. H&E ×50.

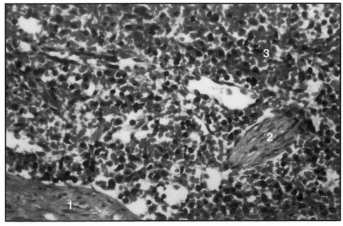

Figure 16.31 Spleen (cat). (1) Part of a trabecula. (2) Ellipsoid. (3) Blood-filled sinusoids of the red pulp. H&E ×400.

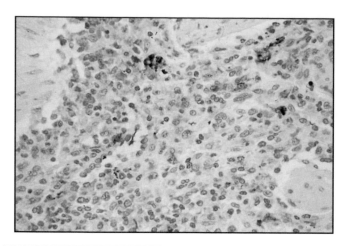

Figure 16.32 Spleen (dog). The site of iron deposits as a result of erythrocyte phagocytosis is blue. Perls' Prussian blue ×250.

response; senescent erythrocytes are also removed from the circulation (**Figure 16.32**).

Red pulp, so called because of the large number of erythrocytes it contains in the reticular frame-work, is a loose arrangement of blood filled, fenestrated sinusoids, opening into venules and draining into the splenic vein to leave at the hilus. The region between the red and white pulp is the marginal zone and is mainly phagocytic (**Figure 16.33**).

Tonsils

The palatine, lingual, and pharyngeal tonsils have a similar histological appearance with differences among species (*see* Chapter 9). Stratified squamous epithelium overlies dense aggregations of lymphoid tissue in the lamina propria (**Figure 16.34**). Small lymphocytes migrate through the epithelium into the pharynx. Primary and secondary lymphatic follicles are present depending upon the activity of the immune response. There are no afferent lymphatics;

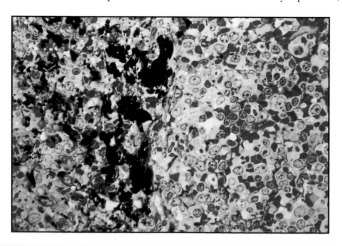

Figure 16.33 Spleen (dog). The marginal zone has a large population of macrophages. These are shown filled with phagocytosed carbon particles. Toluidine blue ×250.

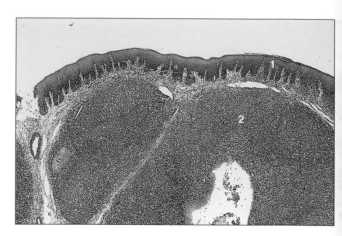

Figure 16.34 Tonsil (dog). (1) Stratified squamous epithelium (2) Dense aggregation of lymphoid tissue in the lamina propria. H&E ×20

the lymph drains into local nodes in the efferent lymphatic vessels. There are also no sinusoids.

Some tonsils contain crypts. Examples include the lingual tonsils of the horse, pig, and oxen; the tubal tonsil of the pig; the paraepiglottic tonsils of the pig, sheep, and goat; and the palatine tonsils of the horse, pig, and ruminants. A crypt with its associated lymphatic tissue constitutes a tonsillar follicle, and several follicles form the tonsil. It is lined with stratified squamous epithelium (*see* Chapter 9). Examples of tonsils without crypts include the tubal tonsil of ruminants, the paraepiglottic tonsil of the cat, and the palatine tonsils of carnivores.

Gastro-Intestinal-Associated Lymphatic Tissue (GALT)

Lymphoid tissue occurs in subepithelial sites in the alimentary tract as well-formed lymphatic nodules or scattered lymphocytes and plasma cells. Many lymphocytes traverse the gut epithelium and are present between the epithelial cells. Most of these lymphocytes are T cells that are involved in initiating the immune response. The large aggregations of lymphoid tissue may flatten the villi in the small intestine and M cells (membranous epithelial cells) can be observed at the "dome" area of the lymphoid follicle (*see* Chapter 9). A microfolding of the luminal plasma membrane of the M cells turns the luminal antigens into endocytes and presents them to the intraepithelial and subepithelial lymphocytes. Large aggregations of lymphocytes often bulge through the muscularis mucosa into the submucosa (**Figures 16.35** and **16.36**). Similar local aggregations of lymphoid tissue occur in the respiratory and urinary tracts.

Lymphatic vessels begin blindly as thin endothelial lined tubes with a minimal amount of supporting connective tissue. Larger lymphatic vessels such as the thoracic duct have a few smooth muscle fibres in their walls. Numerous valves are present.

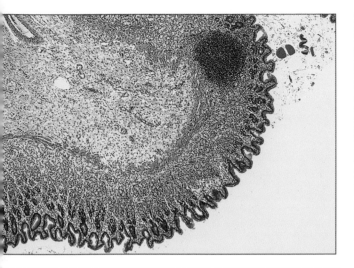

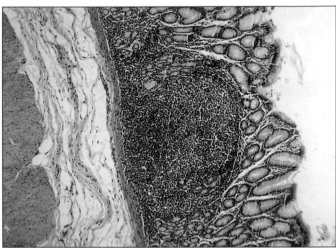

Figure 16.35 Stomach (cat). A lymph nodule is present in the mucosal/submucosal layer of the stomach. H&E ×20.

Figure 16.36 Abomasum (goat). A lymph nodule in the lamina propria of the mucosa. H&E ×62.5.

CLINICAL CORRELATES

The thymus, spleen, bone marrow, and other primary and secondary lymphoid organs are common sites for inflammation because of the phagocytic activity of some of their cellular components. Primary lymphoid tumours such lymphosarcoma are relatively common in many domestic species. Leukaemia, in which abnormal cells circulate in the blood, is less common. Lymphoid tumours may affect almost any site in the body (**Figures 16.37–16.41**), from lymphoid organs to visceral tissues and even skin. Secondary (metastatic) tumour deposits are often found in the lymphatic tissues, the nodes, and ducts, draining the site of the primary neoplasm.

Lymphoid tissues are also targeted organs for infectious agents. Many viruses, like the African swine fever virus, infect mainly monocytes and macrophages. Infected macrophages in lymphoid organs, like the spleen, start producing abundant proinflammatory cytokines, starting a so-called 'cytokine storm' that can have a detrimental effect on the neighbouring cells, inducing apoptosis (programmed cell death) in the lymphocytes within splenic follicles (**Figure 16.42**).

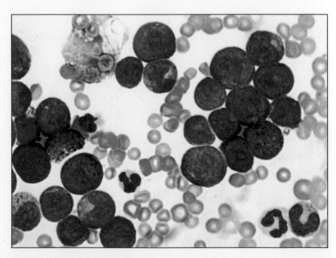

Figure 16.37 Lymphosarcoma (dog). Section of lymph node from a 4-year-old male dog showing dense sheets of large, immature lymphoid cells with round-to-irregular nuclei, and prominent nucleoli. Several mitotic figures are present. H&E ×250.

Figure 16.38 Feline prolymphocytic leukaemia. Note the immaturity of these neoplastic cells, which is characterised by their large nuclei, scanty cytoplasm, and prominent nucleoli. Wright's (smear) ×500.

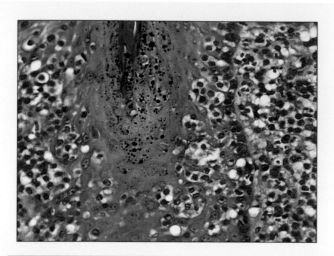

Figure 16.39 Epitheliotrophic lymphosarcoma (dog). In this skin section, taken from a 10-year-old neutered female dog the superficial dermis is occupied by sheets of monomorphic medium-sized lymphoid cells. These cells invade the epithelium in numerous clusters (Pautrier's microabscesses). H&E ×250.

Figure 16.42 Spleen (pig) from an animal suffering acute African swine fever. Apoptosis of lymphocytes can be observed within the splenic follicle within the white pulp. H&E ×100.

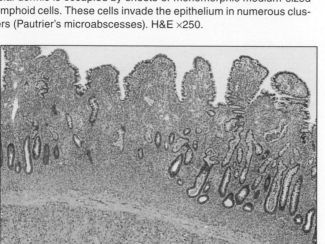

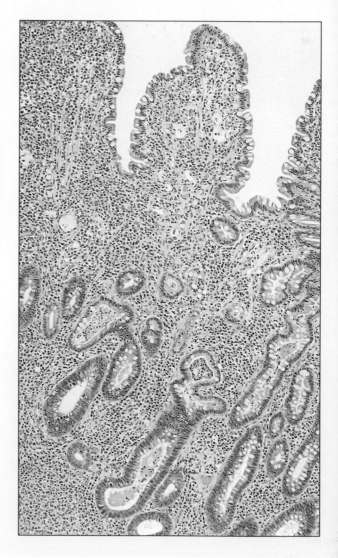

Figure 16.40, Figure 16.41 Alimentary lymphosarcoma (dog). Sections from the small intestine of a 12-year-old Staffordshire Bull Terrier with a history of weight loss and diarrhoea. **Figure 16.40** (lower power) shows shortened villi and crypts distorted by a heavy cellular infiltrate which extends through all layers of the intestinal wall. In **Figure 16.41**, the infiltrating population are large lymphoid cells. In dogs, alimentary lymphosarcoma most commonly affects the small intestine and is usually primary but may be associated with multicentric lymphosarcoma. H&E **Figure 16.40**, ×25; **Figure 16.41**, ×125.

Avian Lymphatic System

The lymphoid system in the bird consists of the spleen, thymus, local nodules in the wall of the lymphatic vessels and the mucosae, and the cloacal bursa (of Fabricius). There are no lymph nodes as such, although diffuse lymphatic tissue and lymphatic follicles may be present.

Thymus

The thymus has a fine connective tissue capsule extending into the parenchyma and dividing it into a variable number of lobes. The lobes are divided into lobules as in mammals. Each lobe has a cortex and medulla of thymocytes. These cells are the source of T lymphocytes. The reticular cells form islands of vacuolated cells, the equivalent of the mammalian thymic corpuscles.

Bursa

The cloacal bursa is an oval sacculated organ dorsal to the cloaca and communicates with it by a small opening. It is a central lymphoid organ that seeds B lymphocytes to the germinal centres of peripheral lymphoid deposits and the spleen and is the primary site for the synthesis of immunoglobulin in the young bird before involution commences from 3 to 4 months. The mucosal wall is thrown into folds covered by a pseudostratified columnar epithelium continuous with that of the cloaca. The folds are subdivided by connective tissue trabeculae into lobules (**Figure 16.43**). Each fold consists of a densely populated outer cortex of lymphocytes and an inner sparsely populated medulla separated by a layer of undifferentiated epithelial cells (**Figure 16.44**). Lymphoid tissue and the overlying mucous membrane are in close apposition, and lymphocytes migrate through the epithelium.

Mucosal-Associated Lymphoid Tissue

The mucosal-associated lymphoid tissue is present in the tubular digestive tract. The densest aggregation is found in the narrow proximal region of the caecum, the so-called

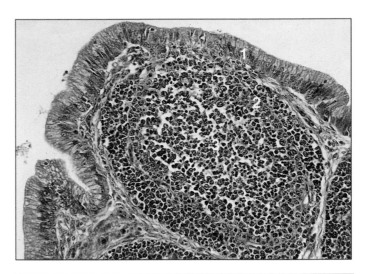

Figure 16.44 Bursa (bird). (1) Simple columnar epithelium of the cloaca. (2) Cortical area of densely packed lymphocytes. (3) Sparsely populated medulla. H&E ×125.

polycryptic caecal tonsil (*see* Chapter 9). Both T and B lymphocytes are present.

Spleen

The capsule is fibromuscular as in the mammal, but thinner. The trabeculae are poorly defined. The white pulp is lymphoid tissue sheathing an artery, and the red pulp is a loose arrangement of fenestrated blood sinusoids. The function of the spleen is phagocytosis of senescent erythrocytes, lymphopoiesis, and antibody production as in the mammal.

CLINICAL CORRELATES

Young pigeons can be infected by the pigeon circovirus (PiC), causing lymphoid depletion of the bursa and, on many occasions, secondary bacterial infections due to the lymphoid depletion caused by the destruction of the bursa (**Figure 16.45**).

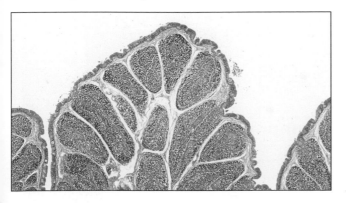

Figure 16.43 Bursa (bird) Follicles separated by connective tissue trabeculae and lined by a simple columnar epithelium. H&E ×50.

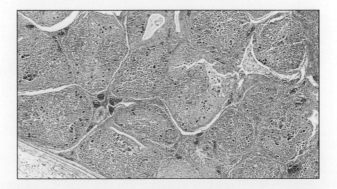

Figure 16.45 Bursa from a pigeon (*Columba livia*) infected with PiC, showing severe necrosis of the lymphoid tissue. H&E ×50.

Reptilian Lymphatic System

Although differences exist between various lymphoreticular and immune systems within the class Reptilia, these systems have many similarities both in form and in function. The thymus may be retained as a functional organ throughout life, or it may involute, either seasonally or because of age and become inactive. Haemopoiesis normally occurs in the bone marrow. It can occur within extramedullary sites, such as the liver, spleen, or kidneys, if the animal experiences acute blood loss caused by traumatic haemorrhage, chronic blood loss, or anaemia. Pre-existing mature erythrocytes and even pluripotential thrombocytes can participate in the formation of erythrocytes by being recruited and transformed into erythrocytes through mitotic and amitotic division. The reptilian spleen is similar to the mammalian and, together with large bone-marrow macrophages, removes senescent blood cells. During the recycling of haeme pigment from engulfed erythrocytes, reptiles produce the green degradation product biliverdin (**Figure 16.46**), instead of the bilirubin that is produced in mammals. Lymph is circulated by contractile lymph hearts, as in amphibians.

The thymus of reptiles is similar to that of mammals (**Figure 16.47**). In many reptiles, the thymus (or thymic lobes) lies adjacent to the parathyroid lobes and the ultimobranchial bodies, which are ventral and very near the internal carotid arteries, jugular veins, and vagus nerves. The lymphoid structure is encapsulated by dense fibrocollagenous connective tissue. It may be lobulated by fine connective tissue trabeculae (in many snakes) or

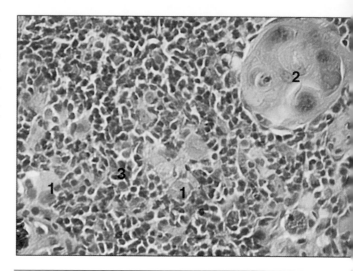

Figure 16.47 Section of cortical part of a thymic lobule from a desert tortoise (*Xerobates agassizi*). (1) Epithelial cells. (2) Concentric (Hassall's) corpuscle. (3) Thymocytes. H&E ×200.

nonlobulated. If lobulated, it may be further arranged into discrete medullary and cortical regions, depending upon the family within the class Reptilia. A hollow cavity may or may not be present. The parenchyma comprises the following: small, round thymic lymphocytes or thymocytes which predominate; epithelial ('epithelioid') cells with pale staining nuclei and prominent nucleoli; elongated pink staining myoid cells; pearl-like, pale pink concentrically lamellated (Hassall's) corpuscles; occasional macrophages; and granulocytes, especially heterophils. Mucoid cysts have been described in some reptilian thymic specimens. Small blood vessels penetrate the thymus at the hilus or enter directly through the capsule and branch out into smaller arterioles, capillaries, and venules before exiting in a parallel manner.

The alimentary, respiratory, cardiovascular, urogenital, and integumentary systems of reptiles contain discrete aggregates of lymphoid tissue. Although reptiles lack a true bursa of Fabricius (similar to that which is present in birds), there are well-delineated lymphoid nodules located within the intestine and cloacal vent of many species.

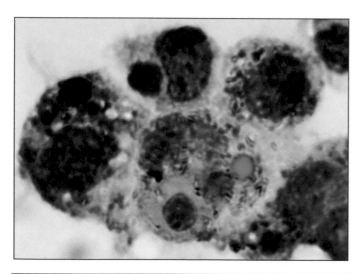

Figure 16.46 Bone marrow cytology specimen from an iguana. This section contains several macrophages with engulfed cellular material and bacteria. Note the green inclusion within the large macrophage which contains biliverdin, the degradation product of haemoglobin in amphibians, reptiles, and birds. H&E ×630.

Amphibian Lymphatic System

The immune system of many amphibians is intermediate between that found in teleost fish and in some reptiles. The thymus is present but variable in size and complexity, depending upon the family of amphibian. In adult mphibians, the thymus is located in the ventral cervical egion, usually just cranial to the heart and great vessels. There may be a discrete spleen or, in some amphibians, a combined splenopancreas that serves as a site for ymphoreticular, haemopoietic, digestive, and endocrine unctions.

Multiple contractile vessels, called 'lymph hearts', circulate lymph throughout the body. Discrete lymph nodes are acking, but lymphoid patches or aggregates are located in he walls throughout the alimentary, respiratory, and urogenital systems. Küpffer cells are arranged within the hepatic inusoids, as they are in higher vertebrates.

Amphibian leukocytes are similar to those observed in birds and reptiles.

Fish Lymphatic System

Generally, the lymphoreticular system of fish is more primitive than that of amphibians and reptiles. Lymph nodes are lacking, but aggregates of lymphoid tissue are located in the walls of the alimentary and urogenital systems. Leukocytes consist of lymphocytes, plasmacytes, large and small monocytic macrophages, heterophil-like granulocytes, eosinophils, azurophils, and basophils. In some species, two types of basophilic granulocytes have been described. The spleen of fish serves as a site of haemopoiesis, although erythrocytes may mature after they have been released into the circulation. It has an immunogenic function, joining with other lymphoid organs in the production of (at least) serum neutralising antibodies in response to antigenic challenge. It removes senescent erythrocytes from the circulating pool of red cells. Stellate (Küpffer) cells line the hepatic sinusoids. The mucus covering the integument of many fish represents the first line of defence against invasion by pathogenic micro-organisms: immunoglobulin A antibodies secreted with the mucus confer a degree of immunity specifically directed against certain antigens. Lymphatics drain erythrocyte-free blood plasma that seeps through capillary walls and returns it to the veins. Lymphatics also drain fat, as chylomicrons, from the intestinal villi.

CLINICAL CORRELATES

Primary lymphoid neoplasms such as thymic lymphosarcoma (**Figure 16.48**), multicentric lymphosarcoma, and lymphatic leukaemia occur in some exotic species. Secondary tumour deposits can also affect the lymphatic system. Other types of lymphatic tumours can infiltrate sites that usually do not contain a substantial lymphoreticular component (**Figure 16.49**).

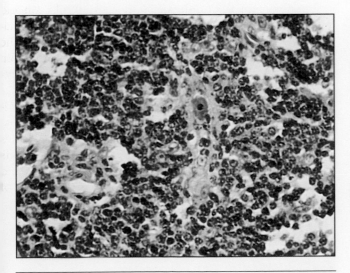

Figure 16.48 Thymic lymphosarcoma in a desert tortoise (*Xerobates agassizi*). The distinction between the cortical and medullary zones is lost as thymocytes proliferate and distort the architecture of the thymic lobule. Scattered pale-staining histiocytes, blood vessels, and small smooth muscle fibres are also present. H&E ×200.

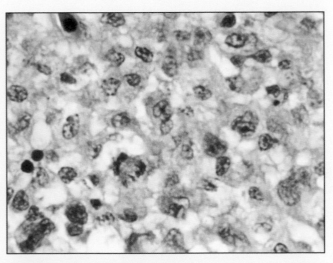

Figure 16.49 Malignant histiocytic lymphoma (lymphosarcoma) in a boa constrictor (*Boa constrictor*). This neoplasm is composed of histiocytic macrophages with vesicular nuclei, prominent nucleoli, and quite abundant pale cytoplasm. The mitotic index is high. H&E ×400.

Cells of the Immune System

The primary function of the lymphatic cells in the lymph nodes, spleen, thymus, and mucosal sites is the protection of the body from infection by organisms such as bacteria and viruses and foreign antigens. This protection, the immune response, constitutes the body's main defence mechanism and operates mainly in two ways: humoral response, the synthesis and release of free antibody into the blood, and cell-mediated immunity, the production of sensitised cells with antibody receptors on the surface. Both of these aspects of the immune response are carried out by lymphoid cells, the commonest being small lymphocytes. These circulate freely between the blood and the lymph, and between the lymphoid organs. They migrate into connective tissue and form diffuse cellular infiltrations.

There are two distinct populations of small lymphocytes: T and B cells (*see* **Figures 16.21** and **16.22**). They are morphologically very similar but functionally distinct and are identified by using special markers. Both lymphocyte subsets first appear in the yolk sac of the embryo, migrating via the blood to the liver in the mammal, and the spleen both in birds and in mammals. From there, they migrate to the bone marrow. In the final migration, the T stem cells migrate to the developing thymus and the B stem cells to the cloacal bursa and equivalent mucosal epithelial related sites in the mammal.

After stimulation by antigens, both cell types proliferate and differentiate to become either memory or effector cells. T cells differentiate in the thymus, and B cells differentiate in the bursa in birds and bone marrow in mammals. Memory cells can mount an immune response after reencounter with an antigen. Effector B cells secrete immunoglobulins and become plasma cells. Effector T cells can be divided into two further subsets: T helper cells, secreting cytokines, and T cytotoxic cells, attaching to antigen on target cells to kill them. NK cells are another type of lymphocytes that do not stain for CD3 or CD19/20, typical of T and B cells, respectively. NK cells can eliminate target cells as T cytotoxic cells or without using the same antigen recognition system

Macrophages are phagocytic cells widely distributed throughout the body (*see* **Figure 16.23**). They form that part of the body's defences, often called the mononuclear phagocyte system, and they originate in the bone marrow. A committed stem cell matures into a monocyte, is released into the circulation, leaves the blood, and migrates into the tissues. There it increases in size, in lysosomal enzyme content and endocytotic activity, to become a macrophage capable of phagocytosis. Macrophages may be free or fixed to other tissue cells. Free macrophages are scattered in the connective tissue as histiocytes (**Figure 16.50**), in body cavities, in pulmonary alveoli as dust cells, and in the spleen and lymph nodes. Fixed macrophages line the blood and lymph sinusoids of the bone marrow, liver (Kupffer cells, **Figure 16.51**), spleen, and lymph nodes, where they are supported by a fine network of reticular fibres (previously known as the reticuloendothelial system). As the blood or lymph moves slowly along the sinusoids, the lining macrophages recognise non-self-antigens (e.g. from virus or bacteria) and altered self (senescent erythrocytes) materials and phagocyte and degrade them. Macrophages also participate in the immune response by processing and presenting antibodies on the cell surface to T cells, triggering activation and proliferation of these cells.

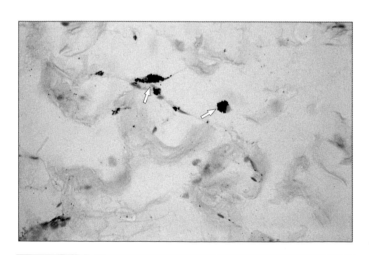

Figure 16.50 Loose connective tissue (dog). The histiocytes have phagocytosed the injected carbon particles (arrowed) ×250.

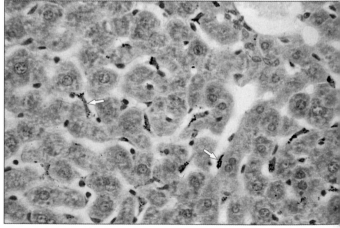

Figure 16.51 Liver (sheep). The macrophages lining the sinusoids have phagocytosed the injected carbon particles (arrowed). Safranin haematoxylin ×250.

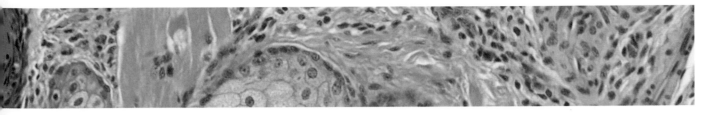

17

INTEGUMENT

The integument includes the skin and its derivatives. The skin is the largest organ in the body. It is a complex organ with many different functions including being the main barrier to the environment. The basic skin structure is quite similar in all mammals, but the thickness of the different layers varies depending on the species and the area of the body, being usually thicker in the dorsal and lateral areas and thinner in the ventral and medial areas.

The skin has three main layers, starting with the outer layer, the epidermis, a specialised epithelium derived from ectoderm, the dermis, a vascular dense connective tissue derived from mesoderm and the hypodermis (subcutis), a layer of loose connective tissue containing a variable amount of fat, which connects the skin to the underlying tissues. Irregular projections of the dermis, the papillae, interdigitate with evaginations of the epidermis, the dermal ridges. The dermis and hypodermis contain the blood vessels, lymphatic vessels, and nerves supplying the skin. Sweat, sebaceous, and mammary glands, as well as hair and feather follicles, are epidermal structures located in the dermis and hypodermis.

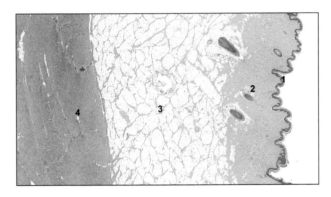

Figure 17.2 Dorsal skin (pig). Epidermis (1), dermis (2) with some hair follicles, hypodermis or subcutis, (3) and underlying muscle (4). H&E ×15.

Skin can be classified as thick or thin. Thick skin occurs in areas of wear and tear such as the footpad (**Figure 17.1**), and thin skin covers the rest of the body (**Figures 17.2** and **17.3**). The epithelium of the epidermis

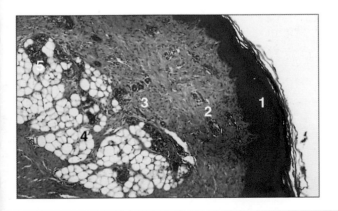

Figure 17.1 Digital pad (cat). (1) Epidermis: stratified squamous keratinised epithelium. (2) Dermis: connective tissue. (3) Hypodermis: loose connective tissue. (4) Adipose tissue. (5) Sweat glands. H&E ×50.

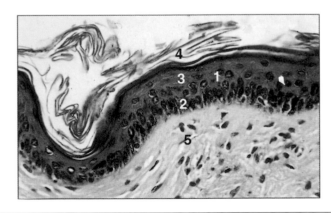

Figure 17.3 Thin hairless skin (horse). (1) The epidermis is three layer deep. The basal germinal layer (2) has melanocytes and clear cells; the middle layer (3) is hexagonal keratocytes; and the surface layer (4) is dead squames in the process of being shed. (5) Dermis: vascular connective tissue. H&E ×200.

DOI: 10.1201/9781003333807-17

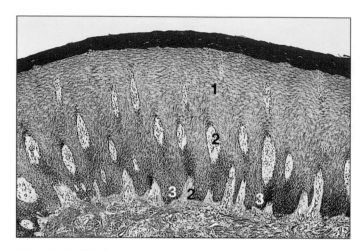

Figure 17.4 Dental pad – thick skin (ox). (1) The epidermis is at least 20 cell deep. (2) Dermal papillae project into the epidermis, necessary for the nutrition of the epidermal cells. (3) Equivalent epidermal pegs. Masson's trichrome ×62.5.

is stratified squamous and contains keratinising epithelial cells: keratinocytes. The number of cell layers varies considerably from 2–4 in thin skin (**Figure 17.2**) to 12–20 in thick skin (**Figures 17.1**, **17.4**, and **17.5**). In less exposed areas, the surface cells are dead squames and are sloughed off to be replaced from the basal layer. The skin of the nose of the horse is thin, with fine hairs, sebaceous, and sweat glands, and occasional sinus hairs. The planum of the nose of the other domestic mammals is covered by a thick, highly keratinised epidermis (**Figure 17.6**).

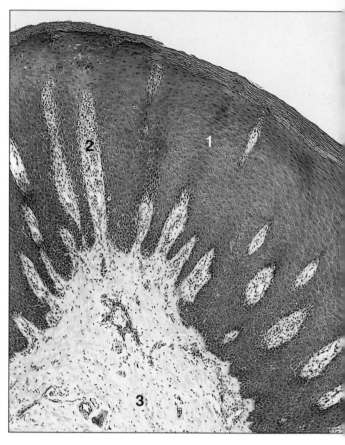

Figure 17.6 Planum nasale – thick skin (ox). (1) The epidermis is at least 20 layers of cells; the surface cells are keratinised. (2) Deep vascular dermal papillae project into the epidermis. (3) Hypodermis. H&E ×62.5.

Epidermis

Epidermis is the outermost layer of the skin and presents different cell layers (from inner to outer).

- The basal or germinal layer (stratum basale, stratum germinativum) consists of a layer of columnar or cuboidal keratinocytes resting on a basement membrane adjacent to the dermis (**Figure 17.7**). The nuclei are oval-shaped and large. Some basal cells have the ability to divide and produce new cells, acting as stem cells.
- The spinous layer (stratum spinosum, 'prickle cell layer') is composed of irregular cuboidal, polygonal, or flattened keratinocytes. Tonofilaments are more prominent in this layer compared to the basal cell layer (**Figure 17.8**).
- The granular layer (stratum granulosum) is formed by several layers of flattened cells migrating toward the surface and accumulating basophilic granules of keratohyalin in the cytoplasm (**Figure 17.9**). Lamellar granules are also present in this layer and composed of mainly lipids and enzymes.

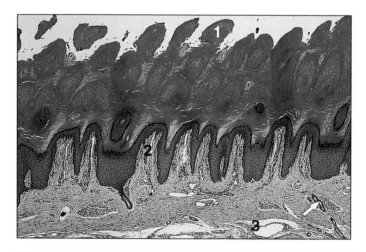

Figure 17.5 Digital pad – thick skin (dog). (1) The epidermis is heavily keratinised. (2) Dermal papillae project into the epidermis. (3) Hypodermis: loose connective tissue with sweat glands (arrowed). H&E ×62.5.

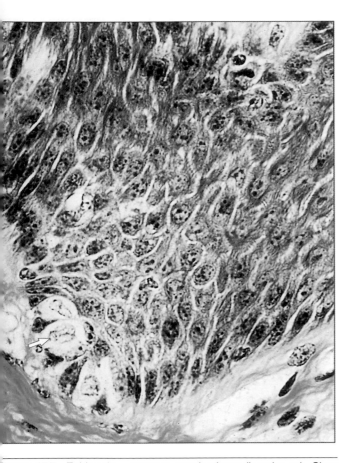

Figure 17.7 Epidermis – stratum germinativum (basale; ox). Clear cells (arrowed) lie in the basal germinal layer of columnar keratocytes. Masson's trichrome ×400.

- The clear layer (stratum lucidum) is only found in some areas of thick skin (e.g., planum nasale or footpad). The keratinised cell loses its clear-cut outline and becomes homogeneous and translucent, losing its nucleus (**Figure 17.10**).

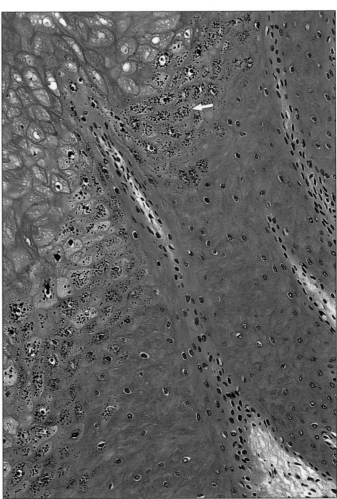

Figure 17.9 Epidermis – stratum granulosum. Hoof (horse). The keratocytes have distinctive deep blue granules in the cytoplasm (arrowed). H&E ×125.

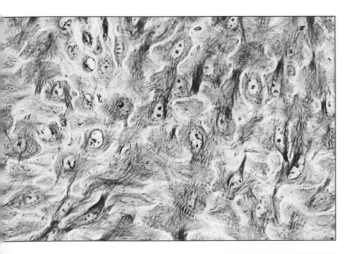

Figure 17.8 Epidermis – stratum spinosum (prickle cell layer; ox). The hexagonal keratocytes have a large clear nucleus. The cytoplasm has extensive spines or prickles. Masson's trichrome ×400.

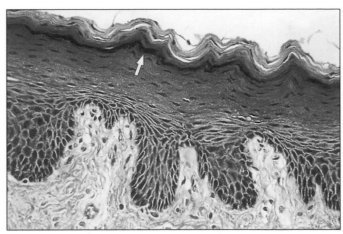

Figure 17.10 Epidermis – stratum lucidum (dog). Digital pad. The bright, thin line (arrowed) marks the clear layer of thick skin. Gomori's trichrome ×100.

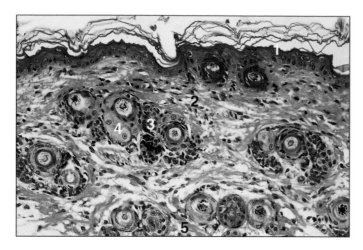

Figure 17.11 Thin skin (cat). (1) Epidermis with two layers of keratocytes. The free surface shows desquamating dead cornified cells. (2) Dermis. (3) Groups of hair follicles. (4) Sebaceous glands. (5) Smooth muscle. H&E ×125.

- The cornified layer (stratum corneum) is the outer layer, consisting of several layers of cells that are non-nucleated, keratinised, and desquamating (**Figures 17.1**, **17.4**, and **17.5**).

In structures composed of hard keratin, such as hooves and claws, both the granular and clear layers are absent. The epidermis of thin skin is composed of relatively few cells, but the number varies with the location. It lacks a clear layer and the granular layer is now always evident (**Figures 17.2** and **17.11**).

Skin colour is determined by the presence or absence of pigment cells, or melanocytes, in the basal layer (**Figure 17.12**). These cells originate from the neural crest.

Langerhans cells are mainly in the stratum spinosum of the epidermis. These are specialised clear intraepidermal macrophages that are involved in the processing of antigen (**Figure 17.7**). Also present in the basal layer are the tactile epithelial cells (Merkel cells), which are similar in appearance to the Langerhans cells but may have a few granules. They have a neuroendocrine function.

Dermis

The dermis consists of loose and dense irregular connective tissue with a mesodermal origin. In thick skin, the superficial, loose tissue of the dermis, the papillary layer, forms projections: the dermal papillae at the dermal–epidermal junction. This increases the surface area for nutrition and may cause surface ridges in the skin. Equivalent epidermal pegs interdigitate with these and anchor the skin. The deep layer of dense irregular tissue is called the reticular layer. Dermal papillae are reduced or absent in thin skin. Smooth muscle fibres (arrector pili muscles) near the hair follicle are present in the dermis. In some specialised areas, smooth (scrotum, penis, or teat) or skeletal (large sinus hairs of the facial area) muscle can be found.

Hypodermis

The hypodermis consists of loose areolar connective tissue with deposits of fat. This may be deposited in well-nourished animals as a continuous layer beneath the skin, the panniculus adiposus, or be confined to places such as the footpad to absorb pressure (*see* **Figures 17.1**, **17.2**, **17.13**, **17.14**, and **17.15**).

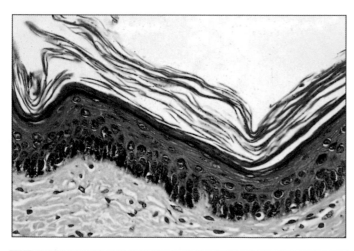

Figure 17.12 Pigmented skin (cat). The epidermis is four cell thick; melanocytes with brown pigment granules are present in the basal layer, the stratum germinativum. H&E ×250.

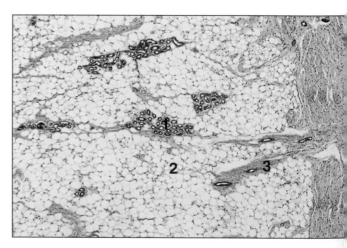

Figure 17.13 Digital pad – hypodermis (dog). (1) Sweat glands. (2) Fat cells or adipocytes. (3) Connective tissue. H&E ×62.5.

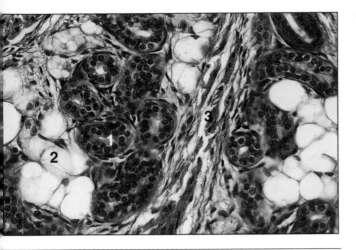

Figure 17.14 Digital pad – hypodermis (cat). (1) Sweat glands. (2) Fat cells or adipocytes. (3) Connective tissue with blood vessels. H&E ×250.

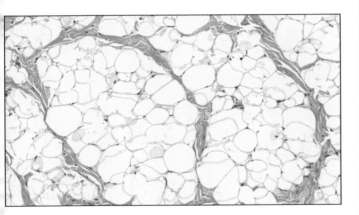

Figure 17.15 Dorsal skin – hypodermis (pig). Abundant adipocytes with connective tissue. H&E ×200.

SKIN APPENDAGES

Hair and Hair Follicles

Hairs are flexible keratinised structures that cover most of the mammals' bodies.

Hairs are produced by hair follicles. The part of the hair within the follicle is the hair root, which has a terminal known the bulb. The free part above the skin surface is called the shaft.

Hair follicles develop in the embryo as localised proliferations of the epidermal epithelium, growing down into the underlying mesenchyme to form a cylinder with an expanded distal end (bulb). The bulb is invaginated by a vascular papilla of dermal connective tissue; the germinal

(matrix) cells of the bulb proliferate to form the hair. A hair near its origin consists of a central medulla of cuboidal cells, a cortex of flattened cells orientated parallel to the long axis of the hair, and an outer cuticle. The germinal cells also form the inner root sheath. This grows from the papilla of the hair bulb to the opening of the sebaceous gland, where the hair becomes a cuticle. The hair cuticle consists of scale-like cells that partially overlap so that their free edges are directed upward. The cells of the sheath cuticle are directed downwards so that the hair and the sheath interlock. The peripheral outer root sheath represents a downward continuation of the epidermis. A dermal sheath abuts the basement membrane of the external root sheath, surrounds the follicle, and blends with the rest of the dermal connective tissue. The arrector pili muscle attaches to the connective tissue sheath of the follicle and the superficial layer of the dermis (**Figures 17.16–17.21**). As the

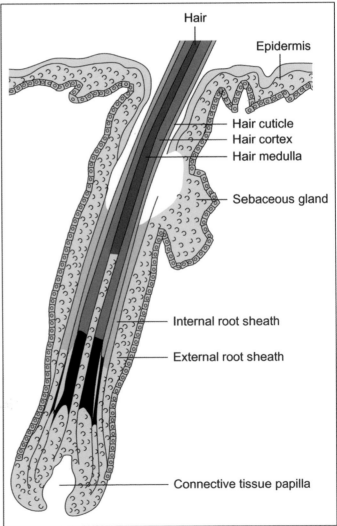

Figure 17.16 Hair follicle.

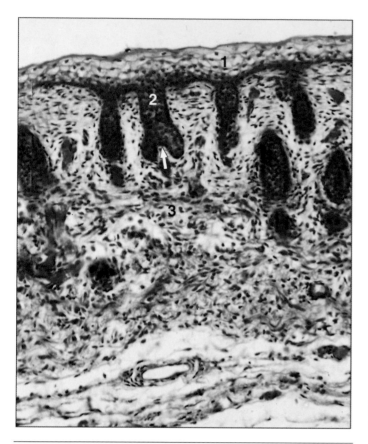

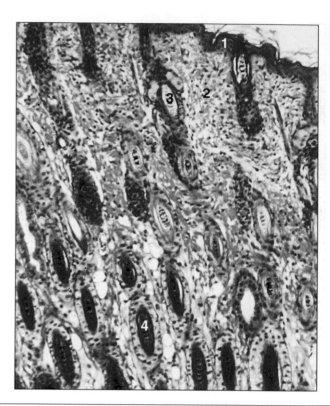

Figure 17.18 Neonatal skin (cat). (1) Epidermis. (2) Dermis. (3) Hair shaft. (4) Root sheath. H&E ×250.

Figure 17.17 Skin. Cat embryo. (1) Ectoderm. (2) The ectodermal cylinder with the invaginated papilla (arrowed) forms the primordium of the hair follicle. (3) Mesoderm. H&E ×250.

hair approaches the surface, both the medullary and the cortical cells shrink and become keratinised, lose their nucleus, and acquire air bubbles. Hair colour is determined by the relative proportions of pigment granules

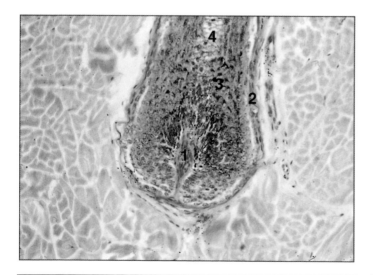

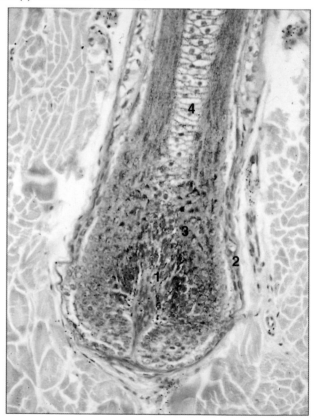

Figure 17.19 Bulb region of the hair follicle. Horse (skin). (1) Dermal papilla. (2) Dermal root sheath. (3) Epidermal root sheath. (4) Hair. Note the melanin pigment in the epidermal sheath. H&E ×125.

Figure 17.20 Hair follicle. Skin (horse). (1) Dermal papilla. (2) Dermal root sheath. (3) Epidermal root sheath with outer and inner sheath. (4) Hair. Note pigmentation. H&E ×250.

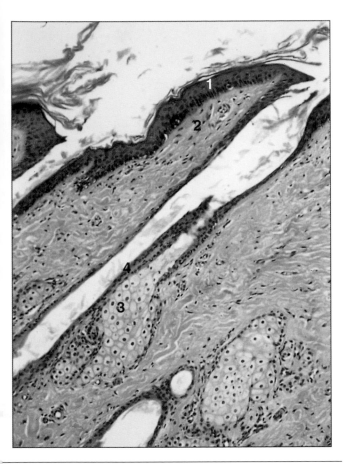

Figure 17.21 Skin (horse). (1) Epidermis. (2) Dermis. (3) Sebaceous gland opening into the hair follicle. (4) Epidermal root sheath is reduced to two layers of cells; the inner layer is thin horny scales. Note the oblique set of the hair follicle. H&E ×125.

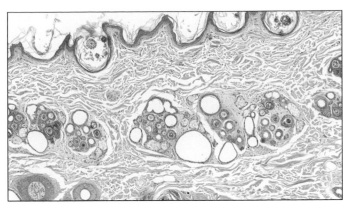

Figure 17.23 Skin (dog). Compound hair follicles in the dermis. H&E ×50.

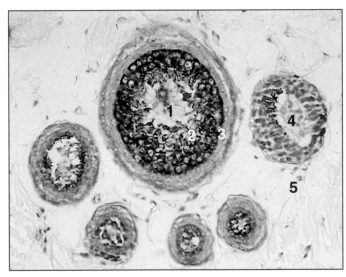

Figure 17.24 Skin. Compound hair follicles (cat). (1) Large cover hair cut in cross-section: (2) epidermal sheath and (3) hair. (4) Fine wool/lanugo hairs. (5) Dermis. H&E ×250.

and air bubbles. Dark hair has more pigment and fewer bubbles (**Figures 17.22** and **17.23**).

Hair follicles are evenly and singly arranged in the horse, ox, and pig (**Figure 17.22**) but occur in groups as compound follicles in the dog and cat (**Figures 17.23** and **17.24**). Hair follicles are set obliquely in the skin except in sheep, where they are vertical (**Figure 17.25**). Compound follicles in the

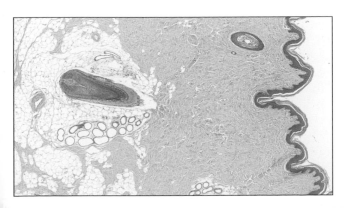

Figure 17.22 Skin (pig). Simple hair follicle within the subcutis and dermis. H&E ×50.

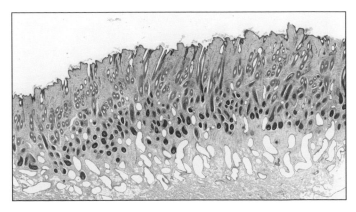

Figure 17.25 Woolly skin. Flank (sheep). The hair follicles are all of the smaller wool/lanugo type, set individually in the dermis and lying vertically. H&E ×20.

sheep consist of a large cover hair and a variable number of fine wool hairs. Hairs are replaced at regular intervals, with the large coarse hairs of the mane and tail lasting throughout life.

Sinus (Tactile) Hairs (Vibrissae)

Tactile hairs are limited to the facial region. The dermal sheath is highly developed and split by a blood sinus into inner and outer layers. In horses, pigs, and ruminants, the sinus is trabeculated throughout its length. In carnivores, the upper region is nontrabeculated, forming an annular sinus. Free sensory nerve endings are associated with the epidermal cells of the hair and the dermal sheath (**Figures 17.26–17.28**).

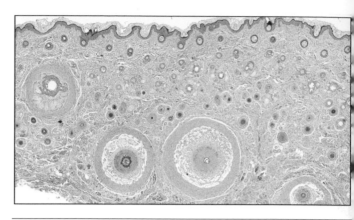

Figure 17.27   Tactile hairs (dog) in the facial region (transversal section). H&E ×25.

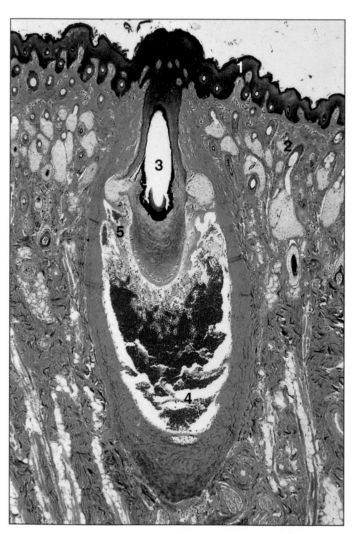

Figure 17.26   Sinus/tactile hair (horse). (1) Epidermis. (2) Dermis. (3) Hair follicle. (4) Blood sinus in the dermal sheath. (5) Connective tissue trabeculae of the dermal sheath. Masson's trichrome. ×50.

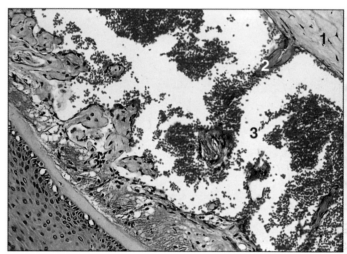

Figure 17.28   Sinus/tactile hair (horse). (1) Dermal sheath. (2) Connective tissue trabeculae of the dermal sheath. (3) Blood sinus. H&E ×125.

Sebaceous Glands

Sebaceous glands are associated with the hair follicles and secrete sebum, an oily substance. The basal layer of each gland is the squamous or cuboidal epithelium. As the cells divide, some are pushed toward the surface away from the basement membrane. Vacuolated secretory cells synthesise lipid; the mature secretory cells degenerate, forming sebum. This is a form of holocrine secretion (**Figures 17.29–17.32**). The smooth muscle of the hair follicle, the arrector pili, contracts and helps to express the sebum onto the skin surface. Sebaceous glands may open directly onto the surface of the skin in the absence of hair follicles in sites such as the tarsal gland of the eyelid (**Figure 17.33**) or lips (**Figure 17.34**).

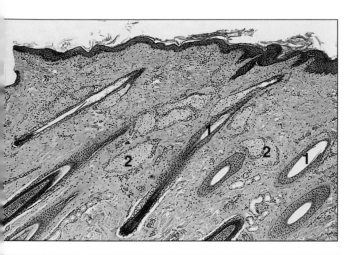

Figure 17.29 Skin (horse). (1) Single hair follicles are present in the dermis. (2) Sebaceous glands. H&E ×62.5.

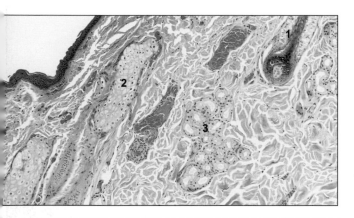

Figure 17.30 Skin (horse). (1) Single hair follicles. (2) Sebaceous and (3) sweat glands. H&E ×100.

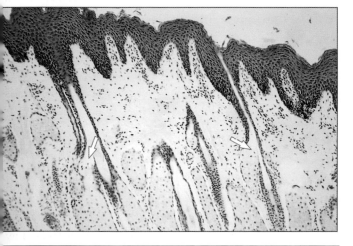

Figure 17.31 Flank skin (horse). The sebaceous glands are numerous and associated with the hair follicles, often opening directly into the follicle (arrowed). H&E ×100.

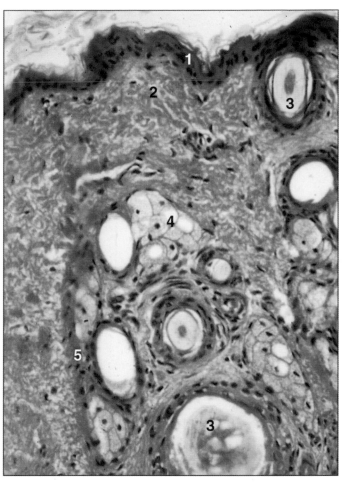

Figure 17.32 Skin (sheep). (1) Epidermis. (2) Dermis. (3) Hair follicle. (4) Sebaceous glands. (5) Arrector pili muscle; contraction assists expression of the sebum. H&E ×250.

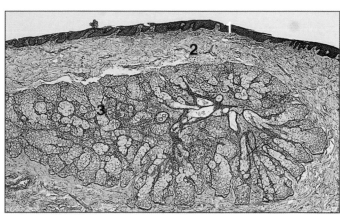

Figure 17.33 Eyelid (horse). (1) Epidermis. (2) Dermis. (3) Tarsal gland, sebaceous secretion. H&E ×62.5.

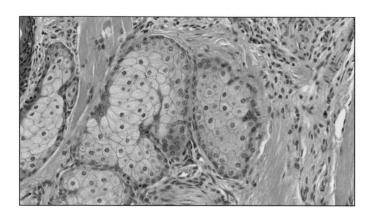

Figure 17.34 Lip (sheep). Large sebaceous glands. H&E ×300.

Sweat Glands

Sweat glands may be winding and highly coiled or tubular and sac-like. There are two types: apocrine and merocrine.

The apocrine gland, which is more common, is situated in the dermis. The simple columnar epithelium has surface blebs of cytoplasm. These pinch off into the lumen, forming a mucoserous secretion. Their function is not completely understood but may involve the secretion of pheromones. Contractile myoepithelial cells lie between the secretory cells and the basement membrane (**Figure 17.35**).

The merocrine (eccrine) sweat gland is situated in the dermis or the hypodermis. The epithelium is cuboidal, and the secretion traverses the luminal cytoplasm without rupturing the membrane. There are few myoepithelial cells (**Figure 17.36**). The excretory ducts of both types of glands open into a hair follicle or directly onto the skin surface.

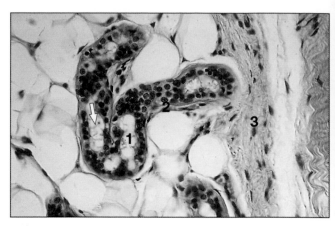

Figure 17.35 Apocrine sweat gland (dog). (1) Gland tubule lined by a simple columnar epithelium; secretory blebs are arrowed. (2) The myoepithelial cells lie between the secretory cell and the basement membrane. (3) Dermal connective tissue. H&E ×400.

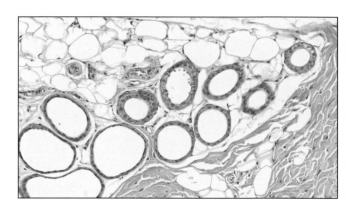

Figure 17.36 Merocrine sweat glands (pig) from the carpal region. H&E ×200.

CLINICAL CORRELATES

In veterinary practice, especially small animal practice, skin disease is one of the most common problems encountered. Horses and production animals also suffer from skin diseases with a relatively high frequency. Genetic factors are important within particular breeds, and environmental factors are responsible in many cases.

Inflammation of the skin is termed dermatitis (**Figure 17.37**). The skin has a fairly limited range of possible response patterns to inflammation; when an area of skin is pruritic, the animal tends to traumatise it, producing secondary changes such as thickening of the epidermis. Over time, hyperpigmentation and scarring of the dermis can develop.

Infection by micro-organisms including bacteria, fungi, viruses, and parasites can cause skin disease. In some occasions, the infection only affects the outer layer of the skin, like in the porcine exudative epidermitis ("greasy skin disease"), produced by the infection

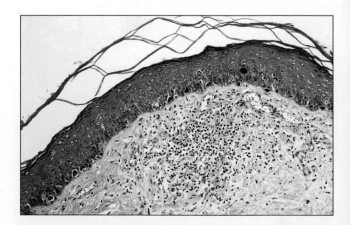

Figure 17.37 Dermatitis (dog). In this section of skin from a dog with dermatitis, there is epidermal hyperplasia with an increased thickness of stratum spinosum (acanthosis) and a mild increase in the depth of the stratum corneum (hyperkeratosis), where the keratin is arranged in a loose, woven pattern. There is slight dermal congestion and oedema and a heavy, mostly perivascular, dermal inflammatory infiltrate of neutrophils. Epithelial pigmentation is quite prominent. H&E ×62.5.

Staphylococcus hyicus and characterised by exfoliation, exudation of a serosebacous fluid, and formation of crust (**Figure 17.38**).

In some cases, the development of the disease may result from a defect in immunity or a breakdown in the host–parasite relationship. In canine demodicosis (**Figure 17.39**), *Demodex canis* mites (also found in low numbers in the hair follicles of normal dogs) pass from dam to pups during suckling and proliferate to cause disease. Species-specific demodex mites are widespread in many species, including humans. Larger parasites, like acari, can also be attached to the epidermis, like *Myobya musculi*, in mice (**Figure 17.40**) or deeper into the dermis.

The integument is also a common site of neoplasms. In some tumours, such as squamous cell carcinoma (**Figure 17.41**) and cutaneous haemangioma and haemangiosarcoma, chronic exposure to solar radiation is a

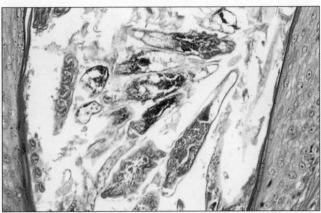

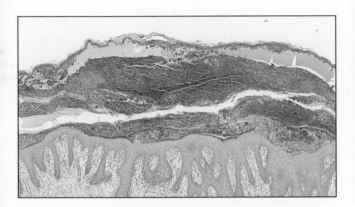

Figure 17.38 Epidermitis (pig) A case of exudative epidermitis. Only the outer layers of the epidermis are heavily affected, and a typical lesion is the presence of a crust containing abundant cell debris on top of the epidermis. H&E ×50.

Figure 17.39 Demodicosis (dog). This high-power micrograph shows a canine hair follicle packed with demodex mites. The mites have elongated, cigar-shaped bodies with four pairs of legs at the head end. H&E ×250.

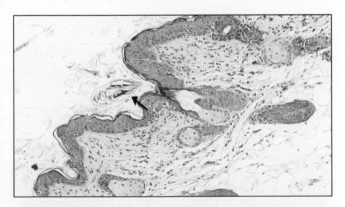

Figure 17.40 *Myobia musculi* (arrow) in the epidermis of an infected mouse. H&E ×200.

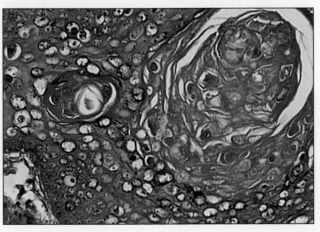

Figure 17.41 Squamous cell carcinoma of the foot of an Indian hedgehog (*Hemiechinus hemiechinus*). Note the marked cellular atypia, squamous metaplasia, and keratin 'pearls'. H&E ×200.

known predisposing factor. Lightly pigmented areas are particularly vulnerable to actinic, or sunlight-induced, lesions. A high incidence of neoplasia, both benign and malignant, is recognised in the dog and, e.g. the quite frequent mast cell tumours (**Figures 17.42–17.43**).

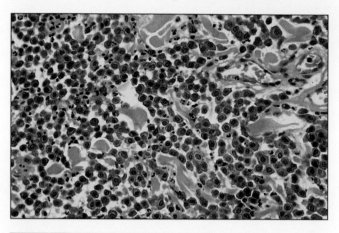

Figure 17.42 Mast cell tumour (dog). From a 5-year-old Boxer. This breed suffers from a high incidence of these and other skin tumours. Irregular sheets and clusters of round cells with round nuclei and well-defined, noticeably granular cytoplasm extend through the connective tissue, accompanied by numerous eosinophils. H&E ×250.

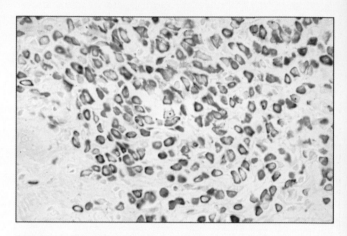

Figure 17.43 Mast cell tumour (dog). In this example of a canine mast cell tumour, the granular cytoplasm of the mast cells is highlighted a blue colour by a special stain. Astra blue ×125.

Avian Skin

Avian skin consists of an epidermis, which is generally thinner than in mammals, and a dermis, which in feathered skin lacks papillae and is nonglandular (**Figure 17.44**).

Feathers are keratinised epidermal derivatives. During initial development, a dermal–epidermal papilla is formed. This sinks beneath the surface and becomes a follicle as in the mammal (**Figure 17.45**). The epidermal cells proliferate to form a cylinder. This becomes the quill, whereas the upper part becomes ridged to form the vane.

Combs and wattles are skin appendages. The dermis contains an extensive network of sinus capillaries and

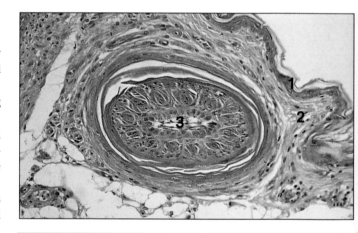

Figure 17.45 Skin (bird). (1) Epidermis. (2) Dermis. (3) Developing feather follicle. H&E ×100.

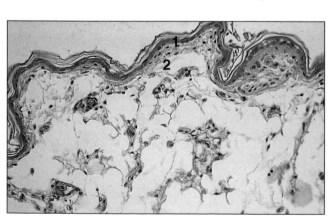

Figure 17.44 Skin (bird). (1) Epidermis. (2) Dermis. H&E ×100.

abundant mucous connective tissue. They are epidermal target organs, highly developed in the male, and are responsive to sex hormones.

The uropygial (preen) gland is the only skin gland in birds and is found at the base of the tail. The bilobed gland, which produces an oily secretion, is bound by a connective tissue capsule and drained by lobar ducts (**Figure 17.46**). Each tubule is divided into a sebaceous zone, a fatty zone, and a glycogen zone; the glycogen zone stains selectively with periodic acid-Schiff (PAS). They are lined with a multilayered epithelium that is very similar to

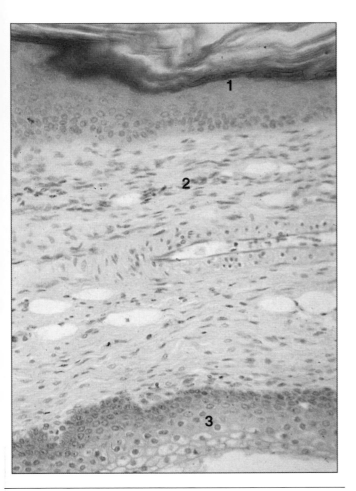

Figure 17.46 Uropygial gland (bird). (1) Epidermis. (2) Dermis. (3) Lobar duct lined by stratified epithelium. H&E ×200.

that of the mammalian sebaceous gland. The basal cells multiply, accumulate a fatty secretion in the cytoplasm, and degenerate. The secretion is passed through each duct to the isthmus and then to the papilla, which opens onto the surface (**Figures 17.47** and **17.48**).

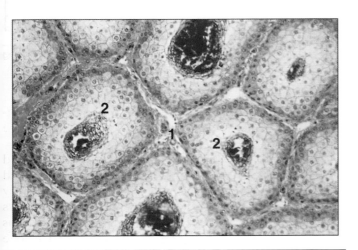

Figure 17.47 Uropygial gland (bird). (1) Connective tissue. (2) Secretory glands units. Masson's trichrome ×100.

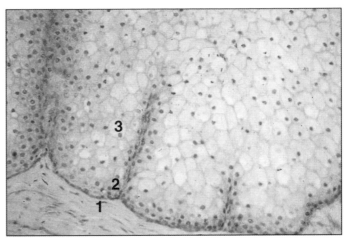

Figure 17.48 Uropygial gland (bird). (1) Connective tissue. (2) Basal layer of the secretory epithelium. (3) Superficial layers of fat-filled cells. H&E ×250.

Reptilian Skin

Depending upon the family within the class Reptilia, there are enormous differences in the integument of snakes, lizards, chelonians, crocodilians, and the tuatara. Snakes and lizards possess skin covered by scales, rounded tubercles, or other epidermal extensions (**Figures 17.49–17.51**). Rattlesnakes have an epidermally derived, loosely segmented, hollow rattle that vibrates when the tail tip is moved rapidly (**Figure 17.52**). A new button-like segment is added each time the rattlesnake moults its skin. Embedded osteoderms are present in the skin of many reptiles: for example, on the dorsal and lateral surfaces of crocodilians and on the limbs of some tortoises. The

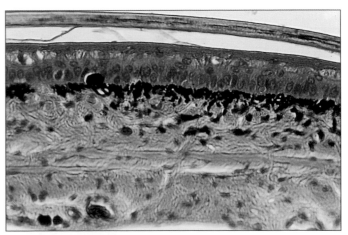

Figure 17.49 Section of pre-ecdysis integument of a mountain kingsnake (*Lampropeltis zonata*). The uppermost keratinised, lightly pigmented, pale pink layer is disengaged from the lightly cornified eosinophilic epithelium that lies immediately above the stratified squamous layer. Numerous black melanophages are present in the upper dermis. H&E ×200.

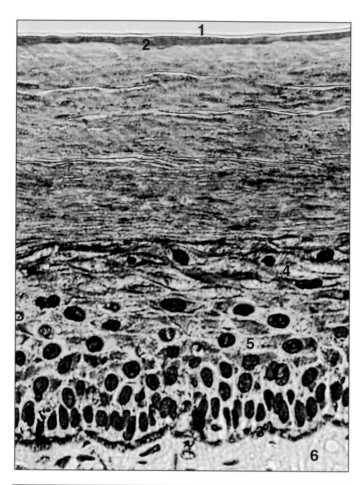

Figure 17.50 Reptile integument. (1) Senescent stratum corneum. (2) Current stratum corneum. (3) Stratum basale. (4) Stratum granulosum. (5) Stratum spinosum. (6) Dermis. Bodian's silver ×400.

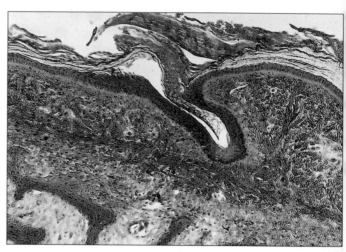

Figure 17.52 Rattle of a small Mexican rattlesnake (*Crotalus enyo*). This keratinised structure forms as a button-like protruberance to which a loosely interlocking segment is added each time the snake moults its integument. The most cranial segment retains a living core of dermis, whereas the tissues of the distal segments are no longer living. H&E ×50.

bony box-like shells of most chelonians are covered with keratinous plates (**Figures 17.53** and **17.54**). Even the soft and leathery shells of soft-shelled turtles are covered with lightly keratinised squamous epithelium. Snakes shed their senescent epidermis periodically as inverted tubes of (usually) unbroken old skin from which they crawl. Lizards, crocodilians, the tuatara, and chelonians moult their old epidermis piecemeal. Aquatic turtles shed one or more of their hard outermost layers of shell plates periodically. Many tortoises merely add more keratin concentrically to their plates (or scutes) as they grow throughout life.

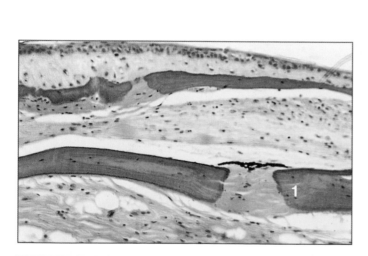

Figure 17.51 The integuments of some reptiles, particularly crocodilians and some lizards, contain multiple bony plaques called osteoderms (1). This section of a blue-tongued skink (*Tiliqua scincoides*) contains several of these ossified flat structures that are attached at their ends by fibrous connective tissue. These hinge-like connections permit flexibility. H&E ×100.

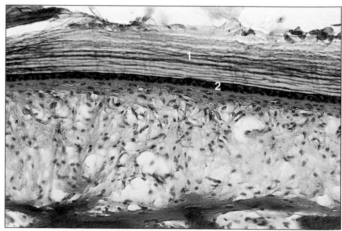

Figure 17.53 The carapacial shell of a red-eared slider turtle (*Trachemys scripta elegans*) is covered by a layer of horn-like keratin (1) and a variable thickness of stratified squamous epithelium (2). The dermis is variable in thickness, depending upon the size and species of the turtle. It covers multiple layers of membranous compact and cancellous bone in which bone marrow fills the cancellous spaces. H&E ×62.5.

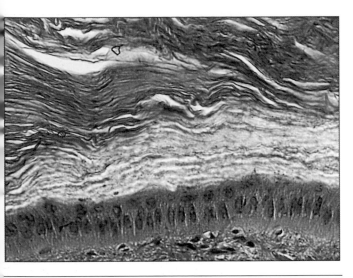

Figure 17.54 Section of the most superficial layers of the keratinised carapace of a terrestrial tortoise (*Xerobates agassizi*). Most terrestrial tortoises possess a much thicker carapacial and plastral shell than aquatic turtles or terrapins of the same size. H&E ×200.

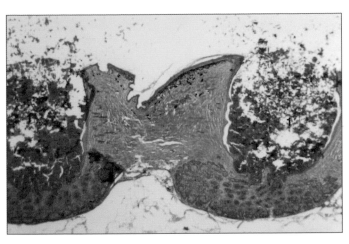

Figure 17.56 A cross-section of the skin of the ventral thigh of a sexually mature male green iguana (*Iguana iguana*), containing two femoral pores (1) The eosinophilic holocrine secretion from these glands is exuded as a waxy substance. H&E ×100.

Reptiles lack sweat glands. However, some lizards have a few sebaceous glands. In some species, modified sebaceous glands secrete holocrine, pheromone-rich waxy substances that are important inducers or releasers of sexual or territorial behaviour (**Figures 17.55–17.57**). Other scent-producing secretions are elaborated by cloacal and hemipenial sheath glandular structures (*see* Chapter 12, **Figure 12.40**) and by the sexual-segment granularity observed in the distal convoluted tubules of some male lizards and snakes (*see* Chapter 10, **Figure 10.21**).

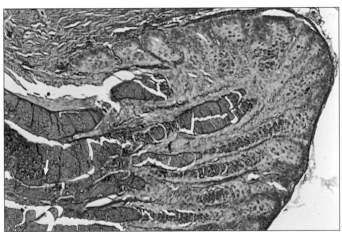

Figure 17.57 Medium-power magnification of a femoral pore from a male iguana. Note the eosinophilic secretion and cellular debris being extruded into the ductal system that empties its contents onto the epidermal surface (arrow). H&E ×100.

Amphibian Skin

Amphibian skin varies depending on the species. It can be either smooth or warty, lacks scales, and is kept variably moist by secretions that are produced by skin glands (**Figure 17.58**). When the skin secretions of many anuran and some caudate amphibians come into contact with a predator's (or human's) mucous membranes or are swallowed or injected, a lethally toxic reaction may result. Skin secretions of some amphibians may also possess potent antimicrobial properties that can be effective against a diverse group of pathogens.

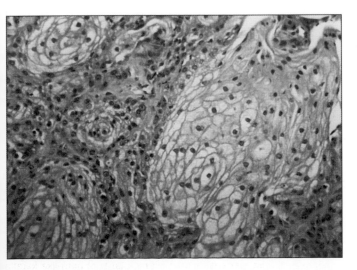

Figure 17.55 Mental gland from a desert tortoise. This modified sebaceous glandular structure is believed to produce a pheromone-rich secretion that initiates and at least partly mediates premating courtship behaviour. H&E ×200.

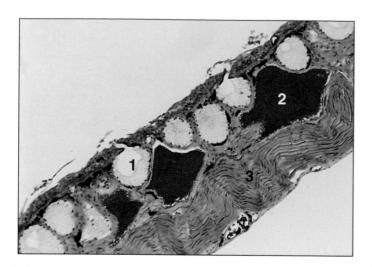

Figure 17.58 Full thickness section of the skin of an African clawed frog (*Xenopus laevis*). The integument contains two types of glandular structures: (1) clear-staining mucus-secreting glands and (2) highly eosinophilic poison glands. The secretory products of both are carried to the skin surface via short ducts. (3) Much of the thickness of the subepithelial tissue is comprised of skeletal muscle fibres. H&E ×50.

Fish Skin

Fish possess a varied spectrum of skin depending on the species. The integument of many elasmobranchs, particularly sharks and rays, is characterised by the presence of embedded tooth-like denticles. Other elasmobranchs lack these mineralised structures. Teleost fish may be either scaled or scaleless, but most of them have an integument containing mucous glands that secrete lubricative products onto the skin and fin surfaces. These mucoid secretions often contain immunoglobulin A antibodies, which provide a protective defence against infection. The mucous glands may be alveolar structures with short epithelium-lined ducts or groups of scattered goblet cells with mucopolysaccharide-rich secretions that flow over and cover the skin. Various barbels or other appendages aid in sensory or visual camouflage, species-specific territorial or courtship recognition, and defence functions.

Specialised Integumentary Structures

Special structures of the integument can be found in different locations in the body. For example, in the external ear (*see* Chapter 15), the pinna is covered on both sides by think skin with numerous sweat and sebaceous glands and

CLINICAL CORRELATES

Inflammation in the skin can be superficial or it can be extensive and involve deeper structures that are located beneath the epidermis and dermis, like in some amphibians (**Figure 17.59**). Snakes of the genus *Pituophis* exhibit a high incidence of pigment cell integumentary tumours (chromatophorous).

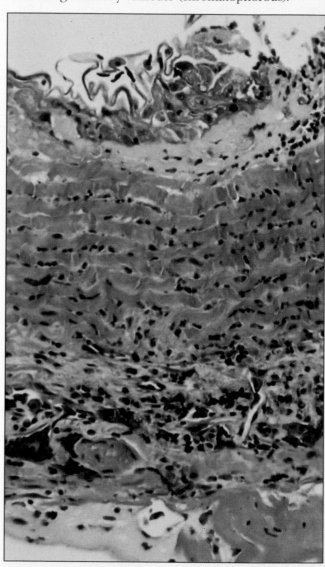

Figure 17.59 Nonsuppurative dermatitis, dermal ulceration, and superficial myositis in the skin of a clawed frog (*Xenopus laevis*). The epidermis is disrupted, and the dermis and skeletal muscular tissues are infiltrated by mixed mononuclear leukocytes. H&E ×250.

a variable number of hair follicles, normally greater in the convex than in the concave surface (**Figure 17.60**).

In the eyelids, typical skin can be found with prominent hairs (cilia or eyelashes) associated with sebaceous

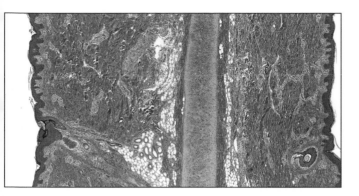

Figure 17.60 External ear (pinna). Outer surface (left), inner surface (right, with the presence of sweat glands and adipose tissue close to the auricular elastic cartilage). H&E ×70.

glands. Tarsal glands are multilobular sebaceous glands better developed in the upper eyelid, well developed in some species like the cat.

Specialised Glands in Mammals

Anal glands are modified tubulosaccular sweat glands opening into the anus. They are present in the dog, cat, and pig (**Figure 17.61**).

Anal sacs are present in carnivores and are located between the internal and external anal sphincters. The walls are lined with sebaceous and apocrine glands in dogs and cats (**Figure 17.62**).

Perianal (circumanal) glands are modified sebaceous glands found in the skin around the anus of dogs, consisting of nonpatent masses of polygonal cells (**Figure 17.63**).

Supracaudal or tail glands are local concentrations of sebaceous glands in the dog and cat. The secretion is used

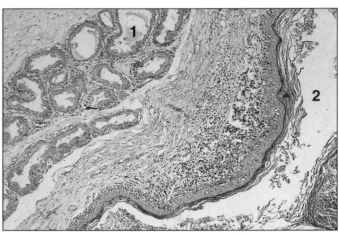

Figure 17.62 Anal sac (dog). (1) Apocrine tubular glands. (2) Anal sac. H&E ×100.

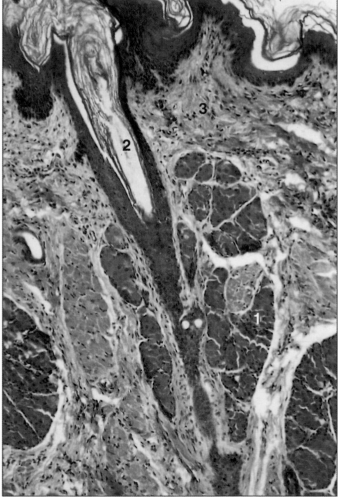

Figure 17.63 Perianal sinus (dog). (1) The glands are a mixture of sebaceous and sweat glands. (2) Hair follicle. (3) Dermal connective tissue. H&E ×250.

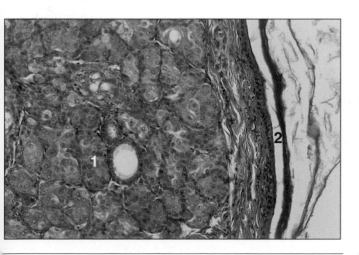

Figure 17.61 Anal gland (cat). (1) Tubular, saccular sweat glands in the circumanal connective tissue. (2) Anus lined by a stratified squamous keratinised epithelium. H&E ×200.

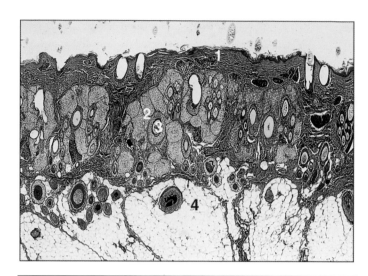

Figure 17.64 Supracaudal organ (cat). The supracaudal organ or tail gland is an area of sebaceous secretory units in the tail region; the secretion is used in grooming. (1) Epidermis. (2) Sebaceous glands. (3) Hair follicles. (4) Hypodermis. Sacpic staining method ×50.

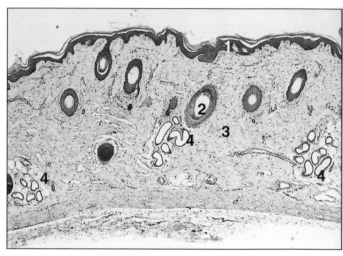

Figure 17.66 Carpal skin (pig). (1) Epidermis. (2) Hair follicles are distributed singly. (3) Dermis. (4) Merocrine carpal sweat glands. H&E ×50.

in grooming (**Figures 17.64** and **17.65**). A similar function in birds is served by the uropygial or preen gland (*see* **Figures 17.46–17.48**).

There are many other small collections of glands serving a variety of purposes. These include porcine carpal glands (**Figures 17.66** and **17.67**); merocrine; the sebaceous scent or horn gland of the goat; the submental sebaceous gland of the cat; and the interdigital glands of the sheep, a mixture of sebaceous and sweat glands (**Figures 17.68** and **17.69**).

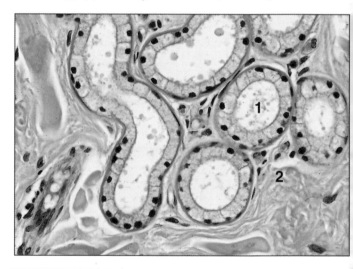

Figure 17.67 Carpal skin (pig). (1) Merocrine sweat glands are lined by a cuboidal epithelium. (2) Dermis. H&E ×400.

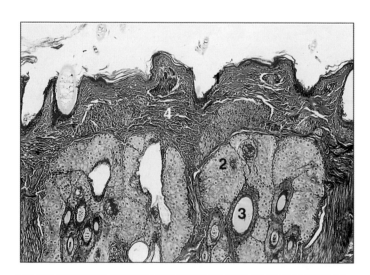

Figure 17.65 Supracaudal organ (cat). (1) Epidermis. (2) Sebaceous glands. (3) Hair follicles. (4) Dermis. Sacpic staining method ×100.

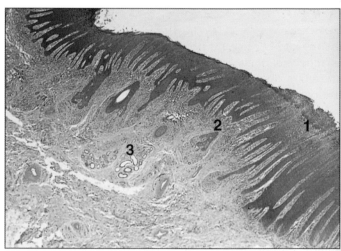

Figure 17.68 Interdigital skin (sheep). (1) Epidermis. (2) Dermis. (3) Sebaceous and sweat glands. H&E ×50.

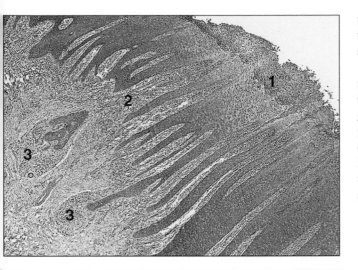

Figure 17.69 Interdigital skin (sheep). (1) Epidermis with epidermal pegs. (2) Dermis with deep dermal papillae. (3) Glandular region of the dermis. H&E ×125.

the bulb (swollen part of the wall behind the frog), and the digital pad where extensive fat deposits form shock-absorbing cushions.

Where the papillae are regular, the overlying epidermis gives rise to a hair-like structure: the horn tubule. Cells at the tip of the papilla form a core or medulla and grow toward the surface. They then shrink and disappear to leave a hollow tube surrounded by columns of cells equivalent to the hair cortex. The cells of the basal layer at the tip of the epidermal peg also grow toward the surface to form an intertubular horn (**Figures 17.70** and **17.71**).

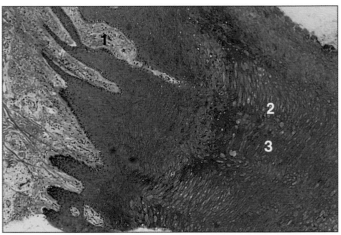

Figure 17.70 Hoof (calf). (1) Deep dermal papillae. Stratified squamous epithelium with (2) tubular and (3) intertubular horn. H&E ×25.

Specialised Glands in Reptiles and Amphibians

Several specialised glands exist in amphibians and reptiles. Many amphibians, especially frogs and toads, possess both mucus-secreting dermal glands and highly toxic poison glands that secrete complex alkaloid and amine-rich venom-like substances. Crocodilians and some chelonians possess paired modified sebaceous mental glands located on the underside of the front of their mandibles. These structures are larger in males than in females. Some lizards have femoral pores or anal pores (*see* **Figures 17.56** and **17.57**) with waxy holocrine secretions that are believed to contain pheromones.

The femoral-pore secretions of some desert-dwelling lizards are highly fluorescent when exposed to ultraviolet illumination. These species are able to see these secretions in reflected ultraviolet light. The femoral and anal pores are more highly developed in male lizards than in female lizards.

Hooves, Horns, and Claws

Hooves, horns, and claws are highly specialised derivatives of the epidermis. The dermis is very vascular and develops deep papillae, which are often sufficient to raise macroscopic ridges. It merges with the periosteum where present. The hypodermis is absent where the skin covers bone but forms a deep layer at the 'frog' (a pad of soft horn between the bars on the sole of a horse's hoof),

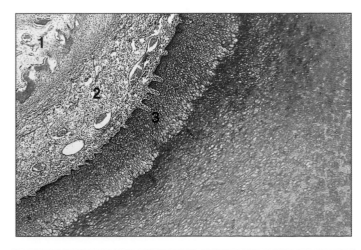

Figure 17.71 Hoof (foal). (1) Developing bone. (2) Vascular dermis. (3) Stratified squamous epithelium. H&E ×100.

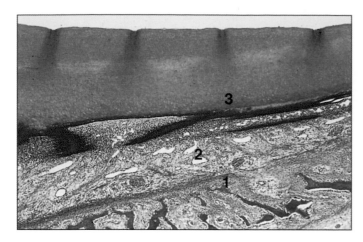

Figure 17.72 Hoof (foal). (1) Phalangeal bone. (2) Vascular dermis. (3) Epidermis with a thick stratum corneum. H&E ×25.

The epidermis is markedly keratinised, and this forms the hard outer surface of the hoof, horn, and claw (**Figures 17.70–17.74**).

Claws are horny plates forming a protective covering on the dorsal surface of the terminal phalanges in the carnivore. The skin is reflected around the third phalanx to form a fold: the bed of the claw. The basal layers of the epidermis form the germinal matrix and grow to become the horny claw (**Figures 17.75–17.77**). A reflection of skin covers the cornual process, and the epidermis proliferates to become the horn sheath composed of tubular and intertubular horn (**Figures 17.78** and **17.79**).

The chestnut and ergot are genetically determined areas of tubular and intertubular horn and are free from hair and glands. A thick pad of adipose tissue lies beneath the ergot.

Figure 17.74 Hoof (goat). The epidermis clearly shows all the layers of a stratified squamous epithelium; note the stratum corneum. H&E ×400.

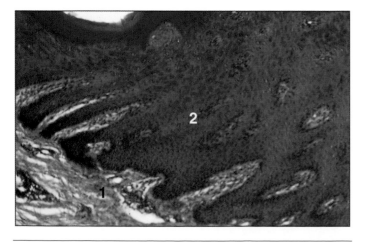

Figure 17.73 Hoof (goat). (1) Dermis. (2) Stratified squamous epidermal epithelium, with a thick stratum corneum. H&E ×125.

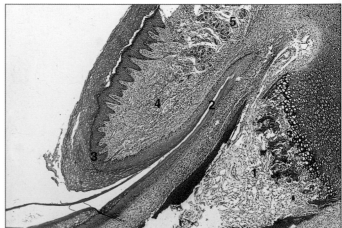

Figure 17.75 Claw (dog). (1) Third phalanx. (2) Claw fold. (3) Stratified squamous epithelium. (4) Dermis. (5) Digital pad. H&E ×50.

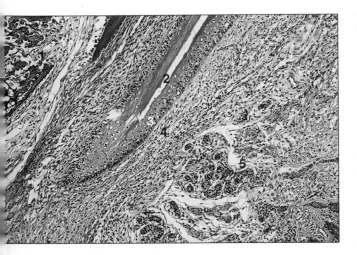

Figure 17.76   Claw (dog). (1) Third phalanx. (2) Claw fold. (3) Stratified squamous epithelium. (4) Dermis. (5) Digital pad. H&E ×62.5.

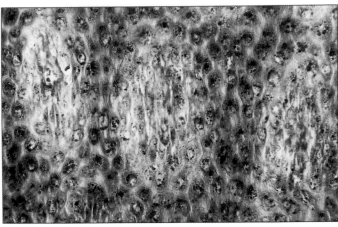

Figure 17.78   Horn (goat). The deep dermal papillae are stained green and surrounded by epidermal cells in columns, the horn tubules. Masson's trichrome ×250.

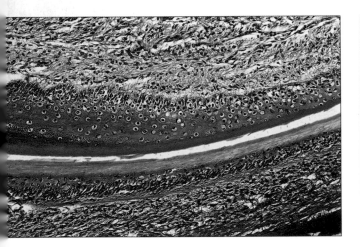

Figure 17.77   Claw (dog). The superficial clear area of the stratified squamous epithelium is the developing claw (equivalent to the human nail). H&E ×125.

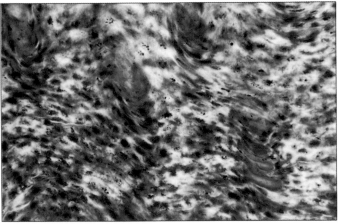

Figure 17.79   Horn (goat). The undulating effect is caused by the alternating tubular and intertubular arrangement of the developing horn. Masson's trichrome ×250.

APPENDICES

Ectoderm	Mesoderm	Endoderm
Central nervous system and eye	Muscle	Epithelium of
Central nervous system and ear	Connective tissue	Pharynx
Hypophysis cerebri	Blood, bone marrow	auditory tube
Epiphysis cerebri	Lymphoid tissue	thyroid, parathyroid,
Chromaffin tissue	Kidney	thymus
Epidermis, hair, nails, skin glands	Gonad and genital ducts	larynx, trachea, lungs
Epithelium of	Suprarenal cortex	digestive tube, liver, and pancreas
oral and	Epithelium of	bladder
nasal cavity	blood vessels,	caudal vagina, vestibule
glands,	lymphatic vessels,	urethra
tooth enamel	body cavities	

Appendix Table 2   Differential White Cell Count: Proportion of White Blood Cells in Domestic Animals, Species Variation

Species	Polymorphonuclear leucocytes (neutrophils) (%)	Eosinophils (%)	Basophils (%)	Lymphocytes (%)	Monocytes (%)
Horse	35–75	2–12	0–3	15–50	2–10
Cattle	15–45	2–20	0–2	45–75	2–7
Sheep	10–50	1–10	0–3	40–75	1–6
Pig	28–45	1–11	0–2	39–62	2–10
Dog	60–77	2–10	Rare	12–30	3–10
Cat	35–75	2–12	Rare	20–50	1–4
Chicken	10–33	1–4	1–3	48–82	1–6
Size	12–15 μm	10–15 μm	10–15 μm	Small, 6–9 μm Large, 9–15 μm	12–18 μm

Kidneys

Figures A1–A7 illustrate the structure of the kidneys.

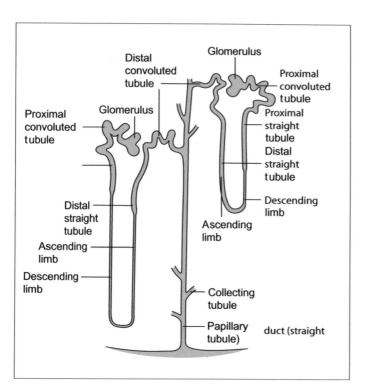

Figure A1 Two uriniferous tubules. On the left is a juxtamedullary nephron with a long loop of the nephron. On the right is an outer cortical glomerulus with a short loop of the nephron.

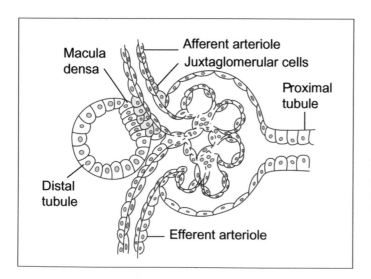

Figure A2 Glomerulus and related structures.

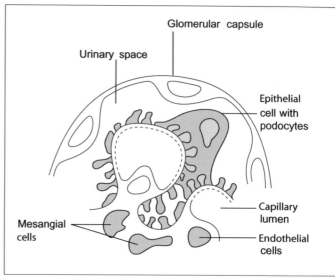

Figure A3 Renal glomerulus.

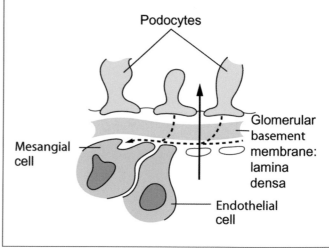

Figure A4 Glomerular filter (filtration slits). The straight vertical arrow indicates the direction of ultrafiltration. The dashed lines show the flow of the basement membrane in the perpendicular direction into the lamina rara interna and mesangium.

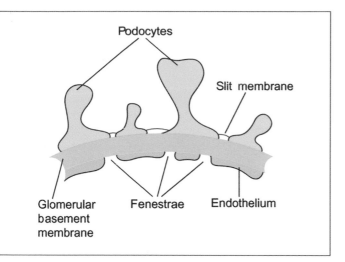

Figure A5 Glomerular capillary wall.

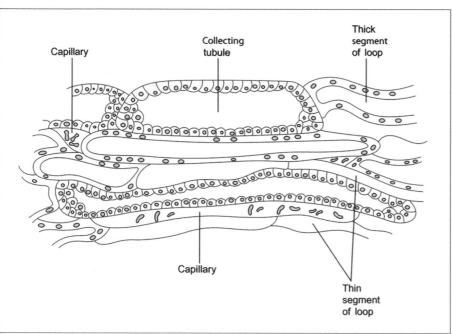

Figure A6 Medulla, longitudinal section.

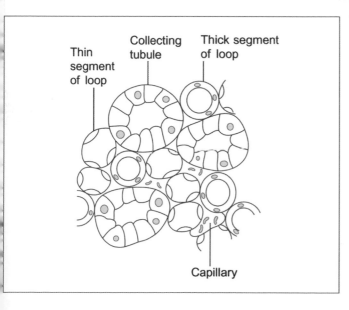

Figure A7 Medulla, transverse section.

Appendix Table 3   Placentation

Definition	Placentation	The intimate apposition of the fetal membranes and parental tissue for the purpose of physiological exchange
Classification	Choriovitelline placenta Chorioallantoic placenta	In the majority of mammals, the chorioallantoic placenta is the definitive one
		The vascular allantoic mesoderm fuses with the chorion and forms the fetal vascular bed as apposed to the maternal vascular bed in the endometrium
		The choriovitelline placenta is transitory
		In the mare and domestic carnivore, the yolk sac precedes the allantois in fusing with the chorion, vascularising and forming the early fetal bed; this is a choriovitelline placenta and remnants may persist until the term in the umbilical cord and form the umbilical vesicle (*see* **Figures 13.39–13.43**)
External configuration	Diffuse placenta	Involves the whole chorionic surface, separation at term is simple with no loss of maternal tissue, nondeciduate (e.g. sow and mare)
	Cotyledonary placenta	Chorionic villi are restricted to the maternal caruncle, and the two together form a placentome, the unit of the ruminant placenta; no loss of maternal tissue at term, nondeciduate (e.g. cow, sheep, and goat)
	Zonary placenta	The invasive zone is confined to a girdle-like band; loss of maternal tissue occurs at term, deciduate (e.g. carnivores)
	Discoidal placenta	Invasive zone is restricted to a single or bilateral disc-shaped area; loss of maternal tissue occurs at term, deciduate (rodents, primates)
Definition	Placentation	The intimate apposition of the fetal membranes and parental tissue for the purpose of physiological exchange
Histological classification	*See* **Figure 13.44**	
Synchoral fusion	Chorionic sacs of adjacent fetuses may fuse in up to 90% of multiple pregnancies	
	Vascular anastomosis between circulations	Rare, except in the cow

INDEX